To Donl[illegible]

a go[illegible] [illegible] and colleague

with my highest regard.

Nezar

Berkely 4/01.

Consuming Tradition, Manufacturing Heritage

Consuming Tradition, Manufacturing Heritage

Global Norms and Urban Forms in the Age of Tourism

edited by

Nezar AlSayyad

Routledge
Taylor & Francis Group

First published 2001 by Routledge
11 New Fetter Lane, London EC4P 4EE

Simultaneously published in the USA and Canada by Routledge, 29 West 35th Street, New York, NY 10001

Routledge is an imprint of the Taylor & Francis Group

Typeset in Sabon and Frutiger by NP Design & Print, Wallingford
Printed and bound in Great Britain by Biddles Ltd *www.biddles.co.uk*

This book was commissioned and edited by Alexandrine Press, Oxford

British Library Cataloguing in Publication Data
A catalogue record for this book is available from the British Library

Library of Congress Cataloguing in Publication Data
A catalog record for this book has been requested

ISBN 0–415–23941–9

Contents

Preface vii

The Contributors ix

Prologue

1. Global Norms and Urban Forms in the Age of Tourism: Manufacturing Heritage, Consuming Tradition 1
   Nezar AlSayyad

Tradition and Tourism: Rethinking the 'Other'

2. Tourism Encounters: Inter- and Intra-Cultural Conflicts and the World's Largest Industry 34
   Mike Robinson

3. Learning to Consume: What is Heritage and When is it Traditional? 68
   Nelson H.H. Graburn

4. Openings to Each Other in the Technological Age 90
   Robert Mugerauer

Imaging and Manufacturing Heritage

5. Colonial Nostalgia and Cultures of Travel: Spaces of Constructed Visibility in Egypt 111
   Derek Gregory

6. Everyday Attractions: Tourism and the Generation of Instant Heritage in Nineteenth-Century San Francisco 152
   J. Philip Gruen

7. Re-Presenting and Representing the Vernacular: The Open-Air Museum 191
   Paul Oliver

Manufacturing and Consuming: Global and Local

8. Making the Nation: The Politics of Heritage in Egypt 212
   Timothy Mitchell

9. The 'New-Old Jaffa': Tourism, Gentrification, and the Battle for Tel Aviv's Arab Neighbourhood 240
Mark LeVine

10. Image Making, City Marketing, and the Aesthetization of Social Inequality in Rio de Janeiro 273
Anne-Marie Broudehoux

Epilogue

11. 'Authentic' Anxieties 298
Dell Upton

Index 307

Preface

This book owes its origins to a discussion I had with Ananya Roy in early 1997. Together, we had just finished organizing the Fifth Conference of the International Association for the Study of Traditional Environments (IASTE), in Berkeley, California, which dealt with the theme 'Identity and the Making of Tradition'. We were both convinced that while the conference had raised the level of awareness and reflected interest in the processes through which identity discourses shape practices of tradition, it had not adequately questioned basic assumptions about the nature of national identities, national heritages, and the commercial dimensions of traditional environments. Ananya challenged me, and then later helped me write the Call for Papers for the following IASTE conference, 'Manufacturing Heritage and Consuming Tradition: Development, Preservation and Tourism in the Age of Globalization'. This conference, held in Cairo, Egypt, in December 1998, was one of the most successful IASTE has ever organized. Participants were asked both to question the impact of tourism on traditional environments, and to consider the ways that the nations of both the First and Third Worlds have resorted to heritage preservation and the reinvention of traditional practices as new forms of resistance against the homogenizing forces of modernity and globalization.

During the 1998 conference it became clear that several of the keynote papers were particularly successful at exploring key conference themes — among them, how nations, regions and cities have utilized and exploited vernacular built heritage to attract international investors at a time of ever-tightening global economic competition, and how the tourist industry has introduced new paradigms of the vernacular and/or traditional, based on the production of entire communities and social spaces that cater almost exclusively to the 'other'. It was then that I decided to pursue the making of this book in its current form, a job that required asking several of the keynote speakers, as well as several presenters at regular sessions at the conference, to rewrite their papers with a set of common themes in mind.

The last few years have witnessed the emergence of a number of good books on the production of cultural landscapes and the making and selling of traditional objects. This book will add a new dimension to this literature. It differs from most previously published works in its primary emphasis on the built environment. The contributing authors provide a rich set of case studies spanning different types of environments, various scales of settlement, and a variety of geographical locales, starting from the late nineteenth century, and culminating with the end of the twentieth century.

While the contributing authors did not originally coordinate their papers, their exposure to each others' work during the conference may have helped smooth the process of putting this book together. I am deeply grateful for their cooperation in reworking their original presentations to meet the larger themes of this volume. In addition, a number of other people deserve special thanks. J.R. Cousineau and Duanfang Lu helped with library research. David Moffat helped edit and prepare the manuscript for publication. Ann Rudkin initiated the idea of the book and handled it in its final stages. I am grateful for her commitment to the project and work on the manuscript. Also, I must thank Caroline Mallinder, who was equally interested, and who kept sending me subtle reminders to finish the book through colleagues she met at recent conferences. I have been delighted with her enthusiasm.

Finally, I have benefited from my association with several research units at the University of California at Berkeley. And I am grateful to my colleagues and students, whose interests have helped shape the book. While some will find answers to their questions here, others may not. It is my hope that a continuing process of critical inquiry around the themes of the book will shape further investigations in this important intellectual arena.

Nezar AlSayyad
Berkeley, October 2000

The Contributors

Nezar AlSayyad is a Professor of Architecture and Planning and Chair of the Center for Middle Eastern Studies at the University of California at Berkeley. He is also the Director of the International Association for the Study of Traditional Environments.

Anne-Marie Broudehoux is an architect from Canada who is currently completing a Ph.D. dissertation in Architecture at the University of California at Berkeley.

Nelson H.H. Graburn is a Professor of Anthropology at the University of California at Berkeley.

Derek Gregory is a Professor of Geography at the University of British Columbia, Vancouver, Canada.

J. Philip Gruen is a Ph.D. candidate in the Department of Architecture at the University of California at Berkeley. His dissertation concerns tourism in the late nineteenth-century urban American West.

Mark LeVine completed his Ph.D. at the Department of Middle Eastern Studies at New York University. He is currently a Mellon Post-Doctoral Fellow at the Department of History, Cornell University, Ithaca, NY.

Timothy Mitchell is an Associate Professor of Politics and Director of the Hagop Kevorkian Center for Near Eastern Studies, New York University.

Robert Mugerauer is Sid Richardson Centennial Professor of Architecture and Planning and also a faculty member in the Departments of Geography, Philosophy, and American Civilization at the University of Texas at Austin.

Paul Oliver is Director of the Oxford Brookes University Centre for Vernacular Architecture Studies.

Mike Robinson is Director of the Centre for Travel & Tourism, University of Northumbria.

Dell Upton is a Professor of Architectural History at the University of California at Berkeley.

Chapter 1

Global Norms and Urban Forms in the Age of Tourism: Manufacturing Heritage, Consuming Tradition

NEZAR ALSAYYAD

The twentieth century has been the century of travel and tourism. Indeed, the inhabitants of the world in the last two decades have met more other people than at any time in known history. As travel around the world has risen to unprecedented levels, the number of tourists visiting certain countries and cities in a given year often exceeds the numbers of those places' native populations. Global travel has encouraged this phenomenal growth of the tourism industry.[1]

Several world travel organizations predict that world tourism will grow at a rate of 4 per cent per annum, reaching a level of more than 700 million international arrivals and more than US$600 billion in revenue in the year 2000. And by 2010 it is predicted that arrivals will reach one billion and revenues will mount to nearly four times the current level. In the late 1970s less than one one-hundredth of a per cent of the world's population took an international trip in any given year.[2] But by the end of the twentieth century this percentage had increased a hundred fold. As the twenty-first century unfolds, people of every class and from every country will be wandering to every part of the planet. This is indeed an age of voyaging on a global scale. Meanwhile, tourist destinations throughout the world find themselves in ever more fierce competition for tourist dollars.[3]

For many parts of the world – especially those marginalized in the global industrial and information economy – tourist development may seem to offer the only hope of surviving in the global era. Yet at the same time that cultural heritage attractions offer income-producing opportunities to some of the poorest (as well as the richest) communities in the world, such mass tourism has often inflamed local and international passions, causing people to decry the irreversible destruction of traditional places and historic sites. In the final years of the twentieth century tourism has been called an 'unstoppable juggernaut, erasing all that is local and particular'.[4] Such anxieties have caused talk about the end of history, the end of geography, and the end of tradition.[5]

In the presence of such trends, and amidst the monotony of global high capitalism, at a time when standardized products and services are increasingly marketed worldwide, there is an increasing demand for built environments that promise unique cultural experiences. Many nations, meanwhile, are resorting to heritage preservation, the invention of tradition, and the rewriting of history as forms of self-definition. Indeed, the events of the last decade have created a dramatically altered global order that requires a new understanding of the role of tradition and heritage in the making of social space and the shaping of city form.

Intersections: Manufacture, Consumption, Heritage and Tradition

This book is about the intersections of four major terms: 'manufacture', 'consumption', 'heritage', and 'tradition'. The standard meanings of these words were primarily established during the modern era, yet these meanings are also inextricably bound up today with the problems or issues they explain. Etymological consciousness of the words is thus essential to understanding the social and intellectual context within which they are used. As 'keywords' in the sense Raymond Williams has pointed to, historical inquiry into their many facets may be used to establish the record of an entire vocabulary of shared meanings.[6] Examination of their dictionary definitions is a useful way to anchor this discussion.

The *Oxford English Dictionary* (*OED*) explains that 'manufacture' comes from two Latin words: *manus*, or hand; and *facere*, to make, and actually means 'the action or process of making by hand'. Yet as a cultural concept in recent times, manufacture has also come to be viewed negatively, as something produced to supply the demands of a market, often without the application of intellect.[7] Likewise, while 'consumption' has always been associated with the notion of using, it has recently come to connote destruction or wasteful expenditure.[8] For its part, 'heritage' derives from the Old French *eritage*, meaning property which devolves by right of inheritance in a process involving a series of linked hereditary successions.[9] Lastly, 'tradition' has been defined as 'the action of transmitting or handing down from one to another a variety of beliefs, rules, and customs'.[10]

These four terms remain very complex concepts that can, have been, and will continue to be used in a variety of ways. The contributors to this volume have not adhered to any specified definitions of them, but instead have explored their many dimensions. In part, this reflects the difficulty of pinning down the meaning of these terms at a time when they have been caught up in larger cultural debates. This has particularly been the case with the notions of heritage and tradition. In fact, it is now possible to distinguish three discrete phases in the change of attitude toward these terms in the last two centuries. The first phase, which roughly corresponded with the end of colonialism, intensified contact between cultures and instituted a period of hybridity. Yet while interest in local indigenous heritage was often initiated during this colonial period, it was only during the second period of postcolonial nationalism that it came to full flower in the demand for historic monuments and symbolic buildings. At this time, invocations of nationalism caused newly established nations to resort to heritage preservation for what they perceived to be a form of resistance against the homogenizing forces of twentieth-century modernity. Today, in the third phase, which has loosely been called globalization, the situation is quite different. As these independent nations compete in an ever-tightening global economy, they find themselves needing to exploit their natural resources and vernacular built heritage to attract international investors. Tourism development has consequently intensified, producing entire communities that cater almost wholly to, or are even inhabited year-round by the 'other'. The new norm appears to be the outright manufacture of heritage coupled with the active consumption of tradition in the built environment. Kenichi Ohmae has even argued that globalization is about the choice between Sony and soil and if given the choice, people will not choose nationalism or soil but satellites and Sony.[11]

Historicizing the New Tourist Landscape

Understanding the connection between heritage preservation and tourism development requires a grounding in both history and political economy. Studies of colonial urbanism have provided valuable insight into the politics of heritage and the discourse of its preservation. Meanwhile, analysis of the macro-economy of global production and investment have afforded a better understanding of the dynamics of tourism. Such an appreciation for history and economics allows one to see how global consumers today seek 'difference' and 'hospitality' as economic goods, and it helps elucidate the role of those producers or suppliers, often in the Third World, who make their living catering to this demand.

What is it that motivates the interest of tourists in others, prompting them to travel to distant lands, sometimes under uncomfortable conditions, often only to see the mundane rituals of daily life? The answer to this complex question may lie in what John Urry has labelled the 'tourist gaze'.[12] After examining the significance of tourism as a major industry in the waning years of the twentieth century, Urry suggested that this gaze is now a core feature of an industry in which the contemporary tourist – like the old-world pilgrim – seeks authenticity and truth in times and places away from his/her own everyday life. But this gaze is not the same everywhere, and its spatial dimension changes from place to place. In fact, the process by which tourists engage the built environment and are engaged by it is one that deserves special identification. I will call this 'engazement', a term I use to mean the process through which the gaze transforms the material reality of the built environment into a cultural imaginary.[13] It is this imaginary that this book attempts to explore.

Here the historical realities and political economy that have marked the development of a global heritage discourse become relevant. Specifically, in looking at the First and Third Worlds, one may notice that, while both possibly possess equal desire to explore the heritage and culture of the 'other', they have fundamentally different motivations for wanting to do so.[14] These differences may be attributed to or explained by earlier relationships of colonialism, political nationalism, and economic dependency. Today, as a result of such historical and economic forces, Third World countries often wish to emulate the 'progress' of the First World and adopt its developmental practices – but only without risking the destabilization of their local cultures. This is clearly a situation of wanting to have one's cake and to eat it too. Thus, as Benjamin Barber has pointed out in the appropriately titled *Jihad vs. McWorld*, such nations want the veil, but they also want the World Wide Web and Coca Cola.[15] Meanwhile, for its part, the First World appears more interested in consuming the cultures and environments of Third World societies. First World nations are often the main advocates for and financial patrons of the preservation of Third World built environments as part of what they define as 'universal' heritage – even when the 'natives' do not recognize its historic value. As a wealthy bloc, which often feels a sense of guilt and responsibility toward its former colonies, the First World has also tried at times to maintain or assist in preserving the dying or disappearing lifestyles and traditions of underdeveloped peoples and places. Yet it is also often the case that First World organizations, foundations and governments have engaged in such efforts while at the same time condemning or rejecting much of the social and political practices of the societies whose traditions they claim to want to preserve – especially when it diverges from Western standards of human rights, gender equality, and environmental sustainability.

As an example of the above dynamic one might consider the island of Bali, Indonesia. Here the First World has come to play the role of guardian of Third World traditions, but only so that would-be First World visitors can continue to appreciate them. In such an environment, the behaviour of the local people becomes fundamentally conditioned by the expectations of tourists.[16] But, unlike Disneyland, where employees are given the title of 'cast members',[17] here tourist industry workers are merely supposed to continue 'being themselves', and act out their supposedly still-genuine culture. Such a theatre, which is globally constructed but locally produced, is, of course, continually in danger of coming apart, since it depends on the willingness of the locals to act for the cameras.

Two other examples of this phenomenon were recently presented as chapters in David Howes's *Cross-Cultural Consumption: Global Markets, Local Realities*. Carol Hendrickson's contribution to this volume looked at how handicrafts heralded as Guatemalan or Mayan may now be purchased through US mail-order catalogues. Not only does this reveal the powerful marketing value of 'cultural difference', Howes pointed out in his introduction, but it also may create a situation of considerable cultural misrepresentation. According to Howes: 'For example, the industrialized cities where many of the Guatemalan artefacts come from are presented as pre-industrial villages in the catalogues. This is so as to agree with the American purchaser's preconceptions about "Mayan life" as well as to foster associations with both "tradition" and "uniqueness".'[18]

In another chapter in Howes's book, Mary Crain examined the employment of native women from Andes villages in the tourist hotels of Quito, Ecuador. According to Howes: 'As is typical of "ethnic tourism" ventures of this kind, "native traditions" are disassembled and rearranged in order to recreate a marketable semblance of "authenticity". In the case cited, this involved the women being required to dress in a gaudy version of their traditional clothing for the purpose of attracting tourism.' But Crain argued that the women have not submitted entirely to such a form of objectification. Thus, according to Howes: '. . . by means of a calculated reconstruction of their gender and ethnic identities, they have actively reshaped the role assigned to them by their employer, and attempted to use it to their own advantage. In other words, they have proceeded to "occupy" and exploit the very stereotypes which were intended to dominate them.'[19]

To be able to understand how such a heritage discourse is indeed an invention of the early modernist empire-building era, one must first frame it within the context of the three phases mentioned earlier.[20] The first phase corresponded principally with the nineteenth century. During this time the world witnessed the rise of modern industrial capitalism and the emergence of organized political dominance represented by colonialism. According to Anthony King, under the colonial paradigm, the world became divided into two kinds of people and two types of societies: powerful, administratively advanced, racially Caucasoid, nominally Christian, and principally European dominant nations; and powerless, organizationally backward, traditionally rooted, and mainly non-White dominated societies.[21] Under this new form of government, a legitimized relationship of unequal cultural and socio-economic exchange was born.

Today one must take this history of political and cultural domination into account if one is to analyse issues related to the invocation of heritage in the built environment. In particular, one must examine the processes during the colonial era by which local identity was violated, ignored, distorted or stereotyped. In terms of the built environment, part of this effort was almost always the introduction of a specifically colonial brand of architecture and urbanism. Thus, in many locales a certain hybrid form of building emerged – one that, at least at a visual level, unified the lands of colonial empires. For example, variations of the bungalow, a hybrid dwelling type first introduced by the British in India, soon appeared all over the British Empire, making it difficult to identify its true origins.[22]

It is interesting to note that at the same time that colonial governments were involved in suppressing indigenous cultural traditions, their fascination with the traditional customs of the 'other' also generated the first impulses toward its preservation. In fact, colonial empires eventually played a central role in maintaining, preserving and restoring much of what is today considered the built heritage of many Third World countries. The different 'World's Fairs' of the nineteenth century were important in this respect. To give an example, the images of a replica of an Egyptian temple built at the Paris Exposition Universelle of 1867 are today the only ones that exist of that destroyed monument. Likewise, the 'Cairo street' built at Chicago's Columbian Exposition of 1893 was one of the few representations ever made of what once constituted a substantial part of a Cairene urban fabric, but which is now in the process of disappearing.[23] Through such cases, one may glimpse how it is becoming increasingly the case now that in the absence of a real thing, the representation has become the thing itself.

When the people of the dominated societies started to rebel against this colonial world order, they had little to cling to in their drive to establish their own sovereignty other than a broad invocation of heritage as an instrument of nation-building. Therefore, in the second phase of cultural transformation an awkward relationship emerged between people and their cultural heritage. Traditions and structures, many of which were no longer appreciated by the native people, were now cast as the prime expressions of a new-found national identity. In some countries, the colonial-influenced abandonment of native heritage had gone so far as to include the discarding of entire and efficient systems of construction because their aesthetic did not fit the modern (colonial) paradigm.[24] But in the second phase, heritage also began to play a different role. It was now invented from the new, not only from the legacy of the past, even when the new was not yet fully developed. As a result, the urban environment of many developing nations was rapidly 'kitschized' or pseudo-modernized.

A good example here is Singapore's massive program to build new public housing to replace its old ethnic neighbourhoods. Responding to the multi-ethnic legacy of the colonial era and attempting to diffuse a potentially explosive cultural issue, the government's intent was to use the new housing programs to forge a postcolonial identity. The programs involved two interlinked objectives: to demolish existing and clearly demarcated ethnic neighbourhoods and the cultures that inhabited them; and to build new large-scale multi-storey housing complexes that would force different groups to live together in what was perceived to be an integrated environment.[25]

This brings me to the third phase in the invocation of heritage by both the dominant and the dominated. In today's climate of global economic and cultural exchange, the search for and reconstruction of identity has become paramount. The reason is that once independence was achieved, the glue that bound these nations together during their independence struggles dissolved, and problems of national and communal difference started to surface. Where it was not resolved, religious and political fundamentalism flourished. Today such a pattern has appeared most violently in such troubled places as Kosovo, Afghanistan, Chechnya, and the Sudan – not to mention the Indian subcontinent and Indonesia. The problem of course originated with the fact that the political units that were formed as nations in the post-World War II era were expected to be homogeneous entities with common cultures. But the reality was otherwise, and eventually, clear internal conflicts emerged based on differences in such essential, though unequal elements of national identity as race, language, religion, history, territory and tradition.

Faced with this problem, Third World governments have resorted to the notion of national identity both to project their image in the international arena and to project the same image internally to the native population. Through their monopoly of policies and resources, many governments have attempted over time to create a national culture, even when they lacked one in the first place. One of the key ingredients of this campaign was often an urban building campaign. Of course, such heavy-handed tactics raised troubling questions about the ethics of one political faction 'designing' national identity from the top down. This question was indeed faced by many fledgling politicians who governed newly independent states, and the architects and planners who worked for them. In some places such issues were affirmatively resolved, as in Singapore, which came later to the nationhood game. Another good example is Bangladesh, where such buildings as Louis Kahn's Dacca Assembly were produced during the height of the modern nationalism movement.[26]

Of course, while it may provide solace against the perceived depredations of foreign domination, simple faith in a myth of traditional origins cannot ultimately provide a stable basis for constructing a true sense of national identity. According to Gwendolyn Wright, 'the past cannot simply mean a retreat to a golden age before the Europeans, before modern industrialization, for these factors have changed us irrevocably'.[27] Furthermore, if one accepts that national identity is a social construct tied to temporal events,[28] it follows that a nation's heritage can only symbolize identity as observed by a single individual or agency at a specific point in time. Of course, the category of identity itself may be problematic; one may even ask if it is possible in the present global era to sustain any coherent, unified sense of the identity. Continuity and historicity of identity will always be challenged by the immediacy and intensity of global cultural confrontations.[29] Furthermore, the problem of national identity today is complicated by the reality of global economic patterns of production and exchange. Thus, not only do many nations now have to mediate between precolonial and colonial legacies, between the traditional and the modern, but they must also deal with the fragmenting effects of globalization. It would be convenient here to adopt Giddens's view that globalization has introduced new forms of world interdependence, 'in which once again there were no "others".'[30] However, it is more likely that, since capitalism thrives on the construction of difference, the present era of economic universalism will only lead to further forms of division, in which culture will become the globally authoritative paradigm for explaining difference and locating the 'other'.[31]

It is important to recognize that the three historical periods mentioned above cannot be read as a simple linear chronology. Indeed, as the examples of this book will demonstrate, the practices and tactics of one period may often resonate with others, or be deliberately deployed for other purposes. The classification of these historical phases then is less a teleology than a cluster of techniques that are often recycled and revisited in interesting ways. Taking this into account, for the rest of this chapter, I present a typology of techniques quite deliberately dissociated from these historical phases. I do not intend thereby to de-historicize my discussion of places; rather, I hope to isolate the clusters of techniques that define various pathways of heritage manufacturing for the consumption of tourism. At times, the coincidence with specific historical phases and associated political economies is obvious. At others, the connections are more open-ended and non-linear. Regardless, I believe that the analytical exercise of typologizing heritage places is as useful as that of historicizing the discussion about heritage, tradition and tourism.

Constructing the 'Other': Toward a Typology

The relation between built form and culture is especially affected in both the exercises of constructing national identity and manufacturing heritage for commercial consumption. One may distinguish three different types of physical environments which are produced today with the planned intent of making them places for the deliberate representation of cultural tradition. Despite the differences between them, it should be emphasized that all three are 'made', in the sense that they embody the clear objective of capturing, reconstructing, manufacturing, and possibly inventing social and built heritage.[32]

The first type is based on the notion of using history to create a dream landscape, a 'Wizard of Oz' land where all conflicts within a given culture are resolved, and where all cultural aspects are reduced to their basic representations. In such a vision, all icons of culture, such as architectural styles, building typologies, and spatial configurations, simply *become* the cultures that they are meant to represent. Authenticity here is desired, and is achieved though the manipulation of images and experiences.

Although such a strategy was often invoked in the building programmes of some reactionary nationalistic regimes aimed at fixing an official ethnic heritage, the ultimate example is, of course, the commercial Disneyland. One must remember, however, that Disney was not the first to pioneer the idea of replicating places of the 'other' for people to experience; for example, the 'World's Fairs' referred to earlier were engaged in just such an activity during the nineteenth century. Disney, however, was the first to recognize the permanent, continuing commercial potential of such installations. At another level, this process has also meshed with present trends, since it is precisely about the manufacture of global cultural products. This phenomenon has been developing for long enough for a certain convergence of consumer preferences and behaviour to have already taken place, as evident in the worldwide appeal of places like Disneyland.[33] Obviously, such places prove that even if the heritage is hyped, history sells. According to Briavel Holcomb: 'Despite critics who argue that the nostalgia industry distorts and commodifies the past, allusions to art and hints of heritage are vital colours in the urban marketer's palette'.[34]

The second type of environment that partakes of these processes of cultural objectification is that with a true claim to history, in the sense that it once was the site of an important historic event but over time has become marginalized. The attempt to resuscitate such environments (which may often be entire cities) by remaking them in their former image may serve one or both of two primary motives: to attract tourists for financial gain; or to serve as 'banks' of national memory and pride to ward off the subversive effects of historical change.

Colonial Williamsburg is a good example of such an environment. A replica of the capital of Revolutionary-era Virginia, it is arguably America's premier public history site. Yet, like other history museums, its legitimacy depends on its claim to 'real' history, as embodied in actual buildings and artefacts. But Colonial Williamsburg itself has long been criticized by historians for many of the same reasons as theme parks. For example, Eric Gable and Richard Handler have called it little more than 'an airbrushed, consumer-oriented, patriotic shrine celebrating an upscale idyll loosely based on the life style of Virginia's Colonial elite'.[35]

Such criticisms have not gone unnoticed. As cultural administrators have sought to keep Colonial Williamsburg at the cutting edge of historical knowledge, a new group of historians, hired in the 1970s, attempted to refashion the site by (among other things) bringing greater prominence to African Americans both in the ranks of its employees and in its narrative of nationhood. Yet the influence of these historians has ultimately remained limited – not, however, because historians are poor contributors to the project of manufacturing an imaginable history, nor because they are reluctant to contribute to the parallel project of facilitating the consumption of heritage. Instead, the lack of 'real' history mainly derives from the concerns of management that visitors will not come, or return, unless their visits are enjoyable.[36] Thus, depicting the harshness of slavery or any of early America's other shortcomings would create a level of discomfort that might ultimately cut into Colonial Williamsburg's popularity or profit. There is a deep irony here. Although tourists generally long to visit 'authentic' places, the authenticity they seek is primarily visual. Thus, their encounter with 'real' history remains marked by distance. And while they may wish to meet the world of the 'other', they also take great pains to limit its influence on them.

There is, however, a third type of environment that seeks to exploit cultural heritage, and in these places any claim to the reality of history is clearly secondary to its potential to generate commercial profit. It is in such places that the loosening of ties between the signs of a culture and their referents may be most apparent. Quite simply, to optimize the desire of the producers to manufacture cultural heritage and the tourists to consume it (all in as pleasant an environment as possible), it is now common for both groups to simply agree to dispense with any pretension to reality altogether. The best case here, of course, is the city of Las Vegas. Unlike the first two types of heritage environment, the sophisticated, themed casino complexes of Las Vegas do not pretend to authenticity. Thus, while the real Doge's Palace does not sit directly on the Piazza San Marco, such an adjustment can easily be made in its desert sister, where the replica of this historic seat of government is the Venetian, a 120,000 square foot gambling casino. Likewise, the Rialto Bridge, which was once the only crossing over the Grand Canal, in Las Vegas is found to connect two powerful gambling institutions. And while the real Bridge of Sighs earned its name by serving prisoners en route to their executions, the only 'sighs' at the Las Vegas version are likely those of gamblers in the process of losing their money.

Thus, unlike real cities, which often resort to the manufacture of heritage for political purposes, or nations which have wilfully allowed the consumption of their traditions by others out of economic necessity, Las Vegas is the ultimate site for the consumption of the heritage of the 'other'. Yet before rushing to dismiss such a project as kitsch, one must consider that in Las Vegas there is no hidden agenda. Las Vegas presents an outrightly manufactured heritage, based on the concept of copying the traditional forms of everywhere for the consumption of everyone. According to Akhil Gupta and James Ferguson, in Las Vegas one may say, 'the local and "exotic" are torn out of place and time to be repackaged for the world bazaar. Time and distance no longer mediate the encounter with "other" cultures.'[37]

Ultimately, one should not forget that the purpose of a categorization such as that I have just made between the three types of manufactured heritage environments is to point to certain social trends. Using such a categorical construction, it is possible to point out certain distinct types of effect on the relationship between tradition and tourist consumption on one hand, and cultural heritage and economic production on the other. Any such division into types may also be criticized because certain cases clearly span between categories, while others cannot be fit into the mould at all. In this regard, Seaside, Florida, and Santa Fe, New Mexico, might be said to fitted between the first and the second categories; while Poundbury, England, and New Gurna, Egypt, might be said to fit between the second and third categories. Meanwhile, Celebration, Florida, might be said to occupy a position by itself. Perhaps most precisely because these cases would appear to be exceptions, it might be most useful to explore their cultural dynamics in greater depth.

It is important to recognize that some of the places mentioned here are the creations of a design movement called New Urbanism (originally known as neo-traditional urbanism). The town of Seaside is perhaps the most well-known icon of this movement. Developed according to a strict zoning/design code known as Traditional Neighborhood Development (TND), it has been a great success in real estate terms. But like much of New Urbanism's output, it has also been criticized as a fake, involved with little more than the selling of nostalgia. Seaside has also been criticized for its exclusionary aesthetics and lack of social diversity. But perhaps the most severe criticism of it has been directed at its particular form of physical determinism, best represented by the belief that 'community' can be created by simply copying historical urban forms.[38]

Still on the margins of the first two categories, Santa Fe provides a slightly different case study. Here, while the built heritage may be real, its meaning has long since been diluted. Thus, although the town's distinctive indigenous adobe forms may be historically inspired, they have been long dissociated from their original cultural and historic context, so that now their consumption operates on an almost purely commercial level. It is well known that much of Santa Fe's authentic-looking adobe structures are in fact cement-plastered wood-frame buildings that give the appearance of adobe. The architecture of Santa Fe has accordingly caused one school of critics to label the town's particular style 'Santa fake'. In this regard, one might contrast the 'fake' authenticity of Santa Fe to the 'authentic' fakery of Las Vegas. One might even say that places such as Santa Fe represent consumed tradition but not manufactured heritage.

A third example comes from Britain. The town of Poundbury originated with Prince Charles's fight with the British architecture establishment. A traditional English village designed by Leon Krier and built on land owned by the Prince, it has attempted to recreate the feeling of a twenty-first-century community that has grown up over time. Of course, the desired effect also includes a long stopover in the golden age of the eighteenth and nineteenth centuries before the advent of architectural modernism. Thus, all services – telephone, electricity, gas and drainage – are buried in channels behind the housing, and the one large satellite dish that serves the entire community is hidden behind a high masonry wall. All that is visible protruding from the roof of a Poundbury house – which generally stands flush with the street so the entire town is just outside the front door – is a stately brick chimney or a polished weather vane.

Like Seaside, when plans for it first emerged in 1989, Poundbury was derided as the product of a kitschy time warp. Yet despite its sentimental pastiche of outmoded styles and small-town concepts, it has increasingly gained favor with its residents, as well as with writers and back-packing day-trippers. As one commentator pointed out, 'the effect is polite, elegant and as English as a vicar's tea party.'[39]

The village of New Gurna near Luxor, Egypt (which is also the subject, in a different sense, of a chapter in this book) provides a counterpoint to Poundbury's story of grudging critical acclaim. The work of Egyptian architect Hassan Fathy, New Gurna was planned in the 1950s as the new home for residents of a settlement the Egyptian government wanted to evict from their houses among the archaeological sites of the ancient Theban necropolis. Fathy designed the village using elaborate mud-brick structures that he imagined represented indigenous traditions. However, in his search for an ideal vernacular, he turned to the geometries and proportions of Islamic styles which had flourished in Cairo several centuries earlier. Among other things, this resulted in the use of unfamiliar forms (domes and vaults) for the project that the local people associated with the tombs and shrines of the dead.

New Gurna was an elegant depiction of an idea, but when the villagers who were meant to live there refused to move in, the attempt to create a new community with no real economic or social justification was revealed as a costly mistake. And in the end it became all too clear that Fathy's true concern was with his reputation among his First World architectural peers. Nevertheless, on account of the publicity his effort to adapt indigenous architectural forms achieved, Fathy came to be considered something of a guru among Third World architects. And today examples of Fathy-like architecture are widespread in the Egyptian landscape. It has only been in hindsight that the full extent of the liberties he took with such forms became clear – as well as his decision to avoid taking into account any of the concerns of the local people.[40]

Finally, there is the case of Celebration, Florida, whose story perhaps best captures the complexities of all the issues discussed so far.

Celebration's success as a place both of the manufacture of heritage and the consumption of tradition was nowhere more evident than in the fact that on November 18, 1995, a lottery was held among 5,000 eager contestants to determine who would be permitted to place a deposit on the first 350 houses and 120 apartments to be built there. Celebration is, in fact, a $2.5 billion real estate development managed and financed by the Disney corporation. At completion, it is expected to have 8,000 dwelling units and up to 20,000 residents. But Celebration's true appeal is that it will actually allow people to inhabit a historical fantasy. According to a recent *New York Times Magazine* report, at the centre of this fantasy is 'a sleepy grid of streets, lined with upscale shops and restaurants'. To emphasize the architectural pedigree of such an idea, however, there will be 'a two-screen movie theater (designed by Cesar Pelli), a bank (by Robert Venturi), a neat toy of a post office (Michael Graves), and a visitors' center (really a sales office designed by the late Charles Moore).'[41]

The idea of such a utopian American community is not new – not even for Disney. In the mid-1960s, the company's founder, Walt Disney, had originally proposed that EPCOT, now a world-cultural theme park, in fact be a high-tech model city of 20,000 residents (the acronym stands for the Experimental Prototype Community of Tomorrow). But this vision was not to be. Shortly after Disney's death in 1966, company executives, no doubt worried about profit margins and the likelihood that a city populated by real people might prove more difficult to manage than a theme park, decided to shelve his utopia. Thus, today Disney executives may glowingly speak of Celebration finally fulfilling their founder's quintessentially American dream of building a 'City on a Hill'. Such a statement, of course, leaves aside more mundane concerns such as the profitability of repackaging what is essentially Florida swampland, bought at $200 an acre in the 1960s, and selling it in quarter-acre lots for more than $250,000 today.[42]

Celebration is perhaps the crowning achievement of New Urbanism's all-enabling invocation of 'community'.[43] It has never been clear what this community is, but it has nevertheless struck a powerful chord for its ability to raise hope and generate profit. According to the *New York Times Magazine* reporter, Michael Pollan, who visited Celebration:

> . . . by the end of the walk the very designed-ness of Celebration had started to weigh on me. Eventually the streetscape began to feel a little too perfect, a little *too* considered. After a while my eye longed for something not quite so orchestrated.
>
> From my research I knew that every last visual detail my eyes had taken in during my two-hour walk, from the precise ratio of lawn to perennials in the front yards to the scrollwork on the Victorian porches, . . . had been stipulated – had in fact been spelled out in the gorgeous and obsessively

detailed 'Pattern Book' that governs every facet of architectural and horticultural life at Celebration. I knew all that, yet now I felt it too, and how it felt packaged, less than real, somewhat more like a theme park than a town.[44]

A cynical reading of New Urbanism would be that such places are simply suburban developments, but with an interesting marketing twist – architectural heritage. An even more suspicious view would question the transfer of design and political control from local governments and citizens to large corporations and the design professionals they hire. It is thus fitting that the town hall in Celebration is privately owned by the Disney Corporation. And it is also appropriate that a Disney executive would proudly describe this building as a one-stop shop for services, the ultimate in private-sector efficiency. One need look no further for the social impacts of such a transfer of control from 'citizens' to 'managers' than the town's design codes, which embody a high level of social control and are exercised according to few of the democratic processes that characterize other American localities. Thus, according to Pollan:

While I was walking around Celebration, I noticed some bright red curtains in the windows of a new Victorian on Longmeadow. Only then did I fully grasp the import of a cryptic little item I'd spotted in [the] monthly newsletter: 'Please refrain from using colored or patterned material in the windows. This can look pretty 'icky' from the street!'

Icky?! So this is the voice of private government in the '90s? It all struck me as fairly creepy, Big Brother with a smiley (Mickey Mouse) face.[45]

It is apparent that the red curtains in the window are clearly symbolic of an excess, a dissonance that New Urbanism cannot contain. Thus, Pollan ended his article by explaining how the banner proclaiming 'Disney's Town of Celebration' had been revised to leave only 'Town of Celebration'.[46] His editors made this connection even more apparent by running the following teaser on the magazine's cover: 'Disney Discovers Real Life: Even with fine design, making the town of Celebration, Fla., turns out to be harder than making entertainment. What follows when people move in? Politics.'[47]

If tradition is about the absence of choice, as Yi-Fu Tuan argued some years ago,[48] heritage then is the deliberate embrace of a single choice as a means of defining the past in relationship to the future. It is clear from the work of Eric Hobsbawm and Terence Ranger, that all traditions are invented.[49] Meanwhile, Benedict Anderson has shown that most nations are imagined communities.[50] Combining the views of these authors, it should become apparent that all heritage is socially manufactured, and that all traditions have the potential to be consumed.

Although the two activities, consuming tradition and manufacturing heritage, are thus produced by different agents, one cannot separate them from each other. In this global era, the consumption of tradition as a form of cultural demand and the manufacture of heritage as a field of commercial supply are two sides of the same coin. And many countries are now actively inventing or re-creating their own heritage, and using tourist revenues to do so. Their design agenda thus has two components: one politically self-serving; the other economically sustaining.

There are many cases around the world today in which the notion of a manufactured heritage has in this sense managed to take over a considerable segment of architectural practice. One might again cite the example of Singapore, alluded to earlier. Here, after government planners had ordered some of the most culturally distinct ethnic quarters to fall to the wrecker's ball, they realized they might actually need such places to compete in the new global tourist marketplace, with its emphasis on heritage sites. The challenge facing these same planners today, therefore, is to recreate new commercial areas that look ethnic enough to recapture some of the city's lost cultural heritage. Besides being patently fabricated, such actions reveal the craven tendency of many governments to bow to both political necessity and economic expediency. As Gupta and Ferguson have written, the new global context does seem to be recreating a sense of place and sense of community in positive ways, 'giving rise to an energetic cosmopolitanism in certain localities.' Yet, in other cases, 'local fragmentation may inspire a nostalgic, introverted and parochial sense of local attachment and identity.' They therefore argue that even if globalization recontextualizes cultural localism, it often does so in ways that are 'equivocal and ambiguous'.[51]

The situation of Hong Kong after its reunification with China in 1997 provides a slightly different twist on the same story. I first became aware of this dilemma at a conference I attended in December 1999, organized by the Hong Kong government to deal with the issues of heritage and tourism. I found the Hong Kong participants at this conference extremely enthusiastic about the prospect of putting their heritage to use for economic gain. However, the questions most participants forgot to ask were 'Whose heritage is to be preserved?' and 'To what end?' At the same time, it occurred to me that the most important topic such people could be beginning to address was how to recast their identity to fit the new reality of Hong Kong's political reunification with the mainland.[52]

What these examples point to are the dangers of heritage professionals who compromise their positions with regard to the tourist industry. After all, the tourist industry is a business, and not a charity, and both its ethics and aesthetics primarily respond to market demands.

The ties of heritage managers to nationalist agendas must also be considered suspect. Nationalism is, and always has been, a divisive governing philosophy because at its core it is exclusionary. The nationalist agenda has always been to set up contrasts: my nation versus yours; my history versus yours; and finally my identity versus yours. Such distinctions have been at the centre of many violent conflicts (which, in turn, may owe their origins to irreconcilable identity positions). To preserve heritage for such exclusionary ends serves little purpose other than to increase the potential for further escalating the sources of intra- or international conflict.

Here is the dilemma of globalization. Because of the importance of the heritage tourism industry in the economy of nations, preserving heritage has become important not only for economic sustenance but also so that nations, regions and cities may position themselves to compete globally. The paradox is that investment in heritage may only stir up further nationalist sentiment that often leads only to invocations of superiority and isolationist tendencies.

As Kevin Robins has written, one must remember that 'Globalization pulls cultures in different, contradictory, and often conflictual, ways. It is about the "de-territorialization" of culture, but it also involves cultural "re-territorialization". It is about the increasing mobility of culture, but also about new cultural fixities.'[53] According to Gupta and Ferguson globalization is also associated with new dynamics of relocalization:

> It is about the achievement of a new global-local nexus, about new and intricate relations between global space and local place. The global-local nexus is about the relation between globalizing and particularizing dynamics in the strategy of the global corporation, and the 'local' should be seen as a fluid and relational space, constituted only in and through its relation to the global . . . Indeed, the very celebration and recognition of 'difference' and 'otherness' may itself conceal more subtle and insidious relations of power . . .[54]

Robins has further written that globalization is about the '. . . increasing transnationalization of economic and cultural life, frequently imagined in terms of the creation of a global space and community in which we shall all be global citizens and neighbours.'[55] The proliferation of common cultural references across the world evokes for some a cosmopolitan ideal. 'There is the sense that cultural encounters across frontiers can create new and productive kinds of cultural fusion and hybridity.'[56]

But the great danger lurking in this new global citizenship, of course, is the erosion of the public sphere. Briavel Holcomb has reported that a high official in the tourist industry once said, 'I can think of no industry other than tourism where the interests of the public and private sectors so closely converge?' Holcomb went on to argue, however, that this observation, 'rings truest when public means government leaders (rather than community) and private means business (not the private citizen).' Holcomb further remarked that the criteria for evaluating the relative costs and benefits of tourism to the public are both volatile and contested.[57] The real issues here involve who manages heritage sites, and how these are in turn managed in relation to both the demands of the tourist market and the goals of the national or municipal governments that control them.[58]

The contributors to this book attempt to address many of the above issues and confront their local/global complexities. Their contributions have been divided into the three sections that comprise the main body of this volume: 'Tradition and Tourism: Rethinking the "Other"'; 'Imagining and Manufacturing Heritage'; and 'Manufacturing and Consuming: Global and Local'.

Tradition and Tourism: Rethinking the 'Other'

The contributors to the first section of the book, Mike Robinson, Nelson Graburn, and Robert Mugerauer, deal with certain conceptual frameworks of encounter with otherness that underlie the activity of tourism. As tradition has increasingly become an object of world tourism, its audience is no longer confined to the members of the cultures that generate it. Instead, the primary consumers of cultural traditions may now be visitors from elsewhere. These outsiders, as well as the local agents who package tours around cultural themes, are no longer willing simply to accept local traditions passively, and have increasingly taken an active role in manipulating and transforming cultures to fit their demands. This local-global process has resulted in the creation of stereotyped notions of 'others', which may be at odds with local people's conceptions of themselves.

The basis for much later discussion in the book is established by Mike Robinson in his chapter 'Tourism Encounters: Inter- and Intra-Cultural Conflicts and the World's Largest Industry'. In the chapter, Robinson discusses the features of current international tourism, its formidable role as a vector of cultural exchange, and the importance of the built environment as the cultural space and place of such encounters. Robinson suggests that tourism is a highly structured and organized form of human activity. He builds his argument on three basic observations: first, there are relatively few nations and cultures that are not affected in some way by tourism and the tourism development process; second, where tourism has emerged as an important economic activity, it is frequently characterized by a rapid and often dramatic expansion in supply; and third, world tourism is the product of a First World ideology that displays fundamental inequalities. Robinson then suggests that cultural conflicts in tourism can be understood on a range of interdependent levels: between individual tourists and representatives of a 'host' culture; between and within host cultures; and between the tourism industry as part of the development process and the host community/culture.

Robinson points out that much of the tourism industry demonstrates no real concern for the cultural dimensions of place or territory. Rather, the challenge is to package, image and transform traditions, rituals, and 'ways of life' into saleable products. Thus, the tourism industry largely conceives of culture(s) in two ways: either as value free, and thus largely as an inconsequential aspect of development; and/or as just another product to be packaged. As a result, culture(s) as embodiments of living traditions are reduced to superficial subjugates of consumerism and lose their active social aspect, political function, and authenticity. Within the short time period of a leisure or business visit, tourists essentially remain strangers and outsiders, with little opportunity or motivation to come to terms with a host culture in any meaningful way. In response to such a dynamic, the development of space, particularly urban and 'inner-city' space, to make it more attractive to tourists, is rarely accompanied by attempts to maintain viable communities of local residents in the same places. On the contrary, such factors as increasing land prices and rents, decreasing security of tenure, heavy competition for business, loss of indigenous control, and the dominance of aesthetics over social function have often created problems. Robinson concludes that in cultural terms, tourism establishes a primarily unequal relationship, since it does not usually take place on the basis of consent and frequently disregards any concern for mutual cultural understanding.

In 'Learning to Consume: What is Heritage and When is it Traditional?' Nelson Graburn next explores the concepts of heritage, tradition and consumption from an anthropological perspective. Using a subjective, personal approach, Graburn constructs a Foucaultian 'genealogy of heritage' which focuses on the way children acquire these concepts, claiming that this is the necessary first step to understanding how adults claim the terms. Graburn argues that heritage is defined as the knowledge of and/or rights over material and non-material things transmitted over time. It is, by definition, owned, and can, with permission from its owners, be consumed. The consumers of a heritage need not be those whose ancestors or immediate predecessors owned it, leaving the concept open to change and manipulation. Graburn's claim that 'all heritage is constructed', just as all environments are culturally constructed, bolsters the idea that heritage can be manufactured. He argues that it can be created and re-created for the purposes of those who claim it. Whether or not the owners of a heritage received it within the bounds of kinship – the original social paradigm of identity – a claim to heritage cannot only give a sense of concrete identity, but it may also demand the preservation and respect of its forms.

One might ask whether it is possible to preserve heritage at all, given that traditions are now often practised in locales far removed from their site of construction. Graburn's definition of tradition is consistent with his view of heritage and consumption. He views tradition as a product of modernity, which itself was the product of change and history. It is the product of a lifetime of experiences by individuals and, having changed much over time, it is bound to change even more as it is passed along to the next generation. Graburn's model can easily be applied to the built environment, where it presumes a set of component parts which once composed the original model, available for deliberate assembly into new variations of it.

Robert Mugerauer's chapter 'Openings to Each Other in the Technological Age' addresses the problems and possibilities that lie within the tension between modernity and traditional environments and peoples. Mugerauer suggests that in its all-consuming drive toward modernization, the West has lost track of its links and debts to other times, cultures, and value systems. By developing scientific attitudes and technologies that have dismissed or subordinated spiritual/cultural values, the West has made material culture and machines the 'measure of mankind'. This has resulted in modernity's negative definition of the 'other' – that is, the traditional – as lacking and disadvantaged. Thus, in the current practices of tourism, a true openness to each other cannot occur. Instead, locals are encouraged to 'dress-up' in traditional costumes and display themselves for touristic consumption, and sometimes for basic subsistence.

Mugerauer points out that if a tradition is vibrant, its members become objectified while unavoidably enacting their tradition. However, in this process personal actions will unavoidably be compared to a preferred version of tradition specified by the tourism industry. Mugerauer also warns that if a tradition is not strong enough to absorb such visitors and their expectations, the local people and places may become a 'fiction' entirely.

Mugerauer notes that, in addition to actually visiting other places as tourists, it is today possible to visit them 'virtually' through film, television, and the World Wide Web. In one sense this may provide an opening for self-articulated and affirmed identities and differences. However, in today's late capitalist global order, both real and virtual visits may often function as little more than instruments of stigmatization and stereotyping. As such, they may propagate the dominant forms of desire and the standard measure of human worth.

In concluding, Mugerauer argues that some aspects of modernist and postmodernist discourse may offer insight into ways to create non-imperialistic opportunities and establish self-determining, differentiated or heterogeneous identities and senses of place, and thus to continually rejuvenate tradition.

Imaging and Manufacturing Heritage

The second section of the book turns to an examination of how heritage environments are manufactured. Derek Gregory, Phil Gruen, and Paul Oliver look at cases of such practices from the nineteenth century to today.

They pay special attention to the interaction between the actual heritage asset, the manufacturing agenda, and consumer demands.

Derek Gregory's chapter 'Colonial Nostalgia and Cultures of Travel: Spaces of Constructed Visibility in Egypt, 1820-2000' is a powerful exposition of the ways in which 'tradition' was manufactured and consumed at several levels in late-nineteenth and early twentieth-century Egypt. He opens by interpreting photographs and texts from travel literature on Egypt that evoke the glory days of late Victorian and Edwardian travel along the Nile. Gregory's project is to ask how these images of colonial nostalgia were constructed and, more importantly, what colonial histories are hidden from the views they presented. Eschewing the term 'postcoloniality', Gregory prefers to explore what he calls the 'colonial present', and to ask why and how, at the end of the twentieth century, people may still be seduced by such stories of colonial power.

Gregory's introduction links his work with that of Eric Hobsbawm on invented traditions as a response to modernity. But he departs from this frame of reference in his use of Edward Said's concept of the 'citationary structure of Orientalism'. Gregory argues that ideas about a 'traditional' Egypt were deliberately constructed through successive, referential travel accounts, and that they were subsequently marketed for the consumption of European travellers who had exhausted the Grand Tour in Italy and were hungry for a new frontier. Yet, at the same time that tourists were viewing 'traditional' Egypt, capitalism was producing a modern Egypt based on tourism and resource exploitation. Tradition, Gregory contends, became at once an indispensable and an irredeemably compromised term.

Key to Gregory's discussion is the idea of 'spaces of constructed visibility'. Drawing on concepts first elaborated by Henri Lefebvre, Michel Foucault and Bruno Latour, he sketches an historical geography of the rationalized, controlled spaces of the Egypt experienced by nineteenth-century tourists. This travellers' Egypt was mainly visual, viewed from the class-bound safety first of leased Nile barges or *dahabeahs* and then from Thomas Cook's steamer ships. Despite the best efforts of colonial powers to regularize and standardize the Nile tour, however, tourists along the river often found themselves negotiating their passage with local individuals, groups and events which disrupted the carefully manufactured view. Thus, Gregory discusses the ways in which Egyptian agents and actors disrupted colonial attempts to dictate images of a 'traditional' or 'authentic' Egypt. Yet he points out that since no written testimonial of the experience of the labourers exists, there is no way of knowing how the tourists' and travellers' gazes were returned or how travel on the Nile was experienced by the colonized people of Egypt. Gregory insists that researchers must seek out such complementary histories of purportedly 'traditional' people and their marginalized existence if the full extent of the cultural exchange is to be understood.

The tourist city is often experienced as a series of free-standing monuments. But in this section's second chapter, 'Everyday Attractions: Tourism and the Generation of Instant Heritage in Nineteenth-Century San Francisco', Phil Gruen argues that tourists in nineteenth-century San Francisco were as interested in the everyday life that made the city whole as they were in the monumental city promoted and manufactured by guidebooks. Gruen starts with an investigation of several key players in the generation of instant heritage in early San Francisco, such as newspaper and magazine articles, photographic panoramas, as well as colour lithographs. Then through a historical survey, he suggests that nineteenth-century San Francisco was not only a source of visitor fascination because of its many 'sights', but it was also a city that offered the visitor multi-layered and multi-sensory experiences: picturesque, exotic, palatial, and 'world-class'. Thus, instead of seeking places of refuge away from the allegedly chaotic and morally decaying nineteenth-century city, tourists revelled in San Francisco's energy, and they considered that very energy a true mark of the city's 'advancement'. Consequently, tourists helped this alleged urban 'other' assimilate among the ranks of America's and the world's 'foremost' cities. Gruen argues that, contrary to the historiography of nineteenth-century tourist encounters, there was more to tourism in the American West than extended visits to the region's natural wonders. He concludes that the manufacture of heritage by official publications and the consumption of this heritage by tourists was a process of selection and exploration, not of direction and slavish obedience.

Lastly in the book's second section, Paul Oliver's chapter 'Re-Presenting and Representing the Vernacular: The Open-Air Museum' illustrates the way in which culture can be both manufactured – in the literal sense – and subsequently consumed, in the sense of being immediately available to those whose desire to experience it in a representational form. Open-air museums, or *skansens*, are assemblages of purportedly redundant buildings which have been relocated to artificially created sites, and to which visitors pay a fee for access. Oliver's examination of the politics of representation shows how history and tradition have been put to use by individual collectors or by national or regional groups searching for a past. In Oliver's view, open-air museums attempt to re-create an image of a past that may have never existed. By peopling their buildings like stage sets, they have destroyed as much cultural heritage as they have preserved. The consumers, in this case, are tourists hungry for authenticity.

Through examining the ways in which buildings have been wrenched from their original sites and repositioned on new ones, Oliver highlights the means by which the late twentieth century is driven by a desire to seek out and consume 'authentic' places. 'Authenticity' may perhaps be an unfortunate term to invoke here for, as Oliver contends, the destructive consumption of any building is continuous from the moment it is built. But rather than preventing the consumption of the historic built environment, open-air museums merely abet its consumption in new ways.

Manufacturing and Consuming: Global and Local

In the third section of the book the discussion turns to examples of how the rise of global culture has had a particular impact upon the dynamics of cultural exchange. As built form is increasingly becoming just another product packaged for the purposes of tourism, it seems to lose its social function and appear value-free. Yet as Timothy Mitchell, Mark LeVine, and Anne-Marie Broudehoux point out, although the global heritage dialogue tends to present the built environment as an empty container, places of heritage remain places where real people live and where real conflicts may arise.

In 'Making the Nation: The Politics of Heritage in Egypt', Timothy Mitchell starts with an assessment of Hassan Fathy's 'model village' of New Gurna. Fathy's account of the events fifty years ago, as told in his *Architecture for the Poor*, relates how the New Gurna project was a story of the progress of ideas impeded by the ignorance of the authorities and lawlessness of the natives.[59] Mitchell's account, however, reveals that Fathy's process of professional re-appropriation of the vernacular came complete with the seeds of its own destruction. The birth of the heritage movement in Egypt inspired by New Gurna was also the moment of its violent demise. For Mitchell, Fathy's failure to place his project into a larger social context makes its content a violent one.

Mitchell explains that today, from the Ministry of Culture and American development experts in Cairo, to the Luxor City Council and local contractors and tourism investors, a new coalition of forces has emerged, working to transform Gurna 'into a site that was clean, well lighted and signposted, with wide roads and ample parking, and people-free: in a word, not just an ancient heritage site but a modern one.' Fifty years after the demise of New Gurna, Mitchell explains how the government is still attempting to evict the population of old Gurna, and still describing them as lawless and unhygienic.

Mitchell claims that the authorities assume that enjoyment of historical treasures can only be secured by their physical separation from the local community. This has resulted in the creation of 'enclave tourism'. Indeed, most Luxor tourists live, eat and sleep in enclave hotels, travel in separate air-conditioned buses, and go to special entertainment sites. According to Mitchell, this process of segregation is being driven not only by the planning of international hotel chains and local entrepreneurs, but also by current Egyptian government policy and World Bank funding. Mitchell argues that such a segregated economic condition can be explained by realizing that the tourist industry, like other conventional industries, relies upon the optimization of resource flows and timetables, and the rearranging of physical space to accommodate it. The twist with the tourist industry is that this process is organized around the maximization not of production but of consumption. To reveal the true nature of this relationship, one must bring the hidden violence of the 'heritage industry' into view.

In 'The "New-Old Jaffa": Tourism, Gentrification and the Battle for Tel Aviv's Arab Neighbourhood', Mark LeVine examines how Arab residents have attempted to re-imagine their 'city' and open up new spaces for agency and empowerment through which they can articulate a more autochthonous synthesis of the city's history and architectural traditions. Jaffa was the economic and cultural capital of pre-1948 Arab Palestine and is now a mixed Arab-Jewish quarter of the city of 'Tel Aviv-Yafo'. During the late 1980s and 1990s Jaffa became an object of 'development' as both a site for tourism and as a new, chic neighbourhood for the burgeoning Jewish elite of 'Global Tel Aviv'. These changes took place in the face of creeping dislocation, and were accompanied (and supported) by daily media and television portrayals of Jaffa as a poor, crime-ridden – and, at the same time, exotic and romantic – place.

Through a study of the ways in which the interplay of discourses on nationalism, modernity, architecture, tourism, and gentrification have influenced the transformation of Jaffa, LeVine argues that the double economy of fixing Jaffa for the Orientalist gaze and developing it according to a changing market economy relies on both the economization and de-politicization of the Arab community. The contested space of Jaffa and Tel Aviv thus epitomizes the complex manner in which architectural movements are inscribed in the politics of national identity in Israel. Both the erasure of 'tradition' through the application of the International Style, and then its reclamation through discourses of heritage promoted by postmodernist professionals, have expressed political idioms inherent in the construction of national identity in Israel.

Through this case study, LeVine also demonstrates that, although it is prevented from expressing itself through the actual planning of its lived environment, Jaffa's Arab population has articulated its identity through a series of 'spatializing social activities'. These have included art festivals, organized protests, and fighting to return streets to their original Arabic names. LeVine draws on one of the key concepts formulated by Henri Lefebvre, regarding 'representational spaces'. He suggests that in studying spaces which are linked to the clandestine or underground side of social life of the inhabitants, as opposed to those created by planners and political authorities), LeVine suggests that in studying the spatial system of Ajami and Arab Jaffa, one is confronted 'not by one social space but many such spaces in which the global does not abolish the local.'

In the final chapter in the book's third section, 'Modernism as Identity: Rio Cidade and the Aesthetization of Social Inequality', Anne-Marie Broudehoux describes and critiques the mechanism of city marketing and image making in a Third World context. In the course of the last few decades, the changing configuration of the global political and economic order has forced cities throughout the world to undergo major restructuring in order to become more competitive in the international market. In their struggle for economic survival, and in search for new sources of employment and revenue, city managers have turned to city marketing and image making to boost local distinctiveness and attract both visitors and capital. With growing awareness of their city's position in the global hierarchy, city officials and local entrepreneurs have often collaborated to exploit their city's image, and to 'sell' their locality by harnessing its actual or perceived attributes.

The late capitalist urban condition is characterized by a trend toward aestheticization, where the primacy of the visual and the centrality of the image have reduced the city to a landscape of visual consumption. Broudehoux points out that, despite a strong economic rationale, there is a social logic to this practice of selling places. She argues that urban image construction through public works and marketing campaigns is often used as a tool of social control, as dominant groups use visual and spatial strategies to impose their views and set the terms for membership in society. In this process some actors are sanctioned as participants, while others are ignored, segregated, and made increasingly invisible.

Broudehoux's study of recent image-making efforts in Rio is a study of just such a relationship between space, power and social justice in a society caught in a free-market frenzy and its concomitant process of socio-economic polarization. With the shift of the national capital to Brasilia in 1960, Rio's economy lost one of its main driving forces. The city was then hit hard by the economic crisis of the 1980s, and the lack of public funds led to massive disinvestment, creating a serious urban crisis. However, since 1993 the city government has conducted a series of public-works programmes to restore the city's image in the hope of retaining investment and making the city competitive on the world tourism market. Through three case studies of specific programmes, Broudehoux demonstrates that the current use of cosmetic solutions as forms of popular pacification is highly unsustainable and may actually be counter-productive.

Urban image construction has increasingly become a means of manipulating public opinion and controlling social behaviour to serve particular social, political or economic interests. Decision-makers use the built environment to manipulate consciousness, and disguise this manipulation in order to reproduce their political ideology and naturalize their power. Yet Broudehoux's case study of Rio shows how such attempts at social control and exclusion are not always met passively, nor do they go unchallenged.

The Image Group: A Concluding Thought

I would now like to return to Disneyland and its Main Street, the street that has captured the imagination of visitors to the park as the most quintessentially 'American' of all places. Recent newspaper reports have indicated that its story may now have come full circle. Marceline, Missouri, and Fort Collins, Colorado, were the two towns that originally inspired the design of Disneyland's Main Street. Marceline was the hometown of Walt Disney, where as a boy he first sketched barnyard animals and fell in love with trains. Fort Collins was the birthplace of Harper Goff, Disneyland's first director in the 1950s. However, as *The New York Times* reported in 'A Tale of Two Main Streets', a reverse flow of cultural capital is now taking place from the copy to the models from which it was derived. According to the article, when Marceline and Fort Collins began to experience economic difficulties, both seized upon the expression of their ties to Disneyland's Main Street as a strategy for survival. In particular, the citizens of Marceline renamed its downtown Kansas Avenue 'Main Street USA' to cement the connection to Disneyland. The town tour now attracts several thousand visitors each year, and there are further plans to transform the train depot into a Walt Disney-Santa Fe Railroad Museum in anticipation of Disney's one-hundreth birthday in 2001. Meanwhile, in Fort Collins, the article reported that the preservation of its downtown had begun to look 'suspiciously like Disneyfiction'.[60]

These examples clearly show how in today's world, where the global heritage industry reigns supreme, the notion of authenticity has sometimes been cut completely loose from its moorings. The image of the thing may now actually replace the thing itself. At times the confused nature of authenticity may border on the absurd. Thus, at EPCOT the Moroccan pavilion was actually subsidized by a foreign government, and Moroccan craftsmen were sent by the King of Morocco to secure the country's place in the new global order.[61] One must ask what kind of authenticity the Moroccan government thought it was buying by investing in such an obviously manufactured environment, especially when none of the wealthier nations represented at the exhibit, such as France or Italy, provided any such funding for their pavilions. Yet in this regard one might also remember the Cairo street built as part of the Paris Exposition of 1889. As part of that exhibit, curators felt compelled to import actual dirt, donkeys and caretakers. The concern for authenticity was so grave that several details of an actual historic structure (a Quranic school for children and a water fountain) were disassembled, shipped to Paris, and installed in the copy.[62] In perhaps a supreme irony, 100 years later, when a foreign preservation team working for the Egyptian government wanted to restore this structure, the only surviving clear and detailed representations of it were those recorded in exhibition publications. Here the copy of the thing became the means by which the thing could continue to exist.

To show just how pervasive such a system of global exchange may be, one may note how such typical movements of 'authenticity' from the Third World to the First, or from 'East' to 'West', may today also operate in reverse. Thus, at the same time that imitations of Western-branded goods such as Nike sneakers clog the shelves of Third World countries, David Howes and Constance Classen described how a Mexican merchant, Fernando Pelletier, actually managed to preempt the authenticity of 'Cartier' products and assert the primacy of the copy over the original. His approach was simple: open a shop and sell accurate 'simulacra' of the original items at cut rates that more accurately reflected the true cost of producing them in Third World environments in the first place.

> When it was discovered that its goods were being counterfeited, the original Cartier company decided to open up its own shop in Mexico. The Mexican Cartier, in response, sent a letter to the President of Mexico denouncing the French Cartier's lack of respect for Mexican industry and government, and suggesting that the French products were fakes. This letter was printed in major Mexican newspapers and soon became the subject of angry editorials against the French invaders. The French Cartier also ran into problems when it tried to register its trademark designs in Mexico, for it found that Pelletier, the Mexican copier, had already registered them and thus had prior rights in them.[63]

I would like to end this introductory chapter with a personal anecdote. At a field trip that was part of a recent conference I attended in Cairo in 1998,[64] I met an American academic on the Giza plateau at the foot of the Pyramids. He was looking down toward the Sphinx. 'Oh, but it is so small', he was saying. He was really disappointed. His comment puzzled me, and it took me a couple of months to figure out what he meant. It turned out the man was a teacher at the University of Nevada, Las Vegas. His city housed the famous Luxor Hotel and Gambling Casino, built as a glass pyramid with a three-times-enlarged Sphinx as its entrance. The professor was used to parking his car in a lot that faced the giant Las Vegas Sphinx. When he was in Giza, he became disappointed, not because the reality did not live up to its image, but because along the way, the reality ceased to be relevant when the image became the principal frame of reference.

This is reminiscent of a tale of mimesis once told by Jean Baudrillard. The cartographers of an empire draw a map that is perfect in every detail and eventually becomes a substitute for 'the real' it represents. The map is slowly rotting in just the parts where the territory in real life becomes desert or occupied by other nations. '[S]imulation is no longer . . . a referential being or a substance. It is the generation by models of a real without origin or reality . . .'[65]

Notes

1. B. Crossette, 'Surprises in the Global Tourism Boom', *The New York Times*, April 12, 1998, p.wk5.

2. *Ibid.*; and review by S. Roaf of J. Jokilehto, *A History of Architectural Conservation*, Oxford, Butterworth Heinemann, 1999, in *Traditional Dwellings and Settlements Review*, vol. 11, no. 1 (Fall 1999), pp. 58-61.

3. K. Robins, 'What in the World's Going On?' in P. du Gay (ed.), *Production of Culture/Cultures of Production,* London, Thousand Oaks, and New Delhi, SAGE Publications, 1997.

4. A. Gupta and J. Ferguson, 'Beyond Culture: Space, Identity, and the Politics of Difference', *Cultural Anthropology*, February 1992, pp. 6-23.

5. See, for example, F. Fukuyama, *The End of History and the Last Man*, New York, Free Press, 1992; R. O'Brien, *Global Financial Integration: The End of Geography*, New York, Council on Foreign Relations Press, 1992; and K. Ohmae, *The End of the Nation State*, Free Press, New York, 1995.

6. R. Williams, *Keywords*, New York, Oxford University Press, 1976.

7. The *Oxford English Dictionary* (*OED*), 2nd ed., CD-ROM, New York, Oxford University Press, 1992, also explains that 'manufacture' first appeared in French, beginning in the sixteenth century. From the seventeenth century onward, however, 'making' was understood to be accomplished either by physical labour (humans) or mechanical power (machines), or both. During Adam Smith's time manufacture was also a branch of productive activity. The *OED* explains how Smith wrote that 'By means of trade and manufactures, a greater quantity of subsistence can be annually imported.' By the nineteenth century, therefore, people had already come to perceive the term in a deprecatory sense as a type of production requiring 'mere mechanical labour', as opposed to work of the intellect – and also as production destined for the market. What is lacking in all of these definitions, of course, is reference to the consumption of ideas and non-material products such as culture, history and propaganda.

8. The *OED* further explains that although 'consumption' is of Latin origin (*consumption-em*), it comes most immediately from the French *consumption*, an early variation of *consomption*. The latter has since been ousted in French by the word *consommation*, which means 'the action or fact of consuming or destroying'. From the thirteenth century, the word was used to refer to the dissipation of moisture by evaporation. Other definitions imply decay, wasting away, and wearing out, specifically of the body by disease. As early as 1535, the word was used to describe the using up of material or food, and, a century later, the products of industry. By the seventeenth century, the word had come to be understood as meaning wasteful expenditure. Present use picks up nuances from many of the above definitions.

9. According to the *OED*, 'heritage' derives from the Old French word *eritage* or *heritage*, meaning any property, especially land, which devolves by right of inheritance. This form of devolution is distinguished from property or realty transfers made by conquest or purchase. This definition was used from the thirteenth century onward. However, beginning in the early seventeenth century, the word began to connote 'that which comes from circumstances of birth; the condition or state transmitted from ancestors'. While these meanings are still implied in the use of the term today, the notion of heritage has been expanded to refer to both the material and non-material world that is passed from ancestors to their direct descendants through indirect means.

10. According to the *OED*, the old French *tradicion* had by the fifteenth and sixteenth centuries come to mean handing down, as in 'a saying handed down' or the handing over of a material object. It was also understood as the giving up or surrender of something, or even the oral delivery of information. By the seventeenth century, 'tradition' had been more substantially defined as 'the action of transmitting or handing down from one to another a variety of beliefs, rules, and customs'. But by the nineteenth century, tradition had again been more vaguely accepted as 'a long established and generally accepted custom or method of procedure, having almost the force of law'. Tradition became the corpus of experiences handed down by predecessors and widely observed.

11. Ohmae, *The End of the Nation State*, 1995.

12. J. Urry, *The Tourist Gaze*, Sage, London, 1990. Another classic text in this context is D. MacCannell, *The Tourist: A New Theory of the Leisure Class*, New York, Schocken Books, 1976.

13. In a recent workshop, Phil Gruen slipped when speaking of 'engagement' with the built environment, and instead said 'engazement'. I have appropriated this accidentally coined term here, because it so aptly describes the engagement of tourists with the built environment under conditions of the gaze.

14. Of course, the First and Third Worlds are not homogeneous entities. The exercise of polarizing them into dualistic categories helps only in fleshing out fundamental differences between the attitudes of the former colonizers and the formerly colonized.

15. B. Barber, *Jihad Vs. McWorld*, New York, Ballantine Books, 1995.

16. There is a significant anthropological literature on Bali by major scholars. See, for example, C. Geertz, *Person, Time and Conduct in Bali: An Essay in Cultural Analysis*, New Haven, CN, Yale University, Southeast Asia Studies, Cultural Report Series, 1966. However, I only draw here upon personal observations as a result of a recent trip.

17. For more on this, see M. Sorkin, 'See You in Disneyland', in Sorkin (ed.), *Variations on a Theme Park*, New York, Hill and Wang, 1992.

18. D. Howes, 'Introduction: Commodities and Cultural Borders', in Howes (ed.), *Cross-Cultural Consumption: Global Markets, Local Realities*, London and New York, Routledge, 1996, p. 12. The chapter referred to is C. Hendrickson, 'Selling Guatemala: Maya Export Products in US Mail-Order Catalogues', pp. 106-124.

19. D. Howes, 'Introduction: Commodities and Cultural Borders', p. 13. The chapter referred to is M. Crain, 'Negotiating Identities in Quito's Cultural

Borderlands: Native Women's Performances for the Ecuadorean Tourist Market', pp. 125-137.

20. N. AlSayyad, 'Urbanism and the Dominance Equation: Reflections on Colonialsm and National Identity', in AlSayyad (ed.), *Forms of Dominance: On the Architecture and Urbanism of the Colonial Enterprise*, Aldershot, Avebury, 1992.

21. A. King, *Culture, Globalization, and the World System*, Houndmills, MacMillan, 1991.

22. A. King, *The Bungalow: Production of a Global Culture*, New York, Oxford University Press, 1995.

23. I. Bierman, unpublished paper submitted to the symposium 'An Authentic City for a Modern World: Cairo in the Nineteenth Century', held at the University of California, Berkeley, in October 1999.

24. N. AlSayyad, 'Urbanism and the Dominance Equation'; 'From Vernacularism to Globalism: The Temporal Reality of Traditional Settlements', *Traditional Dwellings and Settlements Review*, vol. 7, no. 1 (Fall 1995), pp. 13-24; and 'Culture, Identity and Urbanism in a Changing World: A Historical Perspective on Colonialism, Nationalism and Globalization', in M.L. Cohen *et al.* (eds.), *Preparing for the Urban Future: Global Pressures and Local Forces*, Washington, D.C., Woodrow Wilson Center Press, 1996.

25. M. Castells *et al.*, 'Public Housing as Political Strategy', in *The Shep Kip Mei Syndrome: Economic Development and Public Housing in Hong Kong and Singapore*, New York, Pion, 1990, pp. 303-322.

26. L. Vale, *Architecture, Power and National Identity*, New Haven, CN, Yale University Press, 1992.

27. G. Wright, *The Politics of Design in French Colonial Urbanism*, Chicago, University of Chicago Press, 1991.

28. Such a view was first expressed by Louis Snyder in 1954. See L. Snyder, *The Meaning of Nationalism*, Westport, CN, Greenwood Press, 1954.

29. Robins, 'What in the World's Going On?'

30. A. Giddens, *The Consequences of Modernity*, Palo Alto, CA, Stanford University Press, 1990.

31. R. Robertson, 'Social Theory, Cultural Relativity, and the Problem of Globality', in King (ed.), *Culture, Globalisation and the World-System*.

32. There are, of course, many other classifications produced by others that may be helpful in this regard. See, for example, S.S. Fainstein and D.R. Judd, 'Cities as Places to Play', in Judd and Fainstein (eds.), *The Tourist City*, New Haven and London: Yale University Press, 1999. In this chapter, Fainstein and Judd sort out three basic types of tourist cities: resort cities, which are places created expressly for consumption by visitors; tourist-historic cities, which have been sites of tourism for a long time and have transformed themselves into tourist sites through conscious promotion and reconstruction of heritage; and converted cities, which have built an infrastructure for the purpose of attracting visitors, but where the tourist space is insulated from the larger urban milieu within a process of uneven development (see pp. 262-267).

33. Gupta and Ferguson, 'Beyond Culture'.

34. B. Holcomb, 'Marketing Cities for Tourism', in Judd and Fainstein (eds.), *The Tourist City*, p. 65.

35. E. Gable and R. Handler, 'In Colonial Williamsburg, the New History Meets the Old', *Chronicle of Higher Education*, October 30, 1998, pp. B10-B12.

36. *Ibid.*

37. Gupta and Ferguson, 'Beyond Culture', pp. 6-23.

38. D.D. Hall, 'Community in the New Urbanism: Design Vision and Symbolic Crusade', *Traditional Dwellings and Settlements Review*, vol. 9, no. 2 (Spring 1998), pp. 23-36.

39. W. Hoge, 'In Stone, a Prince's Vision of Britain', *The New York Times*, June 11, 1998, pp. B1, B6.

40. N. AlSayyad, 'From Vernacularism to Globalism', pp. 13-24.

41. M. Pollan, 'Town Building is no Mickey Mouse Operation', *New York Times Magazine*, December 14, 1997, p. 62. I have relied heavily on this article for the section on Celebration. Other viewpoints are provided by D. Frantz and C. Collins, *Celebration USA: Living in Disney's Brave New Town*, New York, Henry Holt and Co., 1999; and A. Ross, *The Celebration Chronicles: Life, Liberty and the Pursuit of Property Values in Disney New Town*, New York, Ballantine Books, 1999.

42. Pollan, 'Town Building is no Mickey Mouse Operation', p. 59.

43. Hall, 'Community in the New Urbanism'.

44. Pollan, 'Town Building is no Mickey Mouse Operation', p. 62.

45. *Ibid.*, p. 80.

46. *Ibid.*, p. 88.

47. *Ibid.*, cover page.

48. Y.-F. Tuan, 'Traditional: What Does it Mean?' in N. AlSayyad and J.-P. Bourdier (eds.), *Dwellings Settlements and Tradition: Cross-Cultural Perspectives*, Lanham, MD, University Press of America, 1989.

49. E. Hobsbawm, 'Introduction: Inventing Traditions', in Hobsbawm and R. Ranger (eds.), *The Invention of Tradition*, London, Cambridge University Press, 1983.

50. B. Anderson, *Imagined Communities*, London, Verso, 1983.

51. Gupta and Ferguson, 'Beyond Culture', pp. 6-23.

52. 'International Conference: Heritage & Tourism', Hong Kong, December 13-15, 1999, sponsored by Antiquities Advisory Board, Lord Wilson Heritage Trust, and Antiquities and the Monuments Office of the Home Affairs Bureau.

53. Robins, 'What in the World's Going On?' p. 33.

54. Gupta and Ferguson, 'Beyond Culture', pp. 6-23.

55. Robins, 'What in the World's Going On?' p. 12.

56. *Ibid.*, p. 38.

57. Holcomb, 'Marketing Cities for Tourism', p. 65.

58. Given such conditions, heritage managers – whether nations, cities, agencies or individuals – carry a responsibility that must be separated both from the possibilities of the market and the desires of national or municipal governments. While this may be unrealistic, an appropriate course of heritage administration for the twenty-first century could involve management by non-profit, non-governmental organizations with philanthropic funding, supervised by local municipalities or regions.

59. H. Fathy, *Architecture for the Poor*, Chicago, Chicago University Press, 1971.

60. J.V. Iovine, 'A Tale of Two Main Streets', *The New York Times*, October 15, 1998, pp. B1, B12.

61. As presented by a guide provided by Disney as part of a guided tour organized for ACSA Conference participants in April 1993.

62. Biermen, unpublished paper.

63. D. Howes and C. Classen, 'Epilogue: The Dynamics and Ethics of Cross-Cultural Consumption', in Howes (ed.), *Cross-Cultural Consumption*, pp. 188-189.

64. The conference, for which I served as a director, was called 'Manufacturing Heritage and Consuming Tradition: Development, Preservation and Tourism in the Age of Globalization'. It was the sixth international conference of the International Association for the Study of Traditional Environments (IASTE), held December 15-19, 1998. Many of the articles in this book were originally presented there.

65. J. Baudrillard, *Selected Writings*, ed., M. Poster, Palo Alto, CA, Stanford University Press, 1988, p. 166.

Bibliography

AlSayyad, Nezar, 'Culture Identity and Urbanism in a Changing World: A Historical Perspective on Colonialism, Nationalism and Globalization', in Michael L. Cohen *et al.* (eds.), *Preparing for the Urban Future: Global Pressures and Local Forces*, Washington, D.C., Woodrow Wilson Center Press, 1996.

AlSayyad, Nezar, 'From Vernacularism to Globalism: The Temporal Reality of Traditional Settlements', *Traditional Dwellings and Settlements Review*, vol. 7, no. 1 (Fall 1995), pp. 13-24.

AlSayyad, Nezar, 'Urbanism and the Dominance Equation: Reflections on Colonialsm and National Identity', in AlSayyad (ed.), *Forms of Dominance: On the Architecture and Urbanism of the Colonial Enterprise*, Aldershot, Avebury, 1992.

Anderson, Benedict, *Imagined Communities*, London, Verso, 1983.

Barber, Benjamin, *Jihad Vs. McWorld*, New York, Ballantine Books, 1995.

Baudrillard, Jean, *Selected Writings*, ed., M. Poster, Palo Alto, CA, Stanford University Press, 1988.

Bierman, Irene, unpublished paper submitted to the symposium 'An Authentic City for a Modern World: Cairo in the Nineteenth Century', held at the University of California, Berkeley, in October 1999.

Castells, Manuel *et al.*, 'Public Housing as Political Strategy', in *The Shep Kip Mei Syndrome: Economic Development and Public Housing in Hong Kong and Singapore*, New York, Pion, 1990.

Crain, Mary M., 'Negotiating Identities in Quito's Cultural Borderlands: Native Women's Performances for the Ecuadorean Tourist Market', in David Howes, (ed.), *Cross Cultural Consumption: Global Markets, Local Realities*, London and New York, Routledge, 1996.

Crossette, Barbara, 'Surprises in the Global Tourism Boom', *The New York Times*, April 12, 1998, p. wk5

Fainstein, Susan S. and Dennis R. Judd, 'Cities as Places to Play', in Judd and Fainstein (eds.), *The Tourist City*, New Haven and London: Yale University Press, 1999.

Fathy, Hassan, *Architecture for the Poor*, Chicago, Chicago University Press, 1971.

Frantz, Douglas and Catherine Collins, *Celebration USA: Living in Disney's Brave New Town*, New York, Henry Holt and Co., 1999.

Fukuyama, Frances, *The End of History and the Last Man*, New York, Free Press, 1992.

Gable, Eric and Richard Handler, 'In Colonial Williamsburg, the New History Meets the Old', *Chronicle of Higher Education*, October 30, 1998, pp. B10-B12.

Geertz, Clifford, *Person, Time and Conduct in Bali: An Essay in Cultural Analysis*, New Haven: Yale University, Southeast Asia Studies, Cultural Report Series, 1966.

Giddens, Anthony, *The Consequences of Modernity*, Palo Alto, CA, Stanford University Press, 1990.

Gupta, Akhil and James Ferguson, 'Beyond Culture: Space, Identity, and the Politics of Difference', *Cultural Anthropology*, February 1992, pp. 6-23.

Hall, Denise D., 'Community in the New Urbanism: Design Vision and Symbolic Crusade', *Traditional Dwellings and Settlements Review*, vol. 9, no. 2 (Spring 1998), pp. 23-36.

Hendrickson, Carol, 'Selling Guatemala: Maya Export Products in US Mail-Order Catalogues', in David Howes, (ed.), *Cross-Cultural Consumption: Global Markets, Local Realities*, London and New York, Routledge, 1996.

Hobsbawm, Eric, 'Introduction: Inventing Traditions', in Hobsbawm and R. Ranger (eds.), *The Invention of Tradition*, London, Cambridge University Press, 1983.

Hoge, Warren, 'In Stone, a Prince's Vision of Britain', *The New York Times*, June 11, 1998, pp. B1, B6.

Holcomb, Briavel, 'Marketing Cities for Tourism', in Dennis R. Judd and Susan S. Fainstein (eds.), *The Tourist City*, New Haven and London: Yale University Press, 1999.

Howes, David and Constance Classen, 'Epilogue: The Dynamics and Ethics of Cross-Cultural Consumption', in Howes, (ed.), *Cross-Cultural Consumption: Global Markets, Local Realities*, London and New York, Routledge, 1996.

Howes, David, 'Introduction: Commodities and Cultural Borders', in Howes, (ed.), *Cross-Cultural Consumption: Global Markets, Local Realities*, London and New York, Routledge, 1996.

Iovine, Julie V., 'A Tale of Two Main Streets', *The New York Times*, October 15, 1998, pp. B1, B12.

King, Anthony D., *Urbanism, Colonialism, and the World Economy: Cultural and Spatial Foundations of the World Urban System*, London, New York, Routledge, 1990.

King, Anthony D., 'Introduction: Spaces of Culture, Spaces of Knowledge', in King (ed.), *Culture, Globalization, and the World System*, Houndmills, Macmillan, 1991.

King, Anthony, *Culture, Globalisation and the World-System*, London, MacMillan, 1991.

King, Anthony, *The Bungalow: Production of a Global Culture*, New York, Oxford University Press, 1995.

MacCannell, Dean, *The Tourist: A New Theory of the Leisure Class*, New York, Schocken Books, 1976.

O'Brien, Richard, *Global Financial Integration: The End of Geography*, New York, Council on Foreign Relations Press, 1992.

Ohmae, Kenichi, *The End of the Nation State*, Free Press, New York, 1995.

Oxford English Dictionary, 2nd ed., CD-ROM, New York, Oxford University Press, 1992.

Pollan, Michael, 'Town Building is no Mickey Mouse Operation', *New York Times Magazine*, December 14, 1997.

Roaf, Susan, review of J. Jokilehto, *A History of Architectural Conservation*, Oxford, Butterworth Heinemann, 1999, in *Traditional Dwellings and Settlements Review*, vol. 11, no. 1 (Fall 1999), pp. 58-61.

Robertson, R., 'Social Theory, Cultural Relativity, and the Problem of Globality', in Anthony King (ed.), *Culture, Globalisation and the World-System*, London, MacMillan, 1991.

Robins, Kevin, 'What in the World's Going On?' in Paul du Gay (ed.), *Production of Culture/Cultures of Production*, London, Thousand Oaks, and New Delhi, SAGE Publications, 1997.

Ross, Andrew, *The Celebration Chronicles: Life, Liberty, and the Pursuit of Property Values in Disney New Town*, New York, Ballantine Books, 1999.

Snyder, Louis L., *The Meaning of Nationalism*, Westport, CN, Greenwood Press, 1954.

Sorkin, Michael 'See you in Disneyland', in Sorkin (ed.), *Variations on a Theme Park*, New York, Hill and Wang, 1992.

Tuan, Yi-Fu, 'Traditional: What Does it Mean?' in Nezar AlSayyad and Jean-Paul Bourdier (eds.), *Dwellings Settlements and Tradition: Cross-Cultural Perspectives*, Lanham, MD, University Press of America, 1989.

Urry, John, *The Tourist Gaze*, London, Sage, 1990.

Vale, Lawrence, *Architecture, Power and National Identity*, New Haven, CN, Yale University Press, 1992.

Williams, Raymond, *Keywords*, New York, Oxford University Press, 1976.

Wright, Gwendolyn, *The Politics of Design in French Colonial Urbanism*, Chicago, University of Chicago Press, 1991.

Chapter 2

Tourism Encounters: Inter- and Intra-Cultural Conflicts and the World's Largest Industry

MIKE ROBINSON

Tourism has emerged as a significant international economic activity. Estimates by the World Tourism Organization suggest that in 1996 some 592 million international trips were made, with forecasts that by the year 2020 the number will almost have trebled to some 1.6 billion international trips.[1] Though it is difficult to speak in terms of global figures due to definitional and classification problems, tourism and tourism-related businesses continue to increase in absolute number and relative importance in both developed and developing economies.

As with any economic endeavour, tourism is a force for social, cultural and environmental change. The development and patterns of consumption that accompany world tourism contribute to physical changes in the natural and created environment and also in the cultural meanings attached to spaces and places. The role of international tourism in terms of economic development is well understood, and this is reflected in the academic literature. In addition, a substantive literature has evolved looking at the environmental implications of tourism. However, until relatively recently tourism as a cultural phenomenon and the dynamics of its relationships with place and host communities have remained largely unexplored. While there is a need for empirical research and detailed case studies regarding the cultural implications of international tourism, there is also a need for theoretical frameworks which emphasize the links between the political economy of tourism, its social and environmental dimensions, and its place in cultures and cultural change. This chapter explores the growth of international tourism, its formidable role as a vector of cultural exchange, and the inter- and intra-cultural encounters it produces and directs.[2] The chapter also discusses the importance of the built environment as the cultural space and place in such encounters.

Hannerz's description of the world as a network of social relationships between which there is a 'flow of meanings as well as of people and goods', indicates how the magnitude of world tourism, its apparent unbounded geographical reach and global inevitability, now provides significant opportunities for both positive and negative cultural encounter.[3] Within the context of so-called 'world' tourism, cultures are traded as physical and cultural frontiers are penetrated. This process takes place visibly, sometimes willingly, sometimes reluctantly, and usually irreversibly. As Rojek and Urry have emphasized, 'all cultures get remade as a result of flows of peoples, objects and images across national borders, whether these involve colonialism, work-based migration, individual travel or mass tourism.'[4] In reality, it is often difficult to disentangle the forces of tourism from those of other globalizing influences, but the premise here is that tourism has become an increasingly significant driver of cultural remaking and reinvention. Within these undeniably complex processes, the built environment often plays an intimate and symbolic role, providing the arena for various levels of interaction between tourists and the host community, and, more importantly, between the tourism industry and host community.

World Tourism: Myths and Realities

In recent decades tourism has emerged as such a highly structured and organized form of human activity that it is now referred to as an 'industry'. In reality, it is a collection of different industries, drawn mainly, but not exclusively, from the service sector. These are brought together both formally and informally to supply and service the needs of society to travel for leisure purposes in four broad areas: attractions, accommodation, transport and distribution.[5] The fragmentation of the tourism industry, involving millions of individual businesses, together with governments, public and voluntary agencies as regulators, owners and shareholders, makes it difficult to speak of tourism as a single industrial sector.[6] However, for the purposes of this chapter I shall adopt the term 'industry' essentially to differentiate the effects of tourism from those of tourists.

Despite dealing with the complexities of social motivations and the caprice of human emotions in juxtapositions of work and leisure contexts, the worldwide tourism industry, with minor exceptions, displays a common set of business characteristics: it is driven by the search for profit; it employs people; it is managed in the functional areas of marketing, finance, personnel, etc.; and it is subject to similar economic, political and environmental externalities.[7] The industry is further distinguished by marked polarization between a relatively small number of dominant and powerful multi-national players, particularly in the airline and hotel sectors, and a vast number of small- to medium-sized businesses, often owner-managed.

In addition, tourism developments have increasingly been fed with capital from non-tourism corporations seeking to diversify. Thus, when speaking in terms of cultural conflicts in tourism, one must not only speak of confrontations between individuals, but also conflicts between systems and structures in the context of the processes and organization of capitalism, and differences existing between capitalist and precapitalist economies.[8]

Aside from the complex structural and functional characteristics of world tourism, three further features are outlined here. First, there are relatively few nations and cultures which are not effected in some way by tourism and the tourism development process. Organized activities such as 'alternative' tours to the Arctic and Antarctic and a myriad of other ecotourism and nature tourism ventures to remote corners of the globe exemplify not only a growing public awareness of the planet, but also seem to indicate increasingly sophisticated patterns of First World consumption.[9] However, as Weaver has suggested, 'ecotourism' may indeed be no more than another form of mass tourism, involving, in the main, passive observers and consumers rather than saviours of environments and cultures.[10] The key points are that the touch of tourism does not need to be 'heavy' or 'mass' to produce conflict in distant and often very fragile destinations, and that the growth of ecotourism markets is underpinned by significant capital. The so-called 'new tourist', seeking more exotic experiences and cultural encounters, is also armed with the financial ability to buy into whatever cultural experiences are available.[11]

Second, where tourism has emerged as an important economic activity, it is frequently characterized by a rapid and often dramatic expansion in supply.[12] Growth in accommodation, transport networks, service infrastructure, and leisure space has transformed natural and built landscapes, moulding them to fit new economic imperatives, usually endorsed (implicitly, if not explicitly) by the state. In the city-state of Singapore, for example, much recent urban development has focused specifically on leisure zones and tourism themes.[13] As one would expect, research has tended to focus upon the economic impacts of tourism development, but accompanying these physical transformations are socio-cultural consequences for host communities which can be equally as dramatic, though not immediate, visible, nor easily measurable.

A third feature of world tourism is that by and large it is a First World ideology, and, as such, it displays fundamental inequalities in the patterns and impacts it demonstrates. Though fraught with problems of measurement and interpretation, such inequalities are borne out by World Tourism Organization (WTO) statistics (figures 2.1, 2.2). Figure 2.1 illustrates the regional percentage share of world tourism arrivals over a 22-year period and how there has been growth in apparently 'exotic' destinations (though within the geographical groupings considerable variations do occur). Of particular note is the three-fold growth of arrivals in East Asia/Pacific countries. In part this reflects vigorous price competition in the long-haul market, but it also indicates qualitative changes in the market, as tourists have widened their horizons and sought the exotic and the different.

Figure 2.2 shows the change in percentage distribution of tourism receipts over the same 22-year period. It reveals how some growth has occurred in receipts by 'developing countries', but how this amount is still less than half the level of receipts by the 'industrialized countries'. The

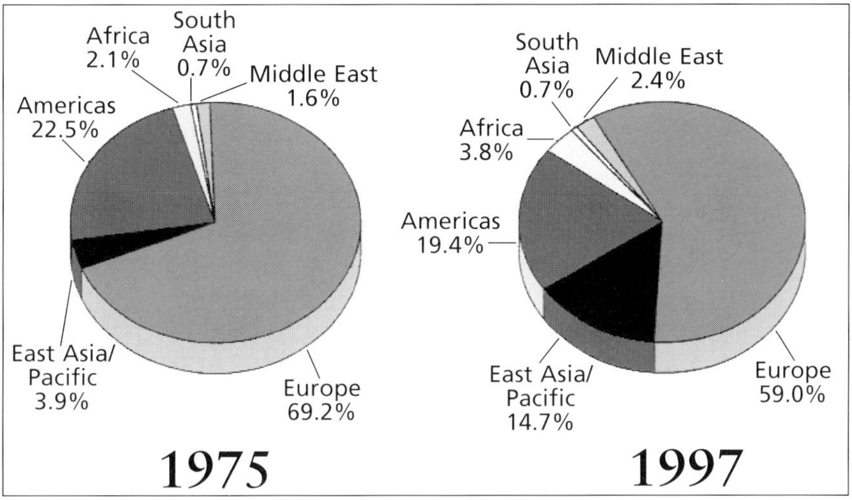

Figure 2.1 Per cent share of World Arrivals – 1975 and 1997. (*Source*: World Tourism Organization (WTO))

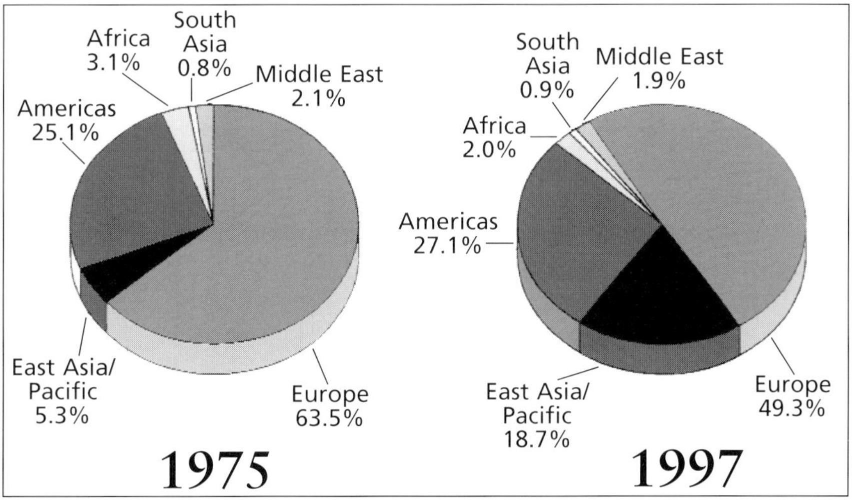

Figure 2.2 Per cent share of world receipts – 1975 and 1997. (*Source*: World Tourism Organization (WTO))

WTO classification of which nations are included or excluded as developing countries is not clear, but figures by Cazes have shown that the low-income nations of the world receive less than 2 per cent of global tourism revenue – a figure which drops to 0.6 per cent if China is excluded.[14] Survival International has estimated that in Thailand, for instance, some 60 per cent of the $40 billion generated through tourism leaves the country.[15] Leaving aside statistics, the essential point is that basic imbalances exist between developed, developing, and lesser-developed nations not only in terms of the spatial distribution of tourist activity, but in terms of the economic benefits such activity generates.[16] Moreover, by far the majority of the world's population do not engage in leisure tourism as participants, nor are they familiar with the social construct of tourism. The culture of tourism remains rooted in the First World as part of the wider consumptive ideologies that developed nations have adopted.[17] The gulf which exists between tourists and non-tourists, the leisured and the working populations, the consumers and the consumed is arguably self-perpetuating.[18] Tourists, by virtue of their ability to 'gaze', effectively reaffirm the cultural dominance of consumption and its capitalist framework. Indeed, one can cynically argue that inequalities – the very presence of poverty, underdevelopment, and the perceived threat of environmental degradation – can add to the tourist experience.

Such fundamental inequalities in world tourism form a backdrop to any discussion of cultural conflict and reveal themselves in the differing ways spaces and places are perceived, used and created.

Tourism and Cultural Conflicts

Cultural conflicts in tourism can be understood on a range of interdependent levels: between individual tourists and representatives of the host culture; between, and within, host cultures themselves; and between the tourism industry as part of the development process and host communities/cultures. Conflicts may manifest themselves in a variety of different forms, ranging from a rather intangible sense of disgruntlement and embarrassment on the part of the host to (in extreme cases) violence against tourists and the component elements of the industry as symbols of external influence and cultural change.

It is tempting to view conflict solely as arising from 'face-to-face' communicative encounters, between tourist and host as they cross over into each other's cultural contexts.[19] Although such encounters can clearly produce benefits such as improved cultural understanding, the erosion of prejudice, and the generation of appreciation and tolerance, they can also result in a variety of negative socio-cultural impacts in line with Gessner and Schade's view that conflict is usually the focus of intercultural communication.[20] However, the interactions, which do take place between individual tourists and hosts, are not straightforward. Nettekoven has noted that within the context of the majority of holiday experiences relatively few tourists actively seek intimate cultural encounters with the host community.[21] Moreover, those encounters which do take place are usually short lived and *ad hoc*, occurring in the context of a limited vacation period and within the restricted space of established resorts. Interactions thus tend to be between tourists and hotel employees, restaurant staff, local shopkeepers, and local tour guides, as both parties respectively conform to established power relations as the consumers of leisure and the providers and servicers of the leisure experience.

Nettekoven has also made the important point that for developing nations, interaction with tourists is seldom the most important driver for cultural change, because contact and acculturation is relatively limited. Given that local tour guides, shopkeepers, and hotel staff will probably already partly share, or at least be in touch with, the value systems of tourists, the extent to which cultural patterns are changed is likely to be limited. However, negative effects of acculturation via the 'demonstration effect', including deviant behaviour to support the imitation of touristic lifestyles, does occur, reflecting the fact that although direct tourist-host encounters may be limited, indirect encounters are far greater and arguably more pervasive.[22] It is difficult to pinpoint cultural conflict to individual tourist-host encounters in a specific cause-and-effect way. Despite possible 'culture shock' and opportunities for misunderstanding between tourist and host, conflict in the form of aggressive behaviour by either party is unlikely to manifest itself in any immediate sense.[23] What is important is the effect of these contact situations in toto over a prolonged period of time.

Tourists (and the industry which supports them) buy into a dynamic of cultural relations, although the degree of exposure and contact between tourists, tourism businesses, and host cultures, clearly varies. At times such internal cultural relations may provide a harmonious context for tourism; yet at others tourism can find itself entangled in a variety of ethnic and cultural clashes within host communities that exist independently of the tourism activity.[24] Such historically conceived clashes invariably politicize culture in order to articulate economic, social and environmental claims. Clearly, when violence erupts, touristic activity will be suspended, and there is unlikely to be much in the way of deliberate contact between tourist and host community. However, tourists and the tourism industry have also been used as targets in conflict situations.[25] And ironically, and often perversely, the physical remnants of such conflict have become absorbed into tourism. Thus, sites of previous cultural and political struggle have drawn their own tourist gaze.[26] For instance, the demilitarized zone of the 38th Parallel, which marks the boundary between North and South Korea, is now a tourist attraction (at least for those visiting South Korea). As was the case with the Berlin Wall, effectively there is nothing much to see, but what tourists seem to feed from is the apprehension of conflict and the emotional responses brought out by the tangible recognition of difference. The conferring of 'heritage' status, commodification, and the marketing of such symbols involves an inherent selectivity, which promotes certain value systems over others and can result in the 'disinheritance' of non-participatory, marginalized groups.

Tourists do not miraculously materialize in a destination. Vacation type and destination choice are influenced by a complex range of 'pushing' factors, and they are also subject to a similarly complex range of 'pulls'. Despite receiving relatively less attention in the literature, pulling factors are significant. Indeed, though tourists may wish to believe otherwise, they are arguably sold holidays to a far greater extent than they buy them. This selling process is undertaken and controlled by the tourism industry and its various components acting individually and collectively. The industry is constantly developing and refining products to sell to the full spectrum of the market, whether this be the still-dominant mass market or emerging niche markets for activity-based tourism or ecotourism. This process locates tourism as an important player in the development process.

For developing economies whose natural resource base is depleted, tourism would appear to provide a rather rapid way of generating hard currency and creating employment. Indeed, utilizing the cultural and ethnic resources of a nation or region for tourism may be the only way to stimulate the economy. Although redolent of the economics of desperation, this course of action holds tremendous appeal. As Smith notes, tourism is seen as a way of achieving political recognition in a competitive world.[27] Compared to the development of manufacturing industry (via state investment, attracting inward investment, or both) it would appear to have multiple benefits: the establishment of tourism infrastructure can be undertaken reasonably quickly; it can serve the wider needs of the population (for example, airport and runway expansion); the environmental costs are perceived to be low; and the core products which lie behind tourism development – attractive natural environments and culture – are assumed to be infinite. For developed economies, too, seeking to restructure and readjust from a manufacturing to a service-sector base, tourism and leisure continues to play a leading role for similar reasons.

However, the development of tourism can generate inter-cultural conflicts broadly centred upon the competition for environmental resources, the commodification of culture, and the extent to which host cultures find themselves economically dependent upon tourism.

Resource Competition and Appropriation

An obvious source of inter- and intra-cultural conflict revolves around competition for physical resources (the natural and built environment) and the ownership and rights of access to these. While at the micro-scale tourists often compete with a host community for access to resources (often unknowingly), the tourists themselves, as transients, have little influence on issues of ownership, planning and management. It is the tourism industry that has the capacity and the power to make major changes to the physical environment.

The tourism industry has long been recognized as a voracious consumer of basic environmental resources. Impacts have ranged from forest clearance or wetland drainage for airport runway extensions and golf courses, to transformations of existing urban and rural spaces into tourist attractions and hotel complexes. Of course, not all such transformations are negative; benefits can accrue for host communities from planning gain, new functions found for old buildings, and new facilities introduced for residents. Negative environmental impacts can also be minimized through anticipatory planning and effective management. This is essentially the *raison d'être* behind the concept of sustainable development – a merging of economic and environmental imperatives. However, to view the utilization of environmental resources by the tourism industry as solely a physical process is to miss the 'intimate interdependencies' which exist with local culture.[28] Various writers have commented on these cultural interdependencies in terms of the culturally constructed natural environment.[29] However, close relationships also exist between host cultures and the built environment too, relationships which the tourism industry can influence.

The international tourism industry, with residual imperialistic flair, is well aware of its power and influence to the extent it remains comfortable in making promises of paradise to the prospective tourist. As Greenwood has written, 'for the moneyed tourist, the tourism industry promises that the world is his/hers to use. All the "natural resources", including cultural traditions, have their price, and if you have money in your hand, it is your right to see whatever you wish.'[30]

The appropriation of natural and cultural resources by the tourism industry in actual and symbolic terms can initiate conflicts between tourist and host, between different cultural groups within a destination, or between the governed and the governing, each not only with different claims to resources, but different interpretations of how they should be used. A good example of how the cultural dimensions of tourism resources can be misunderstood or ignored is the claims made upon Australian Aboriginal culture by the tourism industry.[31] For example, Uluru, as one of the most visited natural attractions in Australia, is also a sacred Aboriginal site, and local Aboriginal communities have requested that tourists not climb the rock. They have attempted to explain this in the Uluru-Kata Tjuta National Park Information Sheet, which reads: 'That's a really important sacred thing that you are climbing . . . You shouldn't climb . . . maybe that makes you a bit sad. But any way, that's what we have to say . . .'[32]

However, the Northern Territory Tourist Brochure offers a different view on the resource of 'Ayer's Rock', advising tourists that if they climb they need to 'take it easy. You're on holiday after all.' However, '. . . once you've negotiated the undulations up on the top . . . there's a wonderful sense of achievement.' The brochure also points out that 'Local Aborigines can't see the logic in climbing the Rock. But when you've come this far it seems the thing to do.'[33]

Such conflict over the utilization of culturally imbued resources has both a moral and legal dimension; ultimately, it represents incommensurable worldviews and, more importantly, the pervasiveness and power of the tourism industry.[34] But the inability of the majority of the tourism industry to identify environmental and cultural resources as something more than merely manageable, tradable products conceived within First World capitalist paradigms is a problem. In terms of addressing environmental issues surrounding resource usage, the tourism industry, driven by self-interest, is increasingly cognizant that with market adjustments and some element of government intervention problems can be managed, and associated cultural conflicts ameliorated. But this reductionist, managerial approach fails to take into account the cultural dimension of the resources, which are at the base of the tourism industry. Even the promise of more sustainable forms of tourism emerging out of a sustainable development framework is an inadequate response. Apart from recognizing the 'value' of indigenous peoples, the discourse of sustainable development actually says very little on culture and the ways it can shape relationships with the environment, and ultimately tourism.[35]

Commodification of Cultures as the Norm

The tourism industry largely conceives of culture(s) in two ways: either as value free, and thus largely an inconsequential aspect of development; or/and as just another product to be packaged. Tourism development in a First World context often progresses in a climate of acceptance or apathy. But this assumption can be, and is, carried into non-Westernized, developing societies with neither tradition nor need for outwardly directed cultural exhibitionism. Entire cultures can be 'showcased' for economic purposes by the tourism industry or for the tourism industry, and such arrangements may be legitimized by the state.[36] As Morris has pointed out:

> Toured communities are increasingly required to live out their manufactured ethnicity for the gaze of the other, with the result that the destruction of some traditions and their replacement by others is required by the state, and then negotiated in various ways by those whose bodies and practices are thus required (but do not necessarily directly consent) to incarnate policy.[37]

Moreover, a deep-seated imperialistic assumption persists that large areas of the world exist solely for the benefit of tourists. In line with Turner and Ash's idea of the 'pleasure periphery', King has noted how the 'paradise' of the Pacific islands has long been viewed as the backyard of Australians, just as the Caribbean has been regarded as a playground for North Americans.[38]

At the heart of such assumptions is a commodification process whereby traditions, rituals, and 'ways of life' are packaged, imaged and transformed into saleable products for tourists.[39] Culture(s), as a living and learning form, together with the idea of culture and its shared meanings, become a superficial subjugate of consumerism and lose their social role, social and political function, and authenticity.[40] However, commodification, in itself, need not generate conflict if it carries the consent of the host culture and the latter can reap the benefits of acceptable commercialization.[41] Indeed, while remaining contentious, the presentation of cultural artefacts and cultural history can be identity affirming, cathartic, and liberating for cultures seeking to explain their traditions and values.[42] The key issue relates to the ability of local cultures to decide for themselves what aspects of culture should be displayed and how they should be presented.

However, conflict may be induced when the commodification process results in trivialization of ethnic groups and their cultural practices and traditions, or when it is controlled by agencies with little insight or understanding of the meanings and historicity of such practices. In non-Westernized developing societies with neither the tradition nor the need for outwardly directed cultural exhibitionism, there has been a long history of such commodification to serve the recreational desires and economic purposes of the developed world.[43] Tourists continue to be offered what MacCannell has termed 'reconstructed ethnicity'.[44] Religious rites, festivals, and ethnic traditions are often reduced and shaped to meet tourist expectations to the point where the host culture loses the deeper meanings and social function of such practices; or to where tensions develop within the community, as some seek preservation of cultural practices while others are happy to provide what the tourists want.[45]

The precise role of the tourism industry in the commodification process is under-researched in terms of the exact points at which decisions are made, but it is clear that such decision-making begins well away from original cultural sites, emphasizing the trans-national and trans-cultural character of tourism, and indicating it is no longer necessary to travel to effect cultural change. Such commodification of distant and complex cultural features often begins with the highly selective and sanitized combinations of words and images within brochures. As Dann has noted, the brochure is recognized as an essential and highly visible aspect of the commodification process.[46] Through glossy photographs and creative prose, unique cultures are effectively reduced and reassembled to appeal to prospective tourists.[47]

For tourists, the tourism industry's selective packaging of culture(s) creates a significant degree of expectation, which emanates from the industry's own value systems, which are carried to chosen holiday destinations.[48] The outcome is that tourists develop contrived expectations of the cultures they visit, which are frequently idealized and inauthentic.[49] Thus, when tourists make contact with the host culture directly, and externally conceived expectations of behaviour and spectacle are not met, dissatisfaction, resentment and conflict can arise between tourist and host, and between tourist and tour operator, in a somewhat exaggerated situation of breach of contract.

Against the backdrop of a First World/Third World development gap and a pervasive imperialist legacy, it is tempting to write off the commodification of cultures as some kind of tacit acceptance of established global power relations and the dominant social paradigm within which international tourism sits. At the same time it is too easy to ignore the involvement of host communities in the commodification of their own cultural traditions in legitimate attempts to secure economic advantages. The slow growth of involvement of indigenous peoples in the ownership and management of tourism businesses is welcome.[50] However, it, too, is not without some degree of selective packaging in order to meet tourist demands, often following dominant First World business models.

However, visited host communities are often far from the homogeneous cultural groups that tourists and the tourism industry takes them to be. Understandably, there are tensions and divisions relating to the extent to which cultural traditions and patterns of behaviour can or should be adjusted to meet important economic goals, resulting in what Greenwood has termed 'conflictual arenas'.[51] Within these arenas intra-cultural struggle may emerge over competing claims regarding the ownership and presentation of cultural assets. Central to such struggles is the state, which may or may not reflect the variety of cultural groups and subcultures within a nation. Culture, in terms of ethnic traditions, language, religious beliefs, and community traditions, together with its symbolic expression in the form of 'cultural capital', are open to political manipulation by the state for both economic and nationalistic reasons. In a majority of cases the international tourism industry negotiates directly with governments and their appointed tourism agencies. Moreover, in much the same way as with other issues of international trade are mediated, 'negotiation' is conducted in the language of neoclassical economics and contemporary capitalism. In the words of Watson and Kopachevsky:

> Tourism by its very nature, is shaped by a very complex pattern of symbolic valuation; and this takes place in a structured social context over which tourists themselves have no immediate control. The essence of modern capitalism is the remanufacture of images, many of which effectively obscure the injuries of class, race and sex.[52]

Tourism Dependency

Various researchers have positioned the phenomenon of international tourism as a manifestation of neocolonialism and imperialism.[53] Similarly, in a neo-Marxist vein it is possible to conceive of the 'pleasure-periphery' idea of tourism as representing the fundamental structural dependency of the developing nations upon the developed nations.[54] The ideas of neocolonialism and global imbalance are borne out not only in terms of the direction of tourist flows from First to Third World, but also by the fact that the necessary enabling elements for world tourism – the means of production, the ideology of consumption, capital, credit, and information – are chiefly located in, and controlled by, the developed nations.

Dependency can also generate its own internal cultural conflicts, creating further negative impacts for the host community. In an examination of tourism development in Anuha in the Solomon Islands, for instance, Sofield identified varying levels of internal conflicts, though he pointed to external tourism development pressures from Australia as the main force initiating these.[55] Different levels of internal cultural conflict can occur between locals and those who work within the tourism industry (perceived as serving the needs of the outsiders first); between competing communities and ethnic groups; and between the masses and the local elite.[56] Given that tourism is a potent economic symbol and a driver for acculturation, particularly in developing countries, active involvement with the industry can create resentment within a passive host community. This would appear to be particularly the case when tourism development provides access to levels of employment and income which may be significantly higher than, for instance, that of local agricultural workers.[57] Compounding the emergence of gaps *vis à vis* access to (in many developing countries) relatively high wage levels, tourism employment opportunities may also be skewed toward certain social and ethnic groups, or labour may be imported from outside the community.[58]

The inability of the host community to 'control' the tourism industry in political and economic terms may exacerbate the potential for resentment and conflict along cultural lines at both a micro- and macro-level. While it would be over-stretching the point to argue that cultural conflicts can be wholly compensated in economic terms, there is nevertheless a trade-off position by which aspects of cultural intrusion and degrees of acculturation can be tolerated in the name of economic development and modernization. Three points may be seen to emerge in this regard. First, the concept of compensation for loss of cultural capital, or the loss of control of that capital, is firmly anchored in the same, 'traditional', First World view which rationalizes the commodification of culture, and has legitimized its trading.[59] Second, the extent of dependency in developing economies does have bearing on the issue of compensation; thus, resentment at the lack of adequacy in terms of 'reward' can manifest itself in cultural tensions. Third, it is worth noting there is nothing wrong *per se* in the utilization of tourism as an agent of development; it could well be the least disturbing of development options, and host communities may be in a position freely to choose tourism as one agent of modernization from among others.[60] Nor should one assume that culture in developing countries is as vulnerable and incapable of adaptation as it is sometimes made out to be.[61]

Cultural Conflicts and Tourism Space and Places

The range of conflicts identified above – between tourists, the tourism industry, and host cultures – takes place within, and interacts with, real space. As Urry has pointed out, 'environments, places and people are being regularly made and remade as tourist objects.'[62] In spatial terms, the processes of construction and reconstruction reflect substantive economic and socio-cultural change. Gottman has noted the clear importance of the economy in shaping the function and form of the built environment.[63] Tourism and leisure, within the context of a dominant service sector economy, may establish themselves as important driving forces in the shaping and reshaping of both urban and rural spaces. However, the concept of space fails to convey the cultural relationships that have developed between host communities and the environment.

Emphasis upon space rather than place, or what Keith and Rogers have termed the 'spatial fetish', fails to acknowledge issues of belonging, 'placeness', and 'territoriality'.[64] Using A.P. Cohen's notion of cultural territories, the natural and built environment can be seen as imbued with cultural meanings and historical contexts and to reflect the values and behaviour of its creators, stewards and inhabitants.[65] Territory in this sense invokes feelings of collective and individual ownership on the part of the local community. In some cases ownership in the legal sense is an issue, but it relates much more to an emotional sense of connectedness with an area, a set of buildings, or streetscape as spatial expressions of cultures. As Cohen has indicated, the boundaries of cultural territories may not be recognizable from the outside, but are learned and recognized from within. Moreover, these boundaries continually shift in both aesthetic and functional terms.

From the perspective of the individual tourist, the act of temporarily leaving his or her own cultural territory to share with (or at least gaze upon) that of another is a large part of the experience. However, there still exists a 'mass' dimension to world tourism whereby tourists are largely defined by their sun-seeking hedonism within the confines of their own transported 'environmental bubble'.[66] While there is evidence that mass tourism is giving way to a more complex postmodern touristic experience, mass tours nevertheless remain the functional core and profit centre of the international tourism market.[67] This arguably partly explains why there is little in the way of emphasis upon, and concern for, the cultural dimensions of place and territory for the majority of the tourism industry. Rather, the challenge is to find (or create) attractive, exciting (and largely value-neutral spaces) that the mass of tourists will want to buy into. As Ringer has noted, 'tourism is essentially about the creation and reconstruction of geographic landscapes as distinctive tourism destinations through manipulations of history and culture.'[68]

For tourists and residents alike expectations are constructed and perceptions shaped as the tourism industry confirms its hegemony through the physical creation and re-creation of place. New leisure-oriented ideologies, derived in the main from the First World, can collide with traditional value systems of host cultures, entailing the view that 'the creation of tourist/recreation places to visit and things worth seeing has a guiding function in [a] community, telling us what is beautiful and worthwhile and what is nice.'[69]

Conflicts and Created Spaces

At this juncture, and broadly following the distinctions between 'mass' and 'niche' tourism, it is important to differentiate between two fundamental types of tourism spaces and places: those which are purpose built to cater to tourist desires, and those which tourists converge upon and actively 'share' with the host community. The former type is perhaps the most straightforward to identify. The tourism industry has been successful in creating its own spaces, such as purpose-built resort complexes and theme parks. These frequently exist within physical boundaries through which the tourist has symbolically to pass in order to experience the leisure product within. The product itself is designed to be total and confined to the one complex.[70] Pearce has referred to such purpose-built centres as tourist 'places', though it is difficult to see them as possessing the same kind of intimate 'placeness' that residents may share with a location.[71] Also, reflecting the postmodernist blurring of leisure boundaries, shopping malls may arguably be added to those spaces that are purposely created with tourists in mind. Large retail outlets, often designed around a series of themes, are clearly part of the landscape of consumption which tourists increasingly inhabit.[72]

Characteristics of such created spaces include functionality and efficiency, self-reliance, and a certain degree of replicabilty. The question of what impacts these designed tourism spaces have upon the built environment is somewhat misplaced, since the built environment overall is likely to be newly designed and constructed so as to meet the requirements of the tourist first and foremost. Rather, issues can arise over the location and environmental positioning of such complexes *vis à vis* such issues as access, visual compatibility with their setting, and the degree of ecological disturbance to which they may or may not contribute.[73]

The 'out of town' location for resorts, shopping malls and theme parks may provide the distance element of the tourist experience – distance both in physical and psychological terms. Such locations allow enough separateness from day-to-day life so that the theme park may, for example, become an 'away' place.[74] But being separate from the ordinariness of the surrounding environment also allows tourist enclaves to develop, ostensibly removing tourists from contact with the host community.[75] It is the dominance and relative isolation of the tourist that is perhaps the most important characteristic of all the types of spaces created by the tourism industry, a self-containment that allows for the development of internal tourist 'communities'.[76] However, whether a theme park which has limited opening hours or a resort which is effectively open all the time, such created spaces seldom allow for meaningful interaction with the host culture. Relationships within created spaces are framed by a two-way process of expectation and tend to be limited to essential encounters between those who serve and those who have paid to be served. Krippendorf has criticized the notion of separate tourist communities as being economically disadvantageous to the host, culturally limiting for both host and holidaymaker, and effectively a sterile experience.[77] This said, although resort complexes, theme parks, and self-contained tourism spaces are highly visible elements of the international tourism industry, by virtue of their designed separateness, prospects for interpersonal 'tourist-tourist' and 'tourist-tourism' conflicts are reduced *de facto*.

Conflicts and Shared(?) Places

A more intangible and yet fundamental characteristic of contemporary tourism is the extent to which it converges on the spaces and places occupied by the host community. Tourists at various times occupy the places which 'belong' to others and which carry cultural meanings for the host community. Within the short time period of a leisure or business visit, tourists essentially remain as Simmel's strangers and outsiders, with little opportunity, nor motivation, to penetrate host cultures in any meaningful way.[78] However, though individual tourists may not stay in one place for very long, the experience for the host community is quite different, consisting of a constant stream of undifferentiated tourists united by their transitoriness, anonymity, and propensity to 'gaze'.

Clearly, the sheer number of tourists to a destination is important, and the host population can find itself overrun and out numbered. This may be particularly the case in rural areas where pressures on communal facilities such as transport, parking, shops, and basic natural resources such as water and air can generate antagonism between tourist and resident. However, tensions can also arise between city 'inhabitants' and 'users', the latter group swelling a city's population and competing for access to its spaces and facilities.[79] Tourists are a key element of the 'user' group, overlapping with such others as shoppers, concert-goers, and sports fans. In world tourism cities such as Venice, for instance, the historic core's population of approximately 80,000 receives nearly 1.5 million tourists each year, together with additional large numbers of excursionists.[80] Such a swollen number of users can induce conflicts with inhabitants, as both groups compete for space and facilities. Moreover, at key points in the tourist season the environmental and social carrying capacity of Venice is likely to be exceeded.

But while the infringement of place by tourists can result in cultural stresses and strains, arguably more significant and lasting cultural impacts on host communities are generated by the tourism industry, as it transforms the built environment to meet and procure tourist demand. On the surface the injection of capital from tourism developers is widely recognized as producing such positive impacts as land reclamation and ecological restoration, preservation of historic buildings, conservation and promotion of vernacular architecture, and introduction of innovative and challenging architectural styles. In the United Kingdom and most large European cities numerous examples exist of old mills and warehouses being restored and providing housing, hotels, offices, shops and art centres. In this way the built environment is often rediscovered, often deliberately with tourism in mind, and remade to fit with new 'symbolic economies'.[81] In other parts of the world, too, tourism has catalysed the revitalization of the built environment. In her studies of Damascus and Aleppo in Syria, for instance, Shackley has pointed out that tourism development has assisted markedly in the conservation of historic buildings and their subsequent conversion to tourist facilities, luxury restaurants, and accommodation.[82]

Having said this, the tourism industry is also open to criticism for its growing legacy and export of 'international' styles of functional, postmodern blandness. Resort areas, waterfront developments, hotels, and attractions have all evolved with little or no concession to environmental setting, local traditions, and the nuances of local culture and ethnic difference. Heng and Chang, for example, have described the redevelopment of the quays along the waterfront of Singapore and pointed to a transformation 'from a historically rich area to one that has the usual restaurants and souvenir shops that can be found virtually everywhere.'[83] Such property-led schemes of physical transformation and regeneration are commonplace in both developed and developing countries, pre- and postindustrial centres, large cities and provincial towns. Yet schemes such as that along the Singapore waterfront, and numerous other examples in the developed and developing world, have been successful in obscuring past economic functions and previous social and community patterns of activity, save perhaps occasional references to the past by way of 'interpretive' heritage centres.

While this process of conversion, gentrification, and effective prohibition from abandonment is usually cited as one of tourism's positive impacts, little attention has been given to the cultural aftermath of such physical transformations. Within a relatively short time the built environments of traditional and non-traditional tourist destinations can be altered in both aesthetic and functional terms to create 'new', or rather different, cultural territories. The issue is not that of change, *per se*, but one of the extent to which a host culture feels a sense of ownership, belonging and participation in the change – especially when investment, development and planning decisions, down to architectural details, are increasingly being shaped by market assessments of visitor potential as much as resident and community use. Furthermore, developing space (particularly urban and inner-city space) to make it more attractive to tourists, has not always been accompanied by attempts to maintain viable communities there. Increasing land prices and rents, decreasing security of tenure, heavy competition for business space, loss of indigenous control, and the dominance of aesthetics over function, have all contributed to changing the social patterns of the host community, and the migration of populations, together with social and economic problems 'out of town'.[84]

Issues of Ownership and Identity

In both developing and developed economies the power of the tourism industry has manifested itself in often-dramatic changes of ownership. Yet the dynamics of ownership in world tourism has attracted relatively little attention in the literature for several reasons. Because of the fluidity of international capital, the low barriers to entry in the tourism business, and the momentum of tourism development, patterns of ownership are difficult to monitor. This is often compounded by the distance and opaqueness of decision-making among corporate players, and between developers and governments. In addition, because tourism is largely measured by its economic success rather than its cultural integrity, the issue of ownership has not commonly been recognized as a problem. However, if one frames ownership in the wider context of territorial belonging, sense of place, and participation in the decisions regarding how places look and function, there are many problems. Cultural territories are contested, and have been, and remain, appropriated from host communities.[85] Arguably in many cases, appropriation is difficult to distinguish from normative processes of economic restructuring and the physical transformations that accompany it. Nonetheless, the impacts upon host cultures remain. As well as the much-referred-to 'tourist gaze', there is also a 'community gaze' which communities experience when they encounter the new, often dramatic, physical and emotional spaces designed for tourists.[86] In skewing the built environment to meet the expectations and preferences of the tourist, the cultural elements of placeness – continuation, evolution, stability and familiarity – are eroded.

In assessing the changes that have taken place on the Victoria and Alfred Waterfront area of Cape Town, Dodson and Killian, for instance, have pointed to its transformation from a functioning port to a place of tourist-oriented consumption.[87] In doing so, they have also highlighted the impact such development and change can have upon a local community, pointing in particular to criticisms of the development as existing in the main for tourists and affluent, White, middle-class residents. Nearby residents of Black townships have been effectively (if unintentionally) excluded from the relatively expensive facilities of the waterfront. And while visitors can now enjoy the cleanliness and safety of the transformed port area, the residents of other areas of Capetown continue to struggle with largely neglected issues of crime and prostitution.

This process of physical transformation to tourism and leisure landscapes, appropriation for tourists, and the tensions and conflicts this can stir among local cultures and subcultures can be accentuated when decisions are taken 'outside' the community – possibly within a wholly different cultural context. In such a process territories can be redefined in accordance with the aspirations, tastes, preferences and budgets of external tourism developers and government agencies, and 'spurious realities' can be created whereby attractions and events bear little in the way of familiarity for the local population.[88] The tourism development process can also highlight different conditions of access to power among local groups and reveal tensions within destination communities.[89] For example, the urban redevelopment and reimaging of Glasgow which led to its being labelled European City of Culture in 1990, revealed a series of tensions between the planners and promoters of the city and the realities of its working-class communities, which were effectively excluded from the high-cultural experiences subsequently offered to tourists.[90]

Identity, as Wearing and Wearing have noted, emerges both from the core of the individual and from within the core of communal culture.[91] Through various formal and informal means of cultural display, tourism development can contribute to the development and reinforcement of social cohesion and cultural identity. Thus, Friedman, in considering the Ainu peoples of Northern Japan, noted that 'tourism production and display have become a central process in the conscious reconstruction of Ainu identity.'[92] Indeed, tourism development can be partly legitimized by its claims to maintain traditions and preserve cultures, particularly in societies where cultural identities reflect long, unbroken histories and powerful continuity.[93] However, the development of tourism and its propensity to change the nature of places may also challenge cultural identities. While it is difficult to argue that tourism is not causally responsible for the loss of cultural identities under pressure for places and their endogenously constructed narratives of reality to become tourist destinations, emphasis has nonetheless shifted from identity to the generation of images.

Bourdieu's view that identity (self-identification) is increasingly shaped through consumptive behaviour and 'lifestyle' may be helpful in explaining the role of the tourist.[94] But the cultural identity of the host community is surely something greater than a collection of individuals seeking to differentiate themselves through distinctive consumption. It is bound up with an intimacy shared with the evolved natural and built environment and is defined in part by its fixedness. Contrast this with the culture of tourism, which advocates movement and a non-fixedness, and the driving force of an industry which has no interest in people staying in one place for very long. The tourism encounter between tourists as the new cosmopolitans, seeking to experience and consume the cultural identity of others, is emblematic of the basic inequality of world tourism referred to earlier – consumers requiring something to consume. Even among the consumed community, whether associated with the provision of tourism (either directly through working in a shop or hotel, or indirectly by just being there), there are conflicts. And seldom are all parts of a destination toured equally; some areas and elements of a host community may be consumed more than others, which itself can contribute to cultural fragmentation involving isolation, poverty, and, in an urban context, the creation of what Sachs-Jeantet has termed 'outlaw zones'.[95]

Conclusion: Uneven Tourism Encounters in a Shrinking World

This chapter has provided in conceptual terms an overview of the inter- and intra-cultural issues and conflicts inherent within the growth of international tourism and the structures that drive and support it. What emerges is a picture in which the cultural implications of tourism cannot be considered remote from the physicality and cultural meaningfulness of place. Destinations, whether purposely designed for the tourist or, more commonly, shared by the tourist, are not value neutral stages for tourism encounters; they are dynamic, culturally conceived theatres of complex interaction. Furthermore, in addition to the potential conflicts at the face-to-face level between tourist and host, the tourism industry often dramatically alters the built environment, and thus the cultural landscape and cultural evolution of a host community. New cultural territories with a focus upon tourism and leisure, often conceived and financed outside the destination culture, create further potentials for conflict between local communities, tourists, and the tourism industry, and can expose power differentials within and between host community groups.

Recognizing that conflict is a necessary – if not sufficient – condition to its resolution, but in identifying specific reasons for inter- and intra-cultural conflicts, there is still a need to address the generic characteristics which tourism as the 'world's largest industry' exhibits. Outstanding are the interrelated issues of tourism's role in the processes of globalization, the dominance of First World capital, and the expansion of the culture and ideology of tourism as a leisure activity.

The concept of globalization, despite its amorphous definition and contestability, nonetheless provides a necessary backdrop to the inter- and intra-cultural relations within tourism. Conceptions of a 'shrinking' world, the New World Order, and the global ideal, are all dominated by progressive, Western, neomodern ideologies in which economic relationships are central.[96] But the 'world as a single place' is conceived largely from an elitist position whereby a unified world without boundaries merely makes for more readily penetrable markets.[97] Conquest, exploitation and imperialism have given way to subtler means of supporting the inevitability of capitalism, whereby globally penetrating technologies now allow the peaceful creation of pseudo-colonial dependencies at a distance. Nevertheless, the telling metaphor of the world marketplace still reflects a neoclassical capitalist belief system that has evolved little in over two hundred years.

In the increasing implementation of global quality standards for tourism businesses and corporate attempts to imitate and assimilate uniform patterns of architecture, social behaviour, language, dress and cuisine, steered by trans-national investment, tourism can be viewed as one of the perpetrators of globalization. Furthermore, as Böröz c has argued, the 'standardisation, normalisation and commercialisation of experience' means the tourism industry will largely seek to maintain the *status quo*.[98] The dramatic economic successes of tourism investment (mainly from the First World) and the increasing ability of the tourism industry to control tourist flows and patterns of development in a remote way has also highlighted how tourism has benefited from globalization. Moreover, it appears that within the shrinking-world perspective, complete with postmodernist compression of time and space, the tourism industry is active in supporting tourists in their search for identity. Quoting Lasch, Bauman has pointed out that identity 'refers both to persons and to things', and that 'both have lost their solidity in modern society, their definiteness and continuity.'[99] Travelling to gaze upon communities which have retained their cultural identity, or which are able to present representations of their identity, and travelling to discover one's own identity, indicate that tourism is, in part at least, a somewhat parasitic search for those things lost.

This gives tourism an ambiguous role in the globalization process. In supply terms, through airline routes and well-developed trans-national distribution and booking systems, tourism engages in a very real sense of interconnectedness. It also continues to partake in the process of 'Coca-Colonisation', and encourages aesthetic and cultural homogeneity, manifest in both form and function of the built environment.[100] Though tourism is only one of a number of catalysts for such changes, it is a significant and symbolic one.[101] At the same time, the market, even within the context of mass tourism, appears to be seeking diversity and cultural difference. The more homogenous the world becomes through the promotion of tourism as a trans-cultural product, the greater the desire to reinvent those values which delineate culture.[102] In this way tourists are being presented with an increasing array of cultural differences to choose from – some authentic, some staged. Relatively few prospective tourists seek total immersion in a different culture, and few host societies wholly seek not to adapt to the needs of tourists. Instead, the tourist seeks safe glimpses of cultural difference, and can often be satisfied with simulacra. This may or may not be accompanied by a desire to understand the culture of the other. In part, tourist safety from overexposure to host cultures is ensured by the short and infrequent duration of contact between themselves and their hosts. Host communities, however, do not go home: they are home, and contact may be continuous.

The problem remains, however, that tourism encounters between and within cultures are shaped by fundamental inequalities. Though, globally, the tourism industry consists of a myriad of small and medium-sized businesses, it is dominated by a relatively small and powerful number of airlines, hotel groups, and leisure developers who are in a position of significant economic, political, and ultimately cultural influence. In market terms too, 80 per cent of international travellers are nationals of just twenty countries. And though there remains a weighting toward mass tourism, with regard to cultural impacts the long-haul, independent travel market is increasingly important.[103] Importantly, the majority of the world's population does not holiday; but this is not a straightforward First/Third World issue, nor a function of economic well-being. It also relates to a lack of tourism culture. The example of the surprising, and relatively small, proportion of the population of the United States who hold passports would seem to indicate that an absence of a cultural framework for tourism (as opposed to leisure) is not automatically linked to economic and social status. The 'learning' of the culture of tourism involves something more than just *how* to consume. It also involves accepting the desire to consume otherness and, implicitly, the need to select, commodify and package the world.

Given the inequities displayed, charges that tourism is merely a different form of colonialism (postcolonialism) and imperialism are difficult to refute. Although forecasts suggest that lesser-developed nations will increasingly become 'sender' countries, they will largely remain as Turner and Ash's 'pleasure periphery' – politically marginalized and economically pocketed by the North and West.[104] Moreover, in a global sense, the pleasure periphery has expanded dramatically, taking in more than the 'Third World' and indigenous populations. Even *within* developed nations pleasure peripheries can be identified, as rural margins are increasingly dependent upon urban markets, city centres have developed as recreational centres for suburbanites, and business travellers with company accounts are served by low-waged peripheral labourers.

The rootedness of world tourism in historical relations, the legacy of imperialism, concentrated ownership of tourism's structures, the tendency to assume a dominant-subordinate relationship, together with still-growing expectations and opportunities among developed countries to engage in tourism, all point to a fundamental inequality, and to a process (however unconsciously articulated) of First World hegemony, reflected in distant, but local tourist-host encounters.[105] But this is not solely an economic process. While representations of tourism as a simplistic and value-neutral exchange in which cultural differences and otherness are traded for tangible economic gain and elusive social well-being still persist, the reality of tourism is much different. Tourism is also usually unequal in cultural terms, does not always take place on the basis of consent, and frequently escapes any notion of mutual cultural understanding.[106]

In any conceptual discussion of world phenomena such as tourism the routes to generalization are open. However, the realities of tourism encounters, understanding the potential for cultural conflicts, the transformations of space and place, and the effects upon the built and natural environment are all framed by this rather uncomfortable feeling of imbalance, of a tourism industry which through its own culture is shaping the world, how it looks, how it functions, and what it means.

Notes

1. *Tourism 2020 Vision*, Madrid, World Tourism Organization, 1997.

2. Proceedings of Round Table, *Culture, Tourism, Development: Critical Issues for the XXIst Century*, Paris, UNESCO/AIEST, 1996.

3. U. Hannerz, 'Cosmopolitans and Locals in World Culture,' in M. Featherstone (ed.), *Global Culture: Nationalism, Globalization and Modernity*, London, Sage Publications Ltd., 1990, p. 238.

4. C. Rojek and J. Urry, 'Transformations of Travel and Theory,' in Rojek and Urry (eds.), *Touring Cultures: Transformations of Travel and Theory*, London, Routledge, 1997, p. 11.

5. K. Tucker and M. Sundberg, *International Trade in Services*, London, Routledge, 1988.

6. S.L.J. Smith, 'Tourism as an Industry: Debates and Concepts,' in D. Ioannides and K.G. Debagge (eds.), *The Economic Geography of the Tourist Industry*, London, Routledge, 1998.

7. M.T. Sinclair, P. Alizadeh, and E.A.A. Onunga, 'The Structure of International Tourism and Tourism Development in Kenya,' in D. Harrison (ed.), *Tourism and The Less Developed Countries*, London, Belhaven Press, 1992.

8. In terms of conflicts between systems and structures, see M. Roche, 'Mega-events and Micro-Modernisation: On the Sociology of the New Urban Tourism,' *British Journal of Sociology*, vol. 43, no. 4, 1992, pp. 563-600. In terms of processes of capitalism, see S. Britton, 'Tourism, Capital and Place: Towards a Critical Geography of Tourism,' *Environment and Planning D: Society and Space*, vol. 9, 1991, pp. 451-578. In terms of the organization of capitalism, see S. Lash and J. Urry, *Economies of Signs and Spaces*, London, Sage Publications Ltd., 1994. In terms of the differences between capitalism and precapitalism, see M. Crick, 'Representations of International Tourism in the Social Sciences,' *Annual Review of Anthropology*, vol. 18, 1989, pp. 307-344.

9. M.E. Johnston, 'Patterns and Issues in Arctic and Sub-Arctic Tourism,' in C.M. Hall and Johnston (eds.), *Polar Tourism: Tourism in the Arctic and Antarctic Regions*, Chichester, John Wiley & Sons, 1995; and C.M. Hall and M. Wouters, 'Issues in Antarctic Tourism,' in *ibid*.

10. D.B. Weaver, *Ecotourism in the Less Developed World*, Wallingford, CAB International, 1998.

11. A. Poon, *Tourism, Technology and Competitive Strategies*, Wallingford, CAB International, 1993.

12. A. Mathieson and G. Wall, *Tourism: Economic, Physical and Social Impacts*, London, Longman, 1982.

13. I.V. Rowe, 'Taiwan: Carrying Capacity in an Industrializing Country,' in F.M. Go and C.L. Jenkins (eds.), *Tourism and Economic Development in Asia and Australasia*, London, Cassell, 1997.

14. G.H. Cazes, 'The Growth of Tourism in the Developing Countries,' in UNESCO/AIEST Proceedings of Round Table, *Culture, Tourism, Development: Critical Issues for the XXIst Century*, Paris, UNESCO/AIEST, 1996.

15. 'Tourism and Tribal Peoples,' Background Sheet, London, Survival International, 1995.

16. D. Harrison, 'Tourism to Less Developed Countries: The Social Consequences,' in Harrison (ed.), *Tourism and The Less Developed Countries*, London, Belhaven Press, 1992.

17. J. Craik, 'The Culture of Tourism,' in Rojek and Urry (eds.), *Touring Cultures*.

18. J. Krippendorf, *The Holiday Makers: Understanding the Impact of Leisure and Travel*, Oxford, Heinemann, 1987.

19. E. Swinglehurst, 'Face to Face: The Socio-Cultural Impacts of Tourism,' in W. Theobald (ed.), *Global Tourism: The Next Decade*, Oxford, Butterworth Heinemann, 1994.

20. V. Gessner and A. Schade, 'Conflicts of Culture in Cross-border Legal Relations: The Conception of a Research Topic in the Sociology of Law,' in M. Featherstone (ed.), *Global Culture: Nationalism, Globalization and Modernity*, London, Sage Publications Ltd., 1990. For an exploration of tourism's benefits, see Y. Amir and R. Ben-Ari, 'International Tourism, Ethnic Contact and Attitude

Change,' *Journal of Social Issues*, vol. 41, no. 3, 1985, pp. 105-115; and Y. Reisinger, 'Social Contact between Tourists and Hosts of Different Cultural Backgrounds,' in A.V. Seaton (ed.), *Tourism: The State of the Art*, Chichester, John Wiley & Sons, 1994.

21. L. Nettekoven, 'Mechanisms of Cultural Interaction,' in E. de Kadt (ed.), *Tourism: Passport to Development?* London, Oxford University Press, 1979.

22. Weaver, *Ecotourism in the Less Developed World*.

23. A. Furnham and S. Bochner, *Culture Shock: Psychological Reactions to Unfamiliar Environments*, London, Routledge, 1986.

24. J.D. Medrano, 'Some Thematic and Strategic Priorities for Developing Research on Multi-Ethnic and Multi-Cultural Societies,' *Management of Social Transformations, Discussion Paper Series*, no.13, Paris, UNESCO, 1996.

25. C. Ryan, *Tourism, Terrorism and Violence: The Risks of Wider World Travel*, Conflict Study 244, London, Research Institute for the Study of Conflict and Terrorism, 1991.

26. J.E. Tunbridge and G.J. Ashworth, *Dissonant Heritage: The Management of the Past as a Resource in Conflict*, Chichester, John Wiley & Sons, 1996.

27. A.D. Smith, 'Towards a Global Culture?' in Featherstone (ed.), *Global Culture*.

28. H.L. Parsons, *Marx and Engels on Ecology*, London, Greenwood, 1977.

29. M. Douglas, *Implicit Meanings*, London, Routledge and Kegan Paul, 1977; S. Vogel, 'Marx and Alienation from Nature,' *Social Theory and Practice*, vol.14, no.3, 1988, pp.367-388; K. Milton, *Environmentalism and Cultural Theory: Explaining the Role of Anthropology in Environmental Discourse*, London, Routledge, 1996; and J. Urry and P. Macnaghten, *Contested Natures*, London, Sage Publications, 1998.

30. D. Greenwood, 'Culture by the Pound: An Anthropological Perspective on Tourism as Cultural Commoditisation,' in V.L. Smith (ed.), *Hosts and Guests: The Anthropology of Tourism*, 2nd ed., Philadelphia, University of Pennsylvania Press, 1989.

31. D. Mercer, 'The Uneasy Relationship between Tourism and Native Peoples: The Australian Experience,' in Theobald (ed.), *Global Tourism*.

32. Uluru-Kata Tjuta National Park, 'Welcome to Uluru-Kata Tjuta National Park,' Information Sheet, undated.

33. Northern Territory Tourist Brochure, Northern Territory Tourist Commission, 1999, pp. 34-35.

34. K. Hollinshead, 'Marketing and Metaphysical Realism: The Disidentifications of Aboriginal Life and Traditions through Tourism,' in R. Butler and T. Hinch (eds.), *Tourism and Indigenous Peoples*, London, Routledge, 1996.

35. World Commission on Environment and Development, *Our Common Future*, Oxford, Oxford University Press, 1987. M. Robinson, 'Collaboration and Cultural Consent: Re-focusing Sustainable Tourism,' *Journal of Sustainable Tourism*, forthcoming.

36. A.H. Walle, 'Habits of Thought and Cultural Tourism, *Annals of Tourism Research*, vol. 23, 1996, pp. 874-890.

37. M. Morris, 'Life as a Tourist Object in Australia,' in M.F. Lanfant, J. Allcock, and E. Bruner (eds.), *International Tourism: Identity and Change*, London, Sage Publications Ltd., 1995.

38. L. Turner and J. Ash, *The Golden Hordes*, London, Constable, 1975. E.M. King, *Creating Island Resorts*, London, Routledge, 1997.

39. E. Cohen, 'Authenticity and Commoditization in Tourism,' *Annals of Tourism Research*, vol. 15, 1987, pp. 371-386.

40. See Greenwood, 'Culture by the Pound'; and D. MacCannell, *Empty Meeting Grounds: The Tourist Papers*, London, Routledge, 1992.

41. P.F. McKean, 'Towards a Theoretical Analysis of Tourism: Economic Dualism and Cultural Involution in Bali,' in Smith (ed.), *Hosts and Guests*; and J. Boissevain, 'Ritual, Tourism and Cultural Commoditization in Malta: Culture by the Pound?' in T. Selwyn (ed.), *The Tourist Image: Myths and Myth Making in Tourism*, Chichester, John Wiley & Sons 1996.

42. S. Macdonald, 'A People's Story: Heritage, Identity and Authenticity,' in Rojek and Urry (eds.), *Touring Cultures*.

43. E. Said, *Orientalism*, New York, Pantheon, 1978; H. Bhabha, *The Location of Culture*, London, Routledge, 1994; and A.E. Coomes, *Reinventing Africa*, New Haven, Yale University Press, 1994.

44. D. MacCannell, 'Reconstructed Ethnicity: Tourism and Cultural Identity in Third World Communities,' *Annals of Tourism Research*, vol. 11, 1984, pp. 361-377.

45. J.C. Altman, 'Tourism Dilemmas for Aboriginal Australians,' *Annals of Tourism Research*, vol. 16, 1988, pp. 456-476; M. Daltabuit and O. Pi-Sunyer, 'Tourism Development in Quintana roo, Mexico,' *Cultural Survival Quarterly*, vol. 14, no. 1, 1990, pp. 9-13; and P.G. Goering, 'The Response to Tourism in Ladakh,' *Cultural Survival Quarterly*, vol. 14, no. 1, 1990, pp. 20-25.

46. G.M.S. Dann, 'Images of Destination People in Travelogues,' in Butler and Hinch (eds.), *Tourism and Indigenous Peoples*.

47. J. Lea, *Tourism and Development in the Third World*, London, Routledge, 1988; I. Silver, 'Marketing Authenticity in Third World Countries,' *Annals of Tourism Research*, vol. 20, 1993, pp. 302-318; and C. Classen and D. Howes, 'Epilogue: The Dynamics and Ethics of Cross-Cultural Consumption,' in Howes (ed.), *Cross-Cultural Consumption: Global Markets Local Realities*, London, Routledge, 1996.

48. R. Parris, 'Tourism and Cultural Interaction: Issues and Prospects for Sustainable Development,' in UNESCO/AIEST Proceedings of Round Table, *Culture, Tourism, Development: Critical Issues for the XXIst Century*, pp. 36-40.

49. G.M.S. Dann, 'The People of Tourist Brochures,' in Selwyn (ed.), *The Tourist Image*.

50. H.D. Zeppel, 'Land and Culture: Sustainable Tourism and Indigenous Peoples,' in C.M. Hall and A.A. Lew (eds.), *Sustainable Tourism: A Geographical Perspective*, London, Longman, 1998; and G. Wall, 'Partnerships Involving Indigenous Peoples in the Management of Heritage Sites,' in M. Robinson and P. Boniface (eds.), *Tourism and Cultural Conflicts*, Wallingford, CAB International, 1998.

51. Greenwood, 'Culture by the Pound.'

52. G.L. Watson and J.P. Kopachevsky, 'Interpretations of Tourism as Commodity,' *Annals of Tourism Research*, vol. 21, 1994, pp. 643-660.

53. Crick, 'Representations of International Tourism in the Social Sciences'; and D. Nash, *Anthropology of Tourism*, Kidlington, Pergamon, 1996.

54. S. Amin, *Unequal Development: An Essay on the Social Formations of Peripheral Capitalism*, New York, Monthly Review Press, 1978; S. Britton, 'The Political Economy in the Third World,' *Annals of Tourism Research*, vol. 9, 1982, p. 331; and M.H. Erisman, 'Tourism and Cultural Dependency in the West Indies,' *Annals of Tourism Research*, vol. 10, 1983, pp. 337-361.

55. T.H.B. Sofield, 'Anuha Island Resort: A Case Study of Failure,' in Butler and Hinch (eds.), *Tourism and Indigenous Peoples*.

56. Crick, 'Representations of International Tourism in the Social Sciences.'

57. J. Cukier-Snow and G. Wall, 'Tourism Employment: Perspectives from Bali,' *Tourism Management*, vol. 14, no .3, 1993, pp. 195-201; and J. Cukier, 'Tourism Employment in Bali: Trends and Implications,' in Butler and Hinch (eds.), *Tourism and Indigenous Peoples*.

58. H. Kakazu, 'Effects of Tourism Growth on Development in the Asia-Pacific Region: The Case of Small Islands,' in UNESCO/AIEST Proceedings of Round Table, *Culture, Tourism, Development: Critical Issues for the XXIst Century*, pp. 7-14.

59. L. Richter, 'Political Instability and the Third World,' in Harrison (ed.), *Tourism and The Less Developed Countries*.

60. C. de Burlo, 'Cultural Resistance and Ethnic Tourism on South Pentecost, Vanuatu,' in Butler and Hinch (eds.), *Tourism and Indigenous Peoples*.

61. Harrison, 'Tourism to Less Developed Countries.'

62. J. Urry, *Consuming Places*, London, Routledge, 1995, p. 192.

63. J. Gottman, *Forces Shaping Cities*, Department of Geography, University of Newcastle upon Tyne, 1978.

64. M. Keith and A. Rogers (eds.), *Hollow Promises? Rhetoric and Reality in the Inner City*, London, Mansell, 1991. Also, E. Cohen, 'Contemporary Tourism – Trends and Challenges: Sustainable Authenticity or Contrived Post-Modernity?' in R. Butler and D. Pearce (eds.), *Change in Tourism: People, Places, Processes*, London, Routledge, 1995; and A.P. Cohen, *The Symbolic Construction of Community*, Chichester, Ellis Horwood Ltd., 1985.

65. Cohen, *The Symbolic Construction of Community*,

66. E. Cohen, 'Towards a Sociology of International Tourism,' *Social Research*, vol. 39, no. 1, 1972, pp. 64-82.

67. Poon, *Tourism, Technology and Competitive Strategies*; and World Tourism Organization, *Tourism 2020 Vision*.

68. G. Ringer, 'Introduction,' in Ringer (ed.), *Destinations: Cultural Landscapes of Tourism*, London, Routledge, 1998, p. 7.

69. J. Lengkeek, 'Collective and Private Interest in Recreation and Tourism – The Dutch Case: Concerning Consequences of a Shift from Citizen Role to Consumer Role,' *Leisure Studies*, vol. 12, 1993, p. 8.

70. S. Zukin, *The Cultures of Cities*, 2nd ed., Oxford, Blackwell, 1997.

71. P.L. Pearce, *The Ulysses Factor: Evaluating Visitors in Tourist Settings*, New York, Springer-Verlag, 1988.

72. R.D Sack, *Place, Modernity, and the Consumer's World*, Baltimore, Johns Hopkins University Press, 1992.

73. H. Ayala, 'Resort Hotel Landscape as an International Megatrend,' *Annals of Tourism Research*, vol. 18, no. 4, 1991, pp.568-587.

74. S.G. Davis, 'The Theme Park: Global Industry and Cultural Form,' *Media, Culture and Society*, vol. 18, 1996, pp.399-422.

75. T.G. Freitag, 'Enclave Tourism Development: For Whom the Benefits Roll?' *Annals of Tourism Research*, vol. 21, no. 3, 1994, pp. 538-554.

76. A.C. Stettner, 'Commodity or Community? Sustainable Development in Mountain Resorts,' *Tourism Recreation Research*, vol. 118, no. 1, 1993, pp. 3-10.

77. J. Krippendorf, *The Holiday Makers: Understanding the Impact of Leisure and Travel,* Oxford, Heinemann, 1987.

78. K.H. Wolffe (ed.), *The Sociology of Georg Simmel*, Glencoe, The Free Press, 1950.

79. G. Martinotti, 'The New Social Morphology of Cities,' *Management of Social Transformations Discussion Papers Series*, no. 16, Paris, UNESCO, 1996, p. 6.

80. J. van der Borg, 'Tourism Management in Venice, or How to Deal with Success,' in D. Tyler, Y. Guerrier, and M. Robertson (eds.), *Managing Tourism in Cities: Policy, Process and Practice*, Chichester, John Wiley & Sons 1998.

81. Zukin, *The Cultures of Cities*, p. 7.

82. M. Shackley, 'Visitors to the World's Oldest Cities: Aleppo and Damascus, Syria,' in Tyler, Guerrier and Robertson (eds.), *Managing Tourism in Cities.*

83. C.K. Heng and V. Chan, 'The Night Zone Storyline: Boat Quay, Clarke Quay, and Robertson Quay,' *Traditional Dwellings and Settlements Review*, vol. 11, no. 2, 2000.

84. J. Edwards, 'Urban Policy: The Victory of Form over Substance?' *Urban Studies*, vol. 34, no. 5, 1997, pp.825-845; and S. Zukin, *Landscapes of Power*, Berkeley, University of California Press, 1991.

85. D. Harvey, *The Urban Experience*, Oxford, Blackwell, 1989; and Britton, 'Tourism, Capital and Place.'

86. J. Urry, *The Tourist Gaze: Leisure and Travel in Contemporary Society*, London, Sage Publications, 1990.

87. B. Dodson and D. Kilian, 'From Port to Playground: The Redevelopment of the Victoria and Alfred Waterfront, Cape Town,' in Tyler, Guerrier and Robertson (eds.), *Managing Tourism in Cities.*

88. S. Papson, 'Spuriousness and Tourism: Politics of Two Canadian Provincial Governments,' *Annals of Tourism Research*, vol. 8, 1981, pp. 220-235.

89. G. Wall, 'Landscape Resources, Tourism and Landscape Change in Bali, Indonesia,' in Ringer (ed.), *Destinations: Cultural Landscapes of Tourism.*

90. M. Boyle and G. Hughes, 'The Politics of 'the Real': Discourses from the Left

on Glasgow's Role as European City of Culture 1990,' *Area*, vol. 23, 1991, pp. 217-228.

91. B. Wearing and S. Wearing, 'Identity and the Commodifcation of Leisure,' *Leisure Studies*, vol. 11, 1992, p. 8.

92. J. Friedman, *Cultural Identity and Global Process*, London, Sage Publications, 1994, p. 110.

93. M.F. Lanfant, 'International Tourism, Internationalization and the Challenge to Identity,' in Lanfant, J.B. Allcock, and E.M. Bruner (eds.), *International Tourism: Identity and Change*, 1995.

94. P. Bourdieu, *Distinction: A Social Critique of the Judgement of Taste*, London, Routledge, 1984.

95. C. Sachs-Jeantet, 'Managing Social Transformations in Cities,' *Management of Social Transformations Discussion Papers Series*, no. 2, Paris, UNESCO, 1996, p. 6.

96. J. Tomlinson, *Cultural Imperialism*, Baltimore, John Hopkins University Press, 1991.

97. A.D. King (ed.), *Culture, Globalization and the World-System*, London, Macmillan, 1991.

98. J. Börözc, *Leisure Migration: A Sociological Study on Tourism*, Kidlington, Pergamon, 1996, p. 194.

99. Z. Baumann, 'From Pilgrim to Tourist – or a Short History of Identity,' in S. Hall and P. DuGay (eds.), *Questions of Cultural Identity*, London, Sage Publications, 1996, p. 23.

100. U. Hannerz, *Cultural Complexity: Studies in the Social Organisation of Meaning*, New York, Columbia University Press, 1992.

101. D. Rowe, 'Leisure, Tourism and Australianness,' *Media, Culture and Society*, vol. 15, 1993, pp. 258-268.

102. J. Naisbitt and P. Auberdene, *Megatrends*, New York, Avon, 1990; and J. Naisbitt, *Global Paradox*, New York, Avon Books, 1995.

103. 'Tourism and Tribal Peoples: The New Imperialism,' Background Sheet, London, Survival, 1998.

104. Turner and Ash, *The Golden Hordes*.

105. D. Nash, 'Tourism as a Form of Imperialism, in Smith (ed.), *Hosts and Guests*.

106. K. Robins, 'Tradition and Translation: National Culture in its Global Context,' in J. Corner and S. Harvey (eds.), *Enterprise and Heritage: Crosscurrents of National Culture*, London, Routledge, 1991; and J. Craik, 'Peripheral Pleasures: The Peculiarities of Post-Colonial Tourism,' *Cultural Policy*, vol. 6, no. 1, 1994, pp. 21-31.

Bibliography

Altman, J.C., 'Tourism Dilemmas for Aboriginal Australians,' *Annals of Tourism Research*, 16, 1988, pp. 456-476.

Amin, S., *Unequal Development: An Essay on the Social Formations of Peripheral Capitalism*, New York, Monthly Review Press, 1978.

Amir, Y. and R. Ben-Ari, 'International Tourism, Ethnic Contact and Attitude Change,' *Journal of Social Issues*, vol. 41, no. 3, 1985, pp. 105-115.
Ayala, H., 'Resort Hotel Landscape as an International Megatrend,' *Annals of Tourism Research*, vol. 18, no. 4, 1991, pp. 568-587.
Baumann, Z., 'From Pilgrim to Tourist – or a Short History of Identity,' in S. Hall and P. DuGay (eds.), *Questions of Cultural Identity*, London, Sage Publications, 1996.
Bhabha, Homi, *The Location of Culture*, London, Routledge, 1994.
Boissevain, J., 'Ritual, Tourism and Cultural Commoditization in Malta: Culture by the Pound?' in T. Selwyn (ed.), *The Tourist Image: Myths and Myth Making in Tourism*, Chichester, John Wiley & Sons 1996.
Börözc, J., *Leisure Migration: A Sociological Study on Tourism*, Kidlington, Pergamon, 1996.
Bourdieu, Pierre, *Distinction: A Social Critique of the Judgement of Taste*, London, Routledge, 1984.
Boyle, M. and G. Hughes, 'The Politics of 'the Real': Discourses from the Left on Glasgow's Role as European City of Culture 1990,' *Area*, vol. 23, 1991, pp. 217-228.
Britton, S., 'The Political Economy in the Third World,' *Annals of Tourism Research*, vol. 9, 1982, pp. 329-331.
Britton, S., 'Tourism, Capital and Place: Towards a Critical Geography of Tourism,' *Environment and Planning D: Society and Space*, vol. 9, 1991, pp. 451-578.
Butler, R. and T. Hinch (eds.), *Tourism and Indigenous Peoples*, London, Routledge, 1996.
Cazes, G.H., 'The Growth of Tourism in the Developing Countries,' in UNESCO/AIEST Proceedings of Round Table, *Culture, Tourism, Development: Critical Issues for the XXIst Century*, Paris, UNESCO/AIEST, 1996.
Classen, C. and D. Howes, 'Epilogue: The Dynamics and Ethics of Cross-Cultural Consumption,' in Howes (ed.), *Cross-Cultural Consumption: Global Markets Local Realities*, London, Routledge, 1996.
Cohen, A.P., *The Symbolic Construction of Community*, Chichester, Ellis Horwood Ltd., 1985.
Cohen, E., 'Contemporary Tourism – Trends and Challenges: Sustainable Authenticity or Contrived Post-Modernity?' in R. Butler and D. Pearce (eds.), *Change in Tourism: People, Places, Processes*, London, Routledge, 1995.
Cohen, E., 'Towards a Scoiology of International Tourism,' *Social Research*, vol. 39, no. 1, 1972, pp. 64-82.
Cohen, E., 'authenticity and Commiditization in Tourism,' *Annals of Tourism Research*, vol. 15, 1987, pp. 371-386.
Coomes, A.E., *Reinventing Africa*, New Haven, Yale University Press, 1994.
Craik, J., 'Peripheral Pleasures: The Peculiarities of Post-Colonial Tourism,' *Cultural Policy*, vol. 6, no. 1, 1994, pp. 21-31.
Craik, J., 'The Culture of Tourism,' in C. Rojek and J. Urry (eds.), *Touring Cultures: Transformations of Travel and Theory*, London, Routledge, 1997.
Crick, M., 'Representations of International Tourism in the Social Sciences,' *Annual Review of Anthropology*, vol. 18, 1989, pp. 307-344.
Cukier, J., 'Tourism Employment in Bali: Trends and Implications,' in R. Butler and T. Hinch (eds.), *Tourism and Indigenous Peoples*, London, Routledge, 1996.
Cukier-Snow, J. and G. Wall, 'Tourism Employment: Perspectives from Bali,' *Tourism Management*, vol. 14, no. 3, 1993, pp. 195-201.

Daltabuit, M. and O. Pi-Sunyer, 'Tourism Development in Quintana roo, Mexico,' *Cultural Survival Quarterly*, vol. 14, no. 1, 1990, pp. 9-13.
Dann, G.M.S., 'Images of Destination People in Travelogues,' in R. Butler and T. Hinch (eds.), *Tourism and Indigenous Peoples*, London, Routledge, 1996.
Dann, G.M.S., 'The People of Tourist Brochures,' in T. Selwyn (ed.), *The Tourist Image: Myths and Myth Making in Tourism*, Chichester, John Wiley & Sons 1996.
Davis, S.G., 'The Theme Park: Global Industry and Cultural Form,' *Media, Culture and Society*, vol. 18, 1996, pp. 399-422.
de Burlo, C., 'Cultural Resistance and Ethnic Tourism on South Pentecost, Vanuatu,' in R. Butler and T. Hinch (eds.), *Tourism and Indigenous Peoples*, London, Routledge, 1996.
Dodson, B. and D. Kilian, 'From Port to Playground: The Redevelopment of the Victoria and Alfred Waterfront, Cape Town,' in D. Tyler, Y. Guerrier, and M. Robertson (eds.), *Managing Tourism in Cities: Policy, Process and Practice*, Chichester, John Wiley & Sons, 1998.
Douglas, M., *Implicit Meanings*, London, Routledge and Kegan Paul, 1977.
Edwards, J., 'Urban Policy: The Victory of Form over Substance?' *Urban Studies*, vol. 34, no. 5, 1997, pp. 825-845.
Erisman, M.H., 'Tourism and Cultural Dependency in the West Indies,' *Annals of Tourism Research*, vol. 10, 1983, pp. 337-361.
Featherstone, M. (ed.), *Global Culture: Nationalism, Globalization and Modernity*, London, Sage Publications Ltd., 1990.
Freitag, T.G., 'Enclave Tourism Development: For Whom the Benefits Roll?' *Annals of Tourism Research*, vol. 21, no. 3, 1994, pp. 538-554.
Friedman, J., *Cultural Identity and Global Process*, London, Sage Publications, 1994.
Furnham, A. and S. Bochner, *Culture Shock: Psychological Reactions to Unfamiliar Environments*, London, Routledge, 1986.
Gessner, V. and A. Schade, 'Conflicts of Culture in Cross-border Legal Relations: The Conception of a Research Topic in the Sociology of Law,' in M. Featherstone (ed.), *Global Culture: Nationalism, Globalization and Modernity*, London, Sage Publications, 1990.
Goering, P.G. 'The Response to Tourism in Ladakh,' *Cultural Survival Quarterly*, vol. 14, no. 1, 1990, pp. 20-25.
Gottman, J., *Forces Shaping Cities*, Department of Geography, University of Newcastle upon Tyne, 1978.
Greenwood, D., 'Culture by the Pound: An Anthropological Perspective on Tourism as Cultural Commoditisation,' in V.L. Smith (ed.), *Hosts and Guests: The Anthropology of Tourism*, 2nd ed., Philadelphia, University of Pennsylvania Press, 1989.
Hall, C.M. and M. Wouters, 'Issues in Antarctic Tourism,' in Hall and M.E. Johnston (eds.), *Polar Tourism: Tourism in the Arctic and Antarctic Regions*, Chichester, John Wiley & Sons 1995.
Hannerz, Ulf, 'Cosmopolitans and Locals in World Culture,' in M. Featherstone (ed.), *Global Culture: Nationalism, Globalization and Modernity*, London, Sage Publications, 1990.
Hannerz, Ulf, *Cultural Complexity: Studies in the Social Organisation of Meaning*, New York, Columbia University Press, 1992.
Harrison, D., 'Tourism to Less Developed Countries: The Social Consequences,' in Harrison (ed.), *Tourism and The Less Developed Countries*, London, Belhaven Press, 1992.

Harvey, David, *The Urban Experience*, Oxford, Blackwell, 1989.
Heng, Chye Kiang and Vivienne Chan, 'The Night Zone Storyline: Boat Quay, Clarke Quay, and Robertson Quay,' *Traditional Dwellings and Settlements Review*, vol. 11, no. 2, 2000.
Hollinshead, K., 'Marketing and Metaphysical Realism: The Disidentifications of Aboriginal Life and Traditions through Tourism,' in R. Butler and T. Hinch (eds.), *Tourism and Indigenous Peoples*, London, Routledge, 1996.
Johnston, M.E., 'Patterns and Issues in Arctic and Sub-Arctic Tourism,' in C.M. Hall and Johnston (eds.), *Polar Tourism: Tourism in the Arctic and Antarctic Regions*, Chichester, John Wiley & Sons, 1995.
Kakazu, H., 'Effects of Tourism Growth on Development in the Asia-Pacific Region: The Case of Small Islands,' in UNESCO/AIEST Proceedings of Round Table, *Culture, Tourism, Development: Critical Issues for the XXIst Century*, Paris, UNESCO/AIEST, 1996, pp. 7-14.
Keith, M. and A. Rogers (eds.), *Hollow Promises? Rhetoric and Reality in the Inner City*, London, Mansell, 1991.
King, Anthony D. (ed.), *Culture, Globalization and the World-System*, London, Macmillan, 1991.
King, E.M., *Creating Island Resorts*, London, Routledge, 1997.
Krippendorf, J., *The Holiday Makers: Understanding the Impact of Leisure and Travel*, Oxford, Heinemann, 1987.
Lanfant, M.F., 'International Tourism, Internationalization and the Challenge to Identity,' in Lanfant, J.B. Allcock, and E.M. Bruner (eds.), *International Tourism: Identity and Change*, 1995.
Lanfant, M.F., J. Allcock, and E. Bruner (eds.), *International Tourism: Identity and Change*, London, Sage Publications Ltd., 1995.
Lash, S. and J. Urry, *Economies of Signs and Spaces*, London, Sage Publications Ltd., 1994.
Lea, J., *Tourism and Development in the Third World*, London, Routledge, 1988.
Lengkeek, J., 'Collective and Private Interest in Recreation and Tourism – The Dutch Case: Concerning Consequences of a Shift from Citizen Role to Consumer Role,' *Leisure Studies*, vol.12, 1993.
MacCannell, Dean, 'Reconstructed Ethnicity: Tourism and Cultural Identity in Third World Communities,' *Annals of Tourism Research*, vol.11, 1984, pp. 361-377.
MacCannell, Dean, *Empty Meeting Grounds: The Tourist Papers*, London, Routledge, 1992.
Macdonald, S., 'A People's Story: Heritage, Identity and Authenticity,' in C. Rojek and J. Urry (eds.), *Touring Cultures: Transformations of Travel and Theory*, London, Routledge, 1997.
Martinotti, G., 'The New Social Morphology of Cities,' *Management of Social Transformations Discussion Papers Series*, no.16, Paris, UNESCO, 1996,
Mathieson, A. and G. Wall, *Tourism: Economic, Physical and Social Impacts*, London, Longman, 1982.
McKean, P.F., 'Towards a Theoretical Analysis of Tourism: Economic Dualism and Cultural Involution in Bali,' in V.L. Smith (ed.), *Hosts and Guests: The Anthropology of Tourism*, Philadelphia, University of Pennsylvania Press, 1989.
Medrano, J.D., 'Some Thematic and Strategic Priorities for Developing Research on Multi-Ethnic and Multi-Cultural Societies,' *Management of Social Transformations, Discussion Paper Series*, no. 13, Paris, UNESCO, 1996.

Mercer, D., 'The Uneasy Relationship between Tourism and Native Peoples: The Australian Experience,' in W. Theobald (ed.), *Global Tourism: The Next Decade*, second ed., Oxford, Butterworth-Heinemann, 1998.

Milton, K., *Environmentalism and Cultural Theory: Explaining the Role of Anthropology in Environmental Discourse*, London, Routledge, 1996.

Morris, M., 'Life as a Tourist Object in Australia,' in M.F. Lanfant, J. Allcock, and E. Bruner (eds.), *International Tourism: Identity and Change*, London, Sage Publications, 1995.

Naisbitt, J. and P. Auberdene, *Megatrends*, New York, Avon, 1990.

Naisbitt, J., *Global Paradox*, New York, Avon Books, 1995.

Nash, D., 'Tourism as a Form of Imperialism, in V.L. Smith (ed.), *Hosts and Guests: The Anthropology of Tourism*, second ed., Philadelphia, University of Pennsylvania Press, 1989.

Nash, D., *Anthropology of Tourism*, Kidlington, Pergamon, 1996.

Nettekoven, L., 'Mechanisms of Cultural Interaction,' in E. de Kadt (ed.), *Tourism: Passport to Development?* London, Oxford University Press, 1979.

Northern Territory Tourist Commission, Northern Territory Tourist Brochure, 1999

Papson, S., 'Spuriousness and Tourism: Politics of Two Canadian Provincial Governments,' *Annals of Tourism Research*, vol. 8, 1981, pp. 220-235.

Parris, R., 'Tourism and Cultural Interaction: Issues and Prospects for Sustainable Development,' in UNESCO/AIEST Proceedings of Round Table, *Culture, Tourism, Development: Critical Issues for the XXIst Century*, Paris, UNESCO/AIEST, 1996, pp. 36-40.

Parsons, H.L., *Marx and Engels on Ecology*, London, Greenwood, 1977.

Pearce, P.L., *The Ulysses Factor: Evaluating Visitors in Tourist Settings*, New York, Springer-Verlag, 1988.

Poon, A., *Tourism, Technology and Competitive Strategies*, Wallingford, CAB International, 1993.

Reisinger, Y., 'Social Contact between Tourists and Hosts of Different Cultural Backgrounds,' in A.V. Seaton (ed.), *Tourism: The State of the Art*, Chichester, John Wiley & Sons, 1994.

Richter, L., 'Political Instability and the Third World,' in D. Harrison (ed.), *Tourism and The Less Developed Countries*, London, Belhaven Press, 1992.

Ringer, G. (ed.), *Destinations: Cultural Landscapes of Tourism*, London, Routledge, 1998.

Ringer, G., 'Introduction,' in Ringer (ed.), *Destinations: Cultural Landscapes of Tourism*, London, Routledge, 1998.

Robins, K., 'Tradition and Translation: National Culture in its Global Context,' in J. Corner and S. Harvey (eds.), *Enterprise and Heritage: Crosscurrents of National Culture*, London, Routledge, 1991

Robinson, Mike, 'Collaboration and Cultural Consent: Re-focusing Sustainable Tourism,' *Journal of Sustainable Tourism*, forthcoming.

Roche, M., 'Mega-events and Micro-Modernisation: On the Sociology of the New Urban Tourism,' *British Journal of Sociology*, vol. 43, no. 4, 1992, pp. 563-600.

Rojek, C. and J. Urry, 'Transformations of Travel and Theory,' in Rojek and Urry (eds.), *Touring Cultures: Transformations of Travel and Theory*, London, Routledge, 1997.

Rowe, D., 'Leisure, Tourism and Australianness,' *Media, Culture and Society*, vol. 15, 1993, pp. 258-268.

Rowe, I.V., 'Taiwan: Carrying Capacity in an Industrializing Country,' in F.M. Go and C.L. Jenkins (eds.), *Tourism and Economic Development in Asia and Australasia*, London, Cassell, 1997.

Ryan, C., *Tourism, Terrorism and Violence: The Risks of Wider World Travel*, Conflict Study 244, London, Research Institute for the Study of Conflict and Terrorism, 1991.

Sachs-Jeantet, C., 'Managing Social Transformations in Cities,' *Management of Social Transformations Discussion Papers Series*, no. 2, Paris, UNESCO, 1996.

Sack, R.D, *Place, Modernity, and the Consumer's World*, Baltimore, John Hopkins University Press, 1992.

Said, E., *Orientalism*, New York, Pantheon, 1978.

Selwyn, T. (ed.), *The Tourist Image: Myths and Myth Making in Tourism*, Chichester, John Wiley & Sons, 1996.

Shackley, M., 'Visitors to the World's Oldest Cities: Aleppo and Damascus, Syria,' in D. Tyler, Y. Guerrier, and M. Robertson (eds.), *Managing Tourism in Cities: Policy, Process and Practice*, Chichester, John Wiley & Sons, 1998.

Silver, I., 'Marketing Authenticity in Third World Countries,' *Annals of Tourism Research*, vol. 20, 1993, pp. 3302-3318.

Sinclair, M.T., P. Alizadeh, and E.A.A. Onunga, 'The Structure of International Tourism and Tourism Development in Kenya,' in D. Harrison (ed.), *Tourism and The Less Developed Countries*, London, Belhaven Press, 1992.

Smith, A.D., 'Towards a Global Culture?' in M. Featherstone (ed.), *Global Culture: Nationalism, Globalization and Modernity*, London, Sage Publications, 1990.

Smith, S.L.J., 'Tourism as an Industry: Debates and Concepts,' in D. Ioannides and K.G. Debagge (eds.), *The Economic Geography of the Tourist Industry*, London, Routledge, 1998.

Smith, V.L. (ed.), *Hosts and Guests: The Anthropology of Tourism*, 2nd ed., Philadelphia, University of Pennsylvania Press, 1989.

Sofield, T.H.B., 'Anuha Island Resort: A Case Study of Failure,' in R. Butler and T. Hinch (eds.), *Tourism and Indigenous Peoples*, London, Routledge, 1996.

Stettner, A.C., 'Commodity or Community? Sustainable Development in Mountain Resorts,' *Tourism Recreation Research*, vol. 118, no. 1, 1993, pp. 3-10.

Survival International, 'Tourism and Tribal Peoples,' Background Sheet, London, 1995.

Survival, 'Tourism and Tribal Peoples: The New Imperialism,' Background Sheet, London, 1998.

Swinglehurst, E., 'Face to Face: The Socio-Cultural Impacts of Tourism,' in W. Theobald (ed.), *Global Tourism: The Next Decade*, Oxford, Butterworth Heinemann, 1994.

Theobald, W. (ed.), *Global Tourism: The Next Decade*, Oxford, Butterworth Heinemann, 1994.

Tomlinson, J., *Cultural Imperialism*, Baltimore, John Hopkins University Press, 1991.

Tucker, K. and M. Sundberg, *International Trade in Services*, London, Routledge, 1988.

Tunbridge, J.E. and G.J. Ashworth, *Dissonant Heritage: The Management of the Past as a Resource in Conflict*, Chichester, John Wiley & Sons, 1996.

Turner, L. and J. Ash, *The Golden Hordes*, London, Constable, 1975.

Tyler, D., Y. Guerrier, and M. Robertson (eds.), *Managing Tourism in Cities: Policy, Process and Practice*, Chichester, John Wiley & Sons 1998.

Uluru-Kata Tjuta National Park, 'Welcome to Uluru-Kata Tjuta National Park,' Information Sheet, undated.

UNESCO/AIEST, Proceedings of Round Table, *Culture, Tourism, Development: Critical Issues for the XXIst Century*, Paris, UNESCO/AIEST, 1996.

Urry, John and P. Macnaghten, *Contested Natures*, London, Sage Publications, 1998.

Urry, John, *Consuming Places*, London, Routledge, 1995.

Urry, John, *The Tourist Gaze: Leisure and Travel in Contemporary Society*, London, Sage Publications, 1990.

van der Borg, J., 'Tourism Management in Venice, or How to Deal with Success,' in D. Tyler, Y. Guerrier, and M. Robertson (eds.), *Managing Tourism in Cities: Policy, Process and Practice*, Chichester, John Wiley & Sons, 1998.

Vogel, S., 'Marx and Alienation from Nature,' *Social Theory and Practice*, vol. 14, no. 3, 1988, pp. 367-388.

Wall, G., 'Landscape Resources, Tourism and Landscape Change in Bali, Indonesia,' in G. Ringer (ed.), *Destinations: Cultural Landscapes of Tourism*, London, Routledge, 1998.

Wall, G., 'Partnerships Involving Indigenous Peoples in the Management of Heritage Sites,' in M. Robinson and P. Boniface (eds.), *Tourism and Cultural Conflicts*, Wallingford, CAB International, 1998.

Walle, A.H., 'Habits of Thought and Cultural Tourism, *Annals of Tourism Research*, vol. 23, 1996, pp. 874-890.

Watson, G.L. and J.P. Kopachevsky, 'Interpretations of Tourism as Commodity,' *Annals of Tourism Research*, vol. 21, 1994, pp. 643-660.

Wearing, B. and S. Wearing, 'Identity and the Commodifcation of Leisure,' *Leisure Studies*, vol. 11, 1992.

Weaver, D.B., *Ecotourism in the Less Developed World*, Wallingford, CAB International, 1998.

Wolffe, K.H. (ed.), *The Sociology of Georg Simmel*, Glencoe, The Free Press, 1950.

World Commission on Environment and Development, *Our Common Future*, Oxford, Oxford University Press, 1987.

World Tourism Organization, *Tourism 2020 Vision*, Madrid, 1997.

Zeppel, H.D., 'Land and Culture: Sustainable Tourism and Indigenous Peoples,' in C.M. Hall and A.A. Lew (eds.), *Sustainable Tourism: A Geographical Perspective*, London, Longman, 1998.

Zukin, Sharon, *Landscapes of Power*, Berkeley, University of California Press, 1991.

Zukin, Sharon, *The Cultures of Cities*, 2nd ed., Oxford, Blackwell, 1997.

Chapter 3

Learning to Consume: What is Heritage and When is it Traditional?

Nelson H. H. Graburn

The aim of this chapter is to take some fundamental concepts in the theme of this volume – heritage, tradition, and consumption – and to relate them to one another in a synthetic model. The concept of heritage requires a sense of ownership, and the consumption of heritage requires a sense of permission. The examination of heritage in relation to 'tradition' as both 'own' and 'owned' makes necessary the consideration of inter-generational cultural continuity, as well as the conception of 'others" and 'alterity'. This essay is therefore about what Foucault might have called the 'genealogy of heritage', but one which focuses on the more restricted but still important sense of the personal acquisition of this set of concepts – their ontogeny rather than their philogeny, which would be the political sense of the historical development and construction of the concept of heritage.[1] It is my claim that one can better understand more accessible adult categories and feelings toward heritage and the built environment by examining their underpinnings and temporal predecessors in earlier life stages.

Using this model, I examine these concepts from the specifically personal and developmental point of view.[2] And I avoid the more commonly discussed matters of political and class interest in heritage, contested exploitation of the past, cultural and nostalgic social uses of the past, and the political geography of space and the past.[3] These approaches to the macro-level relate to the above-mentioned philogeny of concepts of heritage and the past, the work of Bazin and Lanfant.[4]

Furthermore, much as today's anthropology goes beyond poststructural, postmodern and critical theory to focus on the more subjective, experiential, and personal narrative.[5] So in this chapter I turn to the examination of the micro-level, the individual level of the personal story of heritage and tradition. Though political, economic and macro-structural forces shape the cultural categories and social institutions which envelop the developing person, and these are also responsible for the physical, built environment, one must also look at the 'prepolitical' world of the child in which these adult concepts are comprehended, internalized, and eventually 'consumed.'[6]

Kinship, Family and Heritage

I, therefore, use the metaphor of kinship as a model of the relationship between the past and present, alliance and descent, inheritance and appropriation. Most anthropologists and sociologists are familiar with the use of this model and with the intellectual debates which it has generated. Kinship, *per se*, is also deemed cross-culturally to lie at the core of the relationships between the past and present, identity and alterity. In fact, all human beings are raised initially within a kin framework.[7] Although other social frames may be added later, there is some validity to the claim that all relationships are based on paradigms that are an expansion of the original social paradigm of identity, that of kinship.[8]

The metaphorical use of kinship as both folk and analytical model for other genres of cultural structure and wider sets of social relationships is of long standing in the empirical world, as well as within anthropological theory.[9] All students of tradition and the built environment are concerned with models of cultural transmission over time, whether the time is measured in centuries, generations, or tourist seasons. What is transmitted over time can be variously called inheritance, heritage, patrimony (for male-centred societies), tradition, birthright, etc. (see below for concepts in languages other than English). The nature of what is transmitted and what lasts over time is too often narrowly construed as something physical – such as built environment, properties, goods, heirlooms, money, etc. Of course, what is really transmitted is knowledge of and/or rights over those material things and, equally importantly, rights over non-material things such as memories, names, associations, stories, privileges, family traditions, memberships, and so on.

Indeed, the concept of the 'built environment' itself, whether structures or lands, can be examined and elaborated: from a cultural constructivist point of view all environments are 'built' in the sense that their perceived forms and, of course, their meanings are constructed entirely by the culturally productive activities of the local people.[10] Thus, one might say there is no such thing as Fujisan (Mount Fuji) without Japanese people, nor the Cote d'Azure without the French. Later in this chapter I will show that by taking a developmental, child's-eye approach, one is directed to look at a multi-sensory apprehension of the environment, rather than a narrow cognitive, categorical and ocular gaze.

Following the American sociologist Bernard Farber, I have previously used the concept of 'symbolic estate' in my analyses of cultural transmission both within and across cultural (and subcultural, such as ethnic) divisions.[11] Of course, in considering inheritance or heritage, one takes the larger meaning of all that one acquires from one's family or identity-group, not only at death, but, especially, during one's whole lifetime. Thus, one must consider heritage to have a meaning and range comparable to what Edward Tylor called 'culture' in his classic anthropological definition.[12]

The relationship between personal kinship and community heritage has not been unexplored. Indeed, one might expect such an obvious link to be primary in the social sciences. For instance, in their four-volume work *From Family Tree to Family History*, Finnegan and Drake, rather than examining the problematics of this link, assumed its existence and used it as a discovery procedure in examining the ethnography of ethnicity by following personal family trees back until they coincided with the generalities of ethnic history.[13] More to the point, Claude-Marie Bazin has explored how the French concept of *patrimoine* (usually translated as 'heritage' in English) has in the past several decades devolved from the personal legal realm of kinship inheritance to the national or even wider cultural realms of heritage.[14]

This chapter examines the fruitfulness of an analysis, following the metaphor of kinship, which has two key features. First, it concerns the cultural inheritance of a symbolic estate (a set of myths, rights, ownerships, stories and persona) that can be called patrimony (at least in a patrilineal society) – that is, the acquisition throughout one's life of an estate, material and non-material, believed to be one's own by right, by descent, by expectation (that is, by virtue of one's membership in a group with which one identifies and into which one usually enters by birth). Second, it explores how the very common (and perhaps touristic or imperialistic) acquisition and use of the symbolic estates from nearby sources other than birthright and descent (that is, from some nearby 'other') may be likened (in a kinship model) to matrimony – that is, the joining of one familiar or family group to another, whose otherness is defined by the incest taboo or rules of exogamy. More specifically, the parallel is with a man and his family acquiring a supplementary estate at marriage, usually called dowry (or for a woman's family, the bridewealth).

Thus, one may create a the metaphorical model of the acquisition of rights, ownerships and permissions – either from outside the 'symbolic family' or from prior generations within the family (that is, by 'alliance or descent'[15]) if one equates the national or world system with familial terms. Class and ethnic identities, like kinship and family, are also forms of identity and 'belonging' with respect to their own and others' heritage, and they are amenable to similar analyses.

Nostalgia is the ever-unsatisfied yearning to be able to return to a past time – to live and be engaged creatively in that past time, rather than just to inherit it as the 'received truth.' In the familial model, if 'the child is father of the man', nostalgia may be an 'Oedipal' desire, a wish to return to a prior state of creative power, to assume the position of the parent or ancestor for whom one's heritage from the past was their own life.

Just as a child has to be taught what families are and who his/her family is and is not, many authors refer to the fact that all heritage is 'constructed', a cultural model constantly refined by someone (powers that be, educators, or 'cultural producers' in MacCannell's terms[16]) and transmitted to members or segments of a population. One set of questions is about what agency or agencies socialize people into unproblematically claiming, knowing and visiting their 'own heritage', and what agencies strive to modify or add to the heritage that one is brought up to believe in (and for what purposes); or, alternatively, what agencies chose what artefacts be used, refurbished or created as *lieu de memoire* of heritage,[17] and how some peoples or agencies exercise resistance against their own group's heritage being appropriated by the larger hegemonic society.

Heritage, in this sense of a material and symbolic estate, is a large part of identity. Thus defined, heritage may pertain to so-called primordial groups, such as the family, clan, ethnic group, race, and nation or – for some people – all humanity, or even all living creatures. Cross-cutting these primordial categories are others which appear equally primordial to some people, for example, gender, place, space, or even generation. Not only does heritage give a concrete sense of shared identity or belonging, but it also demands responsibility (for preservation, respect, and safety) in the use or enhancement of each of these heritage-identity forms. These feelings can take the form of strong attachments to both forms and spaces in the built environment.

Alterities as Tabooed and Consumed

Based on the above analysis, one might propose the categorical existence of levels of identification with the material and symbolic aspects of heritage: (1) that which is 'naturally' one's own, one's patrimony, one's own 'by birthright' – what one might call 'ascribed' heritage, borrowing a categorical metaphor from sociological role theory; (2) that which may be acquired, appropriated or 'achieved' by an expansion of ownership-identity; and (3) that which is alien or other, or true alterity. This tripartite mental model has a curious parallel with some fundamental categories of kinship and affinity (behaviour towards blood-related relatives and relatives in-law), and with the consumption or avoidance of animal-derived foods. It would not be amiss here to point out that Urry has called the culturally constructed recognition of heritage and the consequent tourist gaze 'the consumption of places', also the title of his recent book.[18]

The late English anthropologist Edmund Leach, in his stimulating paper 'Anthropological Aspects of Language: Animal Categories and Verbal Abuse', pointed out parallels between the ways in which he thought that people (or at least middle-class British people) think about and relate to people at certain culturally defined distances, and to animals at an isomorphically defined set of social 'distances.'[19] One might briefly sum up Leach's contention that people structure their social universe in the same way as their view of nature – and, I am asserting here, as their view of heritage. Briefly, they categorize people into close family (usually living in their house or nearby); distant family and close non-family, who are neighbours and friends who are known and recognized; and strangers, or distant people, who are unknown or not acknowledged. From a young man's point of view with respect to women (and one could make this vice versa just as easily), women who are close – sisters and mothers – are incestuously tabooed and not marriageable; those who are known but more distant, such as distant cousins and friends, and culturally close non-relatives, are the ones among whom sexual partners and potential spouses are selected; and (for the historical classes specified here) complete strangers and the culturally different are not marriageable.

Paralleling these distinctions with relation to animals, those which are very close, those 'of the house and garden', are pets and are not edible at all. Those of the farm (domesticated) and those of the field (wild animals, and birds, which come into the fields) are killed and eaten – although there are usually fairly explicit rules as to the seasons when animals of the field (deer, pheasants, duck) can be eaten and the life stages (or sexual maturity) when domestically reared animals (lamb, ewe, veal, beef, capon, rooster, and so on) are said to be edible. Then there are the animals which are 'strangers' and are not found in people's usual environment, which are truly wild or come from other environments. These animals are known by name but are wild or of the zoo, and are not supposed to be eaten.[20] Leach pointed out that certain animals are anomalous to these categories, such as foxes or pussies; and not only are they not considered to be food, but they are available as 'terms of abuse' when applied to humans.[21]

This tripartite structure – too close, within reach, and too far – is a useful paradigm to look at heritage. One can examine this model not only as a static structured set (the received truth, known to all adults), but ontogenetically, as the model is acquired during the lifecycle. For instance, a baby or toddler does not distinguish between 'own' (too close) and marriageable (indeed, Freudian views insist that the polymorphous feelings of children make close family into sexual objects). It is only when a series of people outside the household becomes well known, and when the 'truth' about the relationship between sexual activity and babies comes along, that such 'adult' distinctions are made. Similarly, children learn about pets very early but may not know that they are eating farm animals until they are older. Wild and zoo animals are also introduced early into the cognitive ken of young children (through picture books, the media, and stuffed animals) but with little or no reference to edibility (figure 3.1).

	Self/Close	Familiar but More Distant	Strange or Very Distant
Social	Close family	Neighbours and friends	Strangers, others
Sex	Tabooed	Permitted	No regular relations
Animals	Pets	Farm and field	Wild or zoo
Edibility	Tabooed	Eaten, with rules	Not usually eaten
Heritage	Family	Cultural or ethnic	Foreign or world
Feelings	Primary	Adult	Optional

Figure 3.1. Proximity and familiarity relationships.

It becomes immediately apparent that the subject matter of this chapter, heritage, the third category above, needs further elaboration and examination. One must also consider the historically specific claim of the particular structural scheme that Leach laid out: that it was limited to English people of a certain place and time. For the past 50 years or so many of the middle classes of the Western world, including the English, have come to claim large parts of the whole world's natural and built environment to be 'world heritage' to be not only visited as tourists but to be protected by international organizations such as UNESCO, or by national membership organizations such as Greenpeace or the Sierra Club.[22] One might also note that, because of increased opportunities for travel and migration, intermarriage with peoples formerly thought strange and distant is becoming much more common. Even more obvious is the recent availability for consumption of all sorts of exotic cuisines and ingredients from all over the world.

Nevertheless, the suggestion is made here that with respect to the feelings or recognition of heritage, perhaps 'ontogeny recapitulates phylogeny'.[23]

Family and Nation

Let me next examine briefly some applications or extensions of the kinship metaphor to ethnicity and nationalism. The extension of kinship-like relationships to other peoples and nations does not always carry positive values (nor should one expect it, considering sibling rivalries and the Oedipus complex). I shall begin with some uses of the metaphor of siblinghood, that is, same-generation relationships. For instance, the

European Portuguese speak of two kinds of 'brothers'. One kind are the Spaniards, who are similar and yet who have historically often been aligned on the other side of geopolitical conflicts and military battles. Portuguese may also refer to Brazilians as *irmaos brasileiros*. Brazil is said to be a brother nation because the eighteenth-century Prince Dom Pedro *segundo* was sent out to serve as regent of Brazil, and while he was there, he elevated himself to 'King' of Brazil, resulting in an untrusted fraternity with Portugal, and eventually in Brazilian national independence.

For the nations of Eastern Europe the concept of 'brother nations', fellow Slavs, was an easy and common call to solidarity during the Communist era, and has been a more sporadic basis for alliance since then. For Egyptians, since World War II and independence under the ascendancy of Gamel Abdul Nasser, calls to solidarity and identity with 'brother nations' usually always referred to Arabs – in Iraq, Palestine, Libya, and so on. But after Anwar Sadat signed a peace treaty with Israel, this alliance was downplayed, and brotherly relations were encouraged with other Muslim nations, including non-Arabs in Pakistan and Indonesia. This particular kind of alliance has also been emphasized more recently with the Muslims of Bosnia and the Albanians of Kosovo – peoples who never before had been seen as sharing heritage. Yet during this same period, familial relationships were spoken of with regard to the 'Israelis, our cousins', supposedly emphasizing the symbolic consanguineal links. Yet this is the same area of the world that is reputed to have originated the saying, 'I and my brother against our cousins . . .'

Thus, metaphorical nationalistic and ethnic relationships of kinship and consanguinity express intrinsic linkages but not necessarily alliances. According to Shakespeare, Henry V of England spoke of 'our sweet sister France' during a period when there were more international wars than centuries. More recently, the English have been forced to admit their cousin-like relationship to the Germans (through historical and linguistic connections as well as the genealogies of royalty) in an era during which there have been two 'world wars' between the nations. Similarly, the Japanese have long acknowledged their sibling-like relationships to the other East Asian Countries, such that they are all *kyodai* together. Yet within that neutral term for siblings they imagine themselves to be *ani*, older brother, to those peoples.

Looking now at the kinship metaphor of the vertical relations of descent rather than horizontal, intra-generational relationships, one may find similar anomalies. For instance, the Egyptians who usually identify themselves as Arabs (invaders of Egypt with the initial expansion of Islam from what is now Saudi Arabia) in fact claim the 'ancient Egyptians' as some kind of ancestors, and they do not call them *jahili*, meaning 'pagans', even though they do call the pre-Muslim ancestors of all other Arab nations *jahili*. But not far back in their history the Egyptians also ignored these pagan ancestors whose built environment so dominated their living space. Perhaps the recent independence of Egypt and its heavy dependence on overseas tourism have forced consideration of the importance and closeness of the pharaohs and other ancients, and at the same time given Egyptians pride as a people previously neglected or ignored.

As another example, the Japanese have looked to their emperor as a father figure, and have even considered the nation as a body, *koka* or *kokutai*, with the emperor as the head – though such a view may have been promulgated by modernizing political forces after the 1868 Meiji Restoration.[24] However, in spite of their enormous debt to Chinese civilization for such things as half their language, literacy, Buddhism, city planning, and the Confucian ethic, Japanese do not normally name China as their mother or father nation.

Among the European nations, another set of patterns may be found. For instance, the British and the Germans, who speak languages not derived from Latin, and who were at the margins of the Roman Empire, have been taught that they have every right to consider the Romans and their empire as their forebears. Both nations have splendid material remains of Roman occupation (all over England and in the southwestern part of Germany, particularly along the Mosel at Trier). Yet other European nations seemingly just as connected to Rome do not include her remains or her occupation as parts of their histories. For instance, neither Bulgaria, which was part of Thrace and has splendid Roman amphitheatres at Plovdiv, nor Portugal, whose Alentejo region (encompassing the town of Evora with its magnificent temple of Diana) was called the 'breadbasket of the Roman Empire', looks back to ancient Rome as an ancestor. The Portuguese see their history starting with Prince Alphonse, who in the thirteenth century started the expulsion of the Moors. Perhaps it was this long intervention of Moorish occupation that cut off the Portuguese (a country with a very close derivative of Latin as its language) from Rome. Likewise, the people of Bulgaria claim that their mythical Finno-Ugric (non-Slavic and non-Roman) Bulgarian ancestors came from the East, and look upon this fifth-century invasion as the beginning of their history – even though the Bulgarian state (like the Portuguese) is quite willing to celebrate its Roman ruins for the sake of attracting and entertaining tourists. Thus, one must conclude that the mere presence of a visible, historical built environment is not enough to determine that it must be considered the heritage of the local inhabitants. On the contrary, it is the specific cultural and historical scope and meanings of the terms heritage and tradition that are most important.

On this note it is telling that under the dictator Salazar the people of Portugal were encouraged to think of themselves, as opposed to the rest of the world, in terms of the trinity: God – Nation – Family. And in this exhortation the word chosen for nation was not the politically accurate but emotionally neutral word *nacion*, but the powerful and kin-based appellation *patria*, 'the fatherland'. Unlike Western Europeans, the Russians and other Slavs are more likely to speak of the motherland, perhaps emphasizing the same non-material, emotional bond that is reflected in the use of *nastledstvo* below. The Japanese too speak of *bokokugo*, the 'mother tongue'. But this is an exception to their generally patriarchal national feelings, and is probably a direct Meiji-era translation of the German term *Mutterzunge*. In most of the nations exemplified here there is a close linguistic, and hence semantic, relationship between family heritage and cultural or national heritage (figure 3.2).

	Family	Cultural	National
Bulgaria	*nastledstvo*	*nastledstvo* or *traditia*	*nastledstvo* or *traditia*
Portugal	*herenca* *patrimonio*	*herenca cultural* or *herenca dos antepassados*	*herenca nacionale*
German	*Erbe(-fall)*	*kultur Erbe*	
	Familien tradition	*Brauch*	*die Brauche*
Turkey	*miras*	*miras kulturel*	*miras milli*
Egypt	*mirath*	*turath*	*turath*
Japan	*issan issan* (material)	(material)	*kokuhosai issan*
	dentoo (traditions: practice/spiritual)		*kokuminteki dentoo*

Figure 3.2. Relationships between family, cultural and national heritages.

Childhood, Heritage and Otherness

Let me next examine some hypotheses about the potentially important relationship between the imagined worlds of family, nature and cultural or built heritage and about the way in which these relationships are acquired and grow, not in historical time but in a person's young life. One possible source of stimulation for this project, as is so often true in the social sciences, is a re-examination of one's own childhood experiences. As Okely has eloquently outlined such a research tactic: 'The ethnographic material for this preliminary enquiry is largely autobiographical, my main informant being myself . . . I deliberately confront the notion of objectivity in research by starting with the subjective, and working from the self outwards.'[25]

A complex set of circumstances which I will not repeat here led me to think in my preschool life in England that I was something called 'Malay', and that I was also 'like a native'.[26] I was brought up by parents who, with their relatives and friends who often visited us, had spent many years or decades of their adult lives in Malaya and sometimes spoke Malay to each other. Their English retirement home was a farmhouse filled with material items and photos of Malaya and copies of the monthly journal *British Malaya*. I thus distinguished myself from other people, children, servants and friends who didn't know anything about Malaya and seemed to be a rather uninteresting bunch of people. Later on I was given specific information about my family's long roots in England: special moral, symbolic and material characteristics (which I would now call family heritage). At school, of course, I also learned about the history and geography of Britain, including Roman Britain (the remains of a Romano-British temple lay in the woods within walking distance of our house). And on endless occasions my attention was directed toward the 'pink bits' (the empire, of which Malaya was one corner) on the world map, toward which my classmates and I were given a sense of ownership and responsibility.

Upon reflection, I discovered that my young worldview was only atypical of English people of my class because of the additional distinctiveness of the Malay connection. But there were plenty of others, including my boarding-schoolmates, who had special connections to other 'pink bits', such as India, South Africa, Kenya, and so on. They too had relatives, family stories, pictures, material objects, and even residences (which I did not have) in these far-off places. They also had stories of other 'natives' who, for the most part, had admired qualities and exotic skills. In my professional career I have run across parallel cases where the children of colonizers have similar complex personal and family heritages. For example, many Portuguese children raised in Angola and Mozambique have identified themselves as Africans, with social lives, built and natural environments different from and, they thought, far superior to the far-off little country of Portugal.

Many such colonial offspring ('colonial brats') have identified in an 'Oedipal' fashion with the native people they knew against the old generation of their own family and those people such as schoolmasters in their home country who stood as authorities in *loco parentis*. My attention to these phenomena was stimulated by Haya Bar-Itzchak, who reported that some young Israeli sabra defended and tried to save aspects of Palestinian culture, particularly agricultural landscapes and certain native trees and plants, against Jewish 'modernization', which their parents were trying to carry out.[27] This 'guilt-like' attitude toward, or emotional identity with, a cultural and material heritage which they or their parents have helped change or destroy is what Renato Rosaldo has defined as 'imperial nostalgia', a common attitude among those responsible for the changes brought about by colonial regimes.[28] Yet for the children brought up in such colonial situations, personal guilt cannot be direct, for, as children, they have destroyed and changed nothing.[29] Thus, the pro-local identifications of the children are probably strongly 'Oedipal' antagonisms, but with a dose of 'inherited' and probably unconscious guilt.

There are very few good studies that link the worldview and experiences of childhood to the consumption of place by adults. Such studies require either an inordinately long period of field research or a theoretical orientation suggesting specific inter-generational lines of enquiry. One brief but exemplary study by an Ethiopian undergraduate student at Berkeley examined the city of Oakland's Black (African-American) peoples' feelings about city areas where they felt most comfortable for their recreation as teenagers and adults.[30] He found that they preferred city parks and street corners, where they could play music, meet their friends, and engage in friendly sports and pick-up games, because those were places where they had been taken by their parents when children, and where a constant cross-generational mix of people was always to be found. This contrasted with their discomfort in visiting the public and much 'wilder' parklands at the top of the Oakland-Berkeley hills where white, middle-class people preferred to go for walks and picnics, and which are valued for the quiet, the sense of nature, and the relatively low density of people. Some Blacks even pointed out the homology between social class and the topography and elevation of residence.

Growing up in Germany after World War II did not provide such inter-generational securities, nor comfort and pride in the built environment, especially for those in Soviet-threatened Berlin and other largely destroyed cities. For those children and young people, identity was precarious. There were the obvious splits between the blameless self and not-so-blameless parents; between those who revolted against their society and had a degree of self-hatred, and those who tried to preserve continuity and pride in the past; between those who admired the all-too-close conquering forces of the United States, as well as France and England, and those who looked forward to their departure. For Berliners, the Americans were particularly seen as saviours (with the Russian presence nearby), and many consciously emulated what they saw as American social, artistic and material values. There were those who upheld the scattered remaining traditionally German houses, *rathausen*, and churches as emblematic of Germany proper. However, a majority felt them to be tainted with Nazism, and they therefore emphasized other family, class or local traditions, whether they were styles of visiting and hospitality, or explorations of foreign food and music such as Jazz and Latin dancing. There were few automatic places of nostalgia or social relaxation. And though monuments such as the Reichstag or the Brandenburg Gate constituted shared heritage, their meanings were neither clear nor sources of pride. The political orientations of different groups – particularly openness to foreign things and ideas as oposed to fear of the non-German – mapped rifts even within the middle classes.

Unlike the neatly structured and isolated English countryside used in Leach's model, Germany had visible fault lines of foreignness all around; the neat tripartite categories of the model, close – near – far, no longer existed. But the Germany of the time did share something with the English countryside – that is, the presence of the 'other inside'. For majority populations the 'other-nearby', be they Jews, Turks (in Germany), Gypsies in many countries, or even those from different regions who may work close by, offer a prime opportunity to define self and own by contrast. The particular aspects of the nature of self and own are enhanced by contrast with this nearby other. For instance, in my own experience Gypsies emphasized my stability vs. their travel, and my house in a garden vs. their caravan in a borrowed field. Okely has shown from the Gypsy point of view that the behaviour and built environment of the Georgios, that is, the mainstream majority population who are the Gypsies' 'other-nearby', provide key identity-markers and negative role models for them too.[31]

For those brought up in Slavic Eastern Europe, parts of which were, like Germany, destroyed in the war, history, continuity and solidarity were important. The recent past was emphasized and its physical remains and locations (the 'destroyed environment') were very meaningful. Things Slavic, particularly Russian, were admired, and for many growing up after the World War II there was an identification between Slavic and Communist. But the outwardly accepted model of the Communist state could only be upheld by the suppression of certain facts from the past, bad class origins, foreign connections, suspicious deeds, even attractions to foreign (that is, Western) styles of art, clothing, buildings, etc. Thus, children were often not told some intimate and important parts of their parents' pasts.

For these Eastern Europeans the 'internal' and the 'external others' were the Tigany (Gypsies) and Westerners, respectively. The latter were objects of suspicion and fear as anti-Communists, whereas the Gypsies were stylistically and materially different, threatening to adults but attractive to curious children. The other major division, however, was between the people of a national capital and the rest of a county's population, such as between Sofia and Plovdiv and the rural areas, between Moscow and Leningrad and outlying areas. The ethnic/racial identity was very strong, particularly against the Germans as recent foes, and Turks as old enemies. In the face of the past and recent conquests, pride in the national and local built environment was high, and rebuilding in 'traditional' styles was a primary aim. Continuity was emphasized in the buildings, but within a socialist framework. On the surface socialism was also quick to emphasize such icons of tradition and continuity as architecture, annual celebrations, dances, and peasant costumes.

Yet these promoted and visible aspects of the constructed environment, these *traditia*, are reminders of Stalin's ethnic and minority policies when he was Commissioner of Nationalities in the early USSR: 'Nationalist in Form and Socialist in Content'. And these physical symbols were never as valued as the inner, spiritual (ethnic and national) heritage, *nastledstvo*, which was inside the person, could not be taken away, and was worth fighting for. In Eastern Europe one characteristically finds that those separable and tangible *traditia* were first preserved and promoted in the service of 'nationalism' as allowed by Communism; and more recently it is exactly the same visible and manmade phenomena which are being promoted for tourism and commercialization – these are not the things one would die for.

But What is Traditional?

This discussion should emphasize that the application of the concept 'traditional' is not only very variable, but it is of lesser concern to young people first experiencing their extra-familial identities. Not only do they not know anything about actual historical time-depth, but they are creatures of the 'here and now' and measure such things on a very personal scale. 'Tradition' like the associated concepts 'heritage' and 'nostalgia' are the products of modernity, which was itself born of a sense of change and history.[32] Children are, if anything, 'premodern', and these modern concepts do not have much salience until the children go through the various life stages and rites of passage implicit in the earlier tripartite formulation discussed above.

For children, tradition has few of its multiple meanings – premodern, unscientific, preindustrial, authentic, non-commercial. But it can express another set of connotations, especially when holidays are spent with grandparents or the parents' older friends and relatives in the countryside. These connotations have to do with being old and surviving, and being different from the ordinary. They consist of connotations such as survival from an earlier age, a reservoir of history and knowledge, being old fashioned – and, especially if the relatives are old, sick, or visibly weakening, connotations such as getting scarcer, and, hence, special or 'sacred', or plainly 'about to disappear'.[33] The actual word 'tradition' (or its equivalents in other languages) may not be used by children, but a set of attitudes, perhaps ambivalent, can easily stem quite unconsciously from such intimate situations where the children are made aware of their youth, inter-generational change, and the impermanence of things and people whom they might have taken for granted.

Most of my European informants have also told me that their families have seemed to make conscious efforts to teach them about heritage and traditions, both at the family and cultural/national levels. At an early age many of them were thus taken to visit the places of their parents' upbringing or old and preserved places described in terms of awe and reverence. These might be the remaining old house, monumental building, or church mentioned above. Or the children may be exposed to more intangible experiences and distinctive physical environments which might seem to counteract the everyday life of modernity. For instance, I have described how in today's Japan, where most children are brought up in big cities in small modern apartments, anxious grandparents and other relatives often take their grandchildren, and quite often their own children too, to *onsen ryokan*, country inns with hot springs.[34] Here the younger generation can learn about 'authentic, traditional' Japanese life, far from the Western-influenced high-rises and emporia of the cities, in romantic buildings which carefully preserve aspects of traditional architecture.[35] They cast off their modern (that is, Western-derived) clothes for a *yukata* and pairs of *geta*, eat high-class *kaiseki* (traditional foods), soak in the 'natural' odiferous waters of the volcanic hot springs, and sleep on straw *tatami* floors. Such activities are *dentooteki*, 'traditional', and help keep up the Japanese belief that they can identify themselves with their heritage as an '*onsen* civilization'.[36]

Discussion and Conclusions

The purpose of this chapter has been to expose the sources and variety of attitudes towards those things called heritage in the modern world, and to examine through people's personal lives how such attitudes develop. This examination has been carried out by inquiring into the topics of family and kinship not only as sources of identity and paradigms of belonging and ownership, but also as the generators of feelings and attitudes toward components of identity as they are modified through life stages. These concepts brought together here do not usually nestle in the same work, yet they stem from a theoretical model which purports to be able to relate the personal, the temporal, and the contextual to the perceived environment.

One can make a few remarks about the consequences of such an approach. First of all, one cannot assume that informants have, or had in their crucial earlier life stages, the same concepts and categories they do as adults. For instance, as Bloch has recently pointed out, it is very difficult to tell if the information in an adult's memory of their childhood experiences is from first-hand experiences or from another adult's telling of those experiences.[37] One does not yet know if there are one, or two, or more kinds of memory. Second, such research, especially cross-cultural applications, further destabilizes such value-laden concepts as authority, tradition and identity. Though undoubtedly these concepts are widespread, if not universal, they may have recently spread from their original historical sources to the leaders of more recently modern societies.[38]

Can one now consider further relations between this model and vernacular architecture and traditional environments? First, I think it is of interest to understand from where adults get their subconscious feelings, their hunches about familiarity, comfort and excitement – their 'habitus' in Bourdieu's terms.[39] It is a topic that has been all but ignored by adult-centred investigators of this field. To get at such issues, one must extend investigations beyond the Western world's characteristically adult-dominant ocular and logo-centric frames. Kenneth Frampton made such a plea to those who study architecture in his essay 'Place, Form and Identity':

> Visual/Tactile: These two alternative modes of experiencing the environment address the way in which the architectural object is open to levels of perception other than the visual stimulus afforded by the object. Architecture possesses the unique capacity for being experienced by the entire sensorium . . . Air movement, acoustics, ambient temperature and smell, all these factors affect our experiences of space.[40]

A similar plea has recently been aimed at anthropologists. 'Recent years have witnessed an increasing number of calls for anthropology to devote more attention to non-visual modes of perception . . . [to] help us escape the supposed malaise of modern "ocular-centrism".'[41] Those who study tourism have perhaps met this criticisms earlier, particularly those who study the relationship between adult vacation needs for the physical rather than cultural (historical and ethnic) sightseeing, or in addition to it.[42] The Finns are elaborately aware of the many aspects of their beloved rural habitus, the great woods, metsaa, where they can hunt mushrooms and smell the piny forests; explorations of such areas is a primary experience even for many urban Finns today. The Japanese are perhaps more aware than most of the human's multi-sensory needs (in spite of their alleged camera-centrism). As I have pointed out in my analysis of the intensely popular Japanese hot spring resort industry, all the senses are catered to, and even the advertisements make their appeals to the senses of touch, smell and sound, indicating that the Japanese may be well aware of the multiple sensory facets of their internal habitus and heritage.[43]

To summarize, I have suggested that the adult world be examined in the light of each adult's earlier life experiences. As the saying goes, 'The child is father of the man.' And if one wishes to understand the man, with all his prejudices, needs and tastes, one must consult the father who gave him the cultural categories, ambivalences and directions for his life's trajectory.

Notes

1. M.-F. Lanfant, 'The Genealogy of Patrimony,' in N. Graburn and Lanfant (eds.), *Tourism, Heritage and Patrimony*, in preparation.

2. This approach was originally presented in my paper 'Tourism, Leisure and Museums,' plenary address to the Canadian Museums Association, Halifax, Nova Scotia, in 1982, and in modified form in my paper 'The Construction of Heritage

and Nostalgia: the Comparative Evidence,' delivered in the session 'Patrimony, Tourism and Nostalgia' at the XIIIth World Congress of Sociology, Bielefeld, Germany in 1994. For this chapter I have interviewed a number of informants on matters of upbringing and identity; most are young middle-class adults. I specifically chose to talk with these unmarried people or young couples without children because I thought they would have a clearer recall of the development of their identities and environmental awarenesses, and particularly because I think that parents reformulate their stories in order to pass them on to their offspring. In addition, I am grateful to Professor Shinji Yamashita of the University of Tokyo and Professor Jerry Eades of Shiga University for their discussion of Japanese materials.

3. In terms of political and class interest in heritage, see G.J. Ashworth and P.J. Larkham, *Building a New Heritage: Tourism, Culture and Identity in the New Europe*, London, Routledge, 1994.; and J.E. Tunbridge and G.J. Ashworth, *Dissonant Heritage: The Management of the Past as a Resource in Conflict*, Chichester and New York, J. Wiley & Sons, 1996. In terms of cultural and nostalgic social uses of the past, see G.J. Ashworth, *On Tragedy and Renaissance: The Role of the Loyalist and Acadian Heritage Interpretations in Canadian Place Identities*, Groningen, Geo Pers, 1993; D. Lowenthal, *The Past is a Foreign Country*, Cambridge, Cambridge University Press, 1982; and D. Lowenthal and M. Binney (eds.), *Our Past Before Us: Why do we Save it?* London, Temple Smith, 1981. On the political geography of space and the past, see D. Crowe (ed.), *Geography and Identity: Living and Exploring Geopolitics of Identity*, Washington, D.C., Maisonneuve Press, 1996; R.D. Sack, *Place, Modernity and the Consumers World*, Baltimore, Johns Hopkins University Press, 1992; and E.W. Soja, *Postmodern Geographies: The Reassertion of Space in Critical Social Theory*, London and New York, Verso, 1989.

4. C.-M. Bazin, 'Industrial Heritage in the Tourism Process in France,' in M.-F. Lanfant, E. Bruner and J. Allcock (eds.), *International Tourism: Change and Identity*, London, Sage Publicaations, 1995; and Lanfant, 'The Genealogy of Patrimony.'

5. On the more subjective, see N.H.H. Graburn, 'Prologue: Southeast Asia in my Mind,' in J. Forshee and S. Cate (eds.), *Convergent Interests: Traders, Travelers and Tourists in Southeast Asia*, Berkeley, Center for Southeast Asian Studies, 1999; and J. Okely, *Own or Other Culture*, London, Routledge, 1996. On the experiential, see S. Lavie, K. Narayan, and R. Rosaldo (eds.), *Creativity/Anthropology*, Ithaca, NY, Cornell University Press, 1993; L. Abu-Lughod, *Veiled Sentiments: Honor and Poetry in Bedouin Society*, Berkeley, University of California Press, 1986; and *Writing Women's Worlds: Bedouin Stories*, Berkeley, University of California Press, 1993. On personal narrative, see J.B. Boddy and V. Lee, *Aman: The Story of a Somali Girl*, New York, Pantheon Books, 1994; and K. Narayan, *Storytellers, Saints, and Scoundrels: Folk Narrative in Hindu Religious Teaching*, Philadelphia, University of Pennsylvania Press, 1989.

6. The child is prepolitical and premodern. These phrases are intended to indicate that the young child, with little or no schooling, does not make distinctions about the world as 'traditional' or 'modern.' The child has little sense of history and rarely thinks about times long 'before I was born,' and hence, like the premoderns, does not see systemic change, directional progress (except for her or his growth), or the possibilities of creating other worlds. Thus, the child is not engaged in efforts to

change 'the system,' and does not (with the possible exception of Oedipal fantasies, which in fact reproduce the system) side with one set of ideological beliefs against another possible narrative of power. In this sense, the child is also prepolitical, which is close to but not necessarily something characteristic of premoderns.

7. J. Goody (ed.), *The Character of Kinship*, Cambridge, Cambridge University Press, 1971; M. Strathern, *Kinship at the Core: An Anthropology of Elmdon, a Village in North-West Essex*, Cambridge, Cambridge University Press, 1981; and M. Strathern, *After Nature: English Kinship in the Late Twentieth Century*, Cambridge, Cambridge University Press, 1992.

8. D. Freeman, 'Kinship, Attachment and the Primary Bond,' in Goody (ed.), *The Character of Kinship*, pp. 109-120.

9. D.M. Schneider, 'What is Kinship All About?' in P. Reining (ed.), *Kinship Studies in the Morgan Centennial Year*, Washington, D.C., Anthropological Society of Washington, 1971, pp. 320-363.

10. D. MacCannell, *The Tourist: A New Theory of the Leisure Class*, New York, Schocken, 1976/1989.

11. B. Farber, *Kinship and Class: A Midwestern Study*, New York, Basic Books, 1971. N.H.H. Graburn, 'Tourism, Leisure and Museums,' Plenary address at the Annual Meetings of the Canadian Museums Association, Halifax, NS, 1982; 'A Quest for Identity,' *Museum International* (UNESCO), vol. 50, no. 3, 1998, pp. 13-18; and B. Assefa, 'Symbolic Estates as Determining Factors in the Leisure Preferences of Minorities in the East Bay (Or Why Blacks Do Not Use Regional Parks),' *Kroeber Anthropological Society Papers*, 67-68, 1988, pp. 7-11.

12. E.B. Tylor, *Primitive Culture*, London, J. Murray, 1871.

13. R. Finnegan and M. Drake (eds.), *From Family Tree to Family History*, 4 vols., Cambridge, Cambridge University Press, 1994.

14. Bazin, 'Industrial Heritage.'

15. Cf. C. Levi-Strauss, *Les structures elementaires de la parente*, Paris, Press Universitaires de France, 1949; G. Homans and D.M. Schneider, *Marriage, Authority and Final Causes*, New York, Free Press, 1995; R. Needham, *Structure and Sentiment*, Chicago, University of Chicago Press, 1962; and L. Dumont, *Introduction à deux theories d'anthropologie sociale; groupes de filiation et alliance de mariage*, Paris, Mouton, 1971.

16. MacCannell, *The Tourist.*

17. P. Nora, *Les Lieux de Memoire*, Paris, Gallimard, 1983; 'Between Memory and History: Les lieux de memoires,' *Representations*, vol. 26, 1989, pp. 7-25; and *Realms of Memory: Rethinking the French Past*, New York, Columbia University Press, 1994.

18. J. Urry, *The Tourist Gaze: Leisure and Travel in Contemporary Societies*, London, Sage, 1990; and *Consuming Places*, London, Routledge, 1995.

19. E. Leach, 'Anthropological Aspects of Language: Animal Categories and Verbal Abuse,' in E.H. Lenneberg (ed.), *New Directions in the Study of Language*, Cambridge, MA, MIT Press, 1964. One valid criticism of Leach's model is that it is most characteristic of a particular historical context and population, i.e., middle-class, property-owning British people in the countryside and small towns, prior to

World War II. Nevertheless, the ideas are so powerful that they suggest some underlying truths that should be tested or adapted for other populations. Leach, himself, one of Britain's most brilliant anthropologists of this century, came from an upper-middle-class wealthy family, but lived a mostly cosmopolitan life in China, Burma, Argentina and Cambridge. I, myself, am British, was brought up in the countryside, and studied under Leach at Cambridge.

20. Readers will realize the culturally and historically limited range of this model. Since the time when it applied, even the English, as well as most Europeans and many North Americans, have come to include the third, more distant category as 'consumable', both in the sense that men are more frequently marrying women outside of their race and nationality (and vice versa), and that the middle classes are becoming culinary explorers. Thus, many foods, including animal foods, which were once distant and tabooed, may be sampled and eaten as parts of home and homeland cuisine (e.g., sashimi, alligator, snake, shark, emu, and so on). And, paralleling these expansions, many middle-class people of the West and East Asia have come to recognize and visit so-called 'world heritage' – buildings and landscapes that are surely parts of other people's heritages. Not only are these admired and celebrated, but the exploratory 'internationalist' class has also tried to take control of them in the name of responsibility for their preservation.

21. Leach, 'Anthropological Aspects of Language'; cf. the work of Mary Douglas on taboos and anomalies, *Purity and Danger*, London, Routledge and Kegan Hall, 1966.

22. A.S. Bartu, 'Reading the Past: The Politics of Cultural Heritage in Contemporary Istanbul,' Ph.D. diss. in Anthropology, University of California, Berkeley, 1997.

23. Although the original meaning of this phrase (Johann Herder) was applied to general evolution, including the biological stages of embryology, its application in socio-cultural anthropology has been to suggest the equation of modern humans' childhood, the adult life of primitive and exotic peoples, and the nature of societies in the early stages of human social evolution (see S. Freud, *Totem and Taboo*, A.A. Brill (trans.), New York, Moffat, Yard & Co., 1918).

24. C. Gluck, *Japan's Modern Myths: Ideology in the Meiji Period*, Princeton, Princeton University Press, 1985; and T. Fujitani, *Splendid Monarchy: Power and Pageantry in Modern Japan*, Berkeley, University of California Press, 1996.

25. As Okely has eloquently outlined such a research tactic in *Own or Other Culture*, p.148.

26. See Graburn, 'Prologue: Southeast Asia in my Mind.'

27. Haya Bar-Itzchak, personal communication, 1996.

28. R. Rosaldo, 'Imperialist Nostalgia,' *Representations*, vol. 26, 1989, pp. 197-222.

29. See, for instance, Elspeth Huxley's reminiscences of her youth in East Africa in her *The Flame Trees of Thika*, London, Chatto and Windus, 1959.

30. Assefa, 'Symbolic Estates as Determining Factors.'

31. Okely, *Own or Other Culture*.

32. N.H.H. Graburn, 'Tourism, Modernity and Nostalgia,' in A. Akbar and Cris Shore (eds.), *The Future of Anthropology: Its Relevance to the Contemporary*

World, London, Athlone Press, University of London, 1995; N.H.H. Graburn, 'Conference Report: Retrospect and Prospect,' *Traditional Dwellings and Settlements Review*, vol. 9, no. 1, 1997, pp. 60-64; and A.E. Horner, 'The Assumption of Tradition: Creating, Collecting and Conserving Cultural Artifacts in the Cameroon Grassfields,' Ph.D. diss., University of California, Berkeley, 1990.

33. M.-F. Lanfant, and N. Graburn, 'International Tourism Reconsidered: The Principle of the Alternative,' in V.L. Smith and W.R. Eadington (eds.), *Tourism Alternatives: Potentials and Problems in the Development of Tourism*, Philadelphia, University of Pennsylvania Press, 1992; and N.H.H. Graburn, 'What is Tradition?' in R. Dobkins (ed.), *The Art of Frank Day: A Symposium, Visual Anthropology Review*, forthcoming.

34. N.H.H. Graburn, 'The Past in the Present in Japan: Nostalgia and Neo-Traditionalism in Contemporary Japanese Domestic Tourism,' in R.W. Butler and D.G. Pearce (eds.), *Changes in Tourism: People, Places, Processes*, London, Routledge, 1995.

35. N.H.H. Graburn, 'Neo-Traditionalism in Japanese Rural Tourist Architecture,' paper presented at the conference 'Value in Tradition: The Utility of Research on Identity and Sustainability in Dwellings and Settlements,' IASTE, Tunis, 1994.

36. Shuzo Ishimori, personal communication, 1989.

37. M. Bloch, *How We Think They Think*, Westwood Press, 1998.

38. Horner, 'The Assumption of Tradition.'

39. C.C. Marcus, *Home-as-haven, Home-as-trap: Explorations in the Experience of Dwelling*, Berkeley, Center for Environmental Design Research, 1986. P. Bourdieu, *Outline of a Theory of Practice*, Cambridge, Cambridge University Press, 1977.

40. K. Frampton, 'Place, Form, and Identity,' in D. Crowe (ed.), *Geography and Identity: Living and Exploring Geopolitics of Identity*, Washington, DC, Maisonneuve Press, 1994, pp. 168-169.

41. N. Bubant, 'The Odour of Things: Smell and Cultural Elaboration of Disgust in Eastern Indonesia,' *Ethnos*, vol. 63, no. 1, 1998, p. 48.

42. S. Veijola and E. Jokinen, 'the Body in Tourism,' *Theory, Culture and Society*, vol. 11, 1994, pp. 125-151; T. Selanniemi, *Matka ikuiseen kesaan: Kulttuuriantropologinen nakokulma suomalaisten etelanmatkailuun [A Journey to Eternal Summer: An Anthropological Perspective on Finnish Sunlust Tourism]*, Helsinki, Suomalaisen Kirjallisuuden Seura, 1994; and N.L. Frey, *Pilgrim Stories: On and Off the Road to Santiago*, Berkeley, University of California Press, 1998.

43. Graburn, 'The Past in the Present in Japan.'

Bibliography

Abu-Lughod, Lila, *Veiled Sentiments: Honor and Poetry in Bedouin Society*, Berkeley, University of California Press, 1986.

Abu-Lughod, Lila, *Writing Women's Worlds: Bedouin Stories*, Berkeley, University of California Press, 1993.

Ashworth, Gregory J. and P.J. Larkham, *Building a New Heritage: Tourism, Culture and Identity in the New Europe*, London, Routledge, 1994.

Ashworth, Gregory J., *On Tragedy and Renaissance: The Role of the Loyalist and Acadian Heritage Interpretations in Canadian Place Identities*, Groningen, Geo Pers, 1993.
Assefa, Brooke, 'Symbolic Estates as Determining Factors in the Leisure Preferences of Minorities in the East Bay (Or Why Blacks Do Not Use Regional Parks),' Kroeber Anthropological Society Papers 67-68, 1988, pp. 7-11.
Bartu, Ayfer S., 'Reading the Past: The Politics of Cultural Heritage in Contemporary Istanbul,' Ph.D. diss. in Anthropology, University of California, Berkeley, 1997.
Bazin, Claude-Marie, 'Industrial Heritage in the Tourism Process in France,' in M-F. Lanfant, E. Bruner and J. Allcock (eds.), *International Tourism: Change and Identity*, London, Sage, 1995.
Bloch, Maurice, *How We Think They Think*, Westwood Press, 1998.
Boddy, Janice B. and Virginia Lee, *Aman: The Story of a Somali Girl*, New York, Pantheon Books, 1994.
Bourdieu, Pierre, *Outline of a Theory of Practice*, Cambridge, Cambridge University Press, 1977.
Bubant, Nils, 'The Odour of Things: Smell and Cultural Elaboration of Disgust in Eastern Indonesia,' *Ethnos*, vol. 63, no. 1, 1998, pp. 48-80.
Crowe, Dennis (ed.), *Geography and Identity: Living and Exploring Geopolitics of Identity*, Washington, D.C., Maisonneuve Press, 1996.
Douglas, Mary, *Purity and Danger*, London, Routledge and Kegan Hall, 1966.
Dumont, Louis, *Introduction a deux theories d'anthropologie sociale; groupes de filiation et alliance de mariage*, Paris, Mouton, 1971.
Farber, Bernard, *Kinship and Class: A Midwestern Study*, New York, Basic Books, 1971.
Finnegan, Ruth and Michael Drake (eds.), *From Family Tree to Family History*, 4 vols., Cambridge, Cambridge University Press, 1994.
Frampton, Kenneth, 'Place, Form, and Identity,' in Dennis Crowe (ed.), *Geography and Identity: Living and Exploring Geopolitics of Identity*, Washington, D.C., Maisonneuve Press, 1994.
Freeman, Derek, 'Kinship, Attachment and the Primary Bond,' in J. Goody (ed.), *The Character of Kinship*, Cambridge, Cambridge University Press, 1973.
Freud, Sigmund, *Totem and Taboo*, A.A. Brill (trans.), New York, Moffat, Yard & Co., 1918.
Frey, Nancy L., *Pilgrim Stories: On and Off the Road to Santiago*, Berkeley, University of California Press, 1998.
Fujitani, Takashi, *Splendid Monarchy: Power and Pageantry in Modern Japan*, Berkeley, University of California Press, 1996.
Gluck, Carol, *Japan's Modern Myths: Ideology in the Meiji Period*, Princeton, Princeton University Press, 1985.
Goody, Jack (ed.), *The Character of Kinship*, Cambridge, Cambridge University Press, 1971.
Graburn, Nelson H. H. 'What is Tradition?' in Rebecca Dobkins (ed.), *The Art of Frank Day: A Symposium, Visual Anthropology Review*, forthcoming.
Graburn, Nelson H.H., 'A Quest for Identity,' *Museum International* (UNESCO) vol. 50, no. 3, 1998, pp. 13-18.
Graburn, Nelson H.H., 'Conference Report: Retrospect and Prospect,' *Traditional Dwellings and Settlements Review*, vol. 9, no. 1, 1997, pp. 60-64.
Graburn, Nelson H.H., 'Neo-Traditionalism in Japanese Rural Tourist

Architecture,' paper presented at the conference 'Value in Tradition: The Utility of Research on Identity and Sustainability in Dwellings and Settlements,' International Association for the Study of Traditional Environments, Tunis, 1994.

Graburn, Nelson H.H., 'Prologue: Southeast Asia in my Mind,' in Jill Forshee and Sandra Cate (eds.), *Convergent Interests: Traders, Travelers and Tourists in Southeast Asia*, Berkeley, Center for Southeast Asian Studies, 1999.

Graburn, Nelson H.H., 'The Past in the Present in Japan: Nostalgia and Neo-Traditionalism in Contemporary Japanese Domestic Tourism,' in Richard W. Butler and Douglas G. Pearce (eds.), *Changes in Tourism: People, Places, Processes*, London, Routledge, 1995.

Graburn, Nelson H.H., 'Tourism, Leisure and Museums,' plenary address at the Annual Meetings of the Canadian Museums Association, Halifax, NS, 1982.

Graburn, Nelson H.H., 'Tourism, Modernity and Nostalgia,' in Ahmed Akbar and Cris Shore (eds.), *The Future of Anthropology: Its Relevance to the Contemporary World*, London, Athlone Press, University of London, 1995.

Homans, George and David M. Schneider, *Marriage, Authority and Final Causes*, New York, Free Press, 1995.

Horner, Alice E., 'The Assumption of Tradition: Creating, Collecting and Conserving Cultural Artifacts in the Cameroon Grassfields,' Ph.D. diss., University of California, Berkeley, 1990.

Huxley, Elspeth, *The Flame Trees of Thika*, London, Chatto and Windus, 1959.

Lanfant, Marie-Francoise and Nelson Graburn, 'International Tourism Reconsidered: The Principle of the Alternative,' in Valene L. Smith and William R. Eadington (eds.), *Tourism Alternatives: Potentials and Problems in the Development of Tourism*, Philadelphia, University of Pennsylvania Press, 1992.

Lanfant, Marie-Francoise, 'The Genealogy of Patrimony,' in Nelson Graburn and Lanfant (eds.), *Tourism, Heritage and Patrimony*, in preparation.

Lavie, Smadar, Kirin Narayan, and Renato Rosaldo (eds.), *Creativity/Anthropology*, Ithaca, NY, Cornell University Press, 1993.

Leach, Edmund, 'Anthropological Aspects of Language: Animal Categories and Verbal Abuse,' in Eric H. Lenneberg (ed.), *New Directions in the Study of Language*, Cambridge, MA, MIT Press, 1964.

Levi-Strauss, Claude, *Les structures elementaires de la parente*, Paris, Press Universitaires de France, 1949.

Lowenthal, David and M. Binney (eds.), *Our Past Before Us: Why do we Save it?* London, Temple Smith, 1981.

Lowenthal, David, *The Past is a Foreign Country*, Cambridge, Cambridge University Press, 1982.

MacCannell, Dean, *The Tourist: A New Theory of the Leisure Class*, New York, Schocken, 1976/1989.

Marcus, Clare Cooper, *Home-as-haven, home-as-trap: Explorations in the Experience of Dwelling*, Berkeley, Center for Environmental Design Research, 1986.

Narayan, Kirin, *Storytellers, Saints, and Scoundrels: Folk Narrative in Hindu Religious Teaching*, Philadelphia, University of Pennsylvania Press, 1989.

Needham, Rodney, *Structure and Sentiment*, Chicago, University of Chicago Press, 1962.

Nora, Pierre, *Les Lieux de Memoire*, Paris, Gallimard, 1983.

Nora, Pierre, 'Between Memory and History: Les lieux de memoires,' *Representations*, vol. 26, 1989, pp. 7-25.

Nora, Pierre, *Realms of Memory: Rethinking the French Past*, New York, Columbia University Press, 1994.
Okely, Judith, *Own or Other Culture*, London, Routledge, 1996.
Rosaldo, Renato, 'Imperialist Nostalgia,' *Representations*, vol. 26, 1989, pp. 197-222.
Sack, Robert D., *Place, Modernity and the Consumers World*, Baltimore, Johns Hopkins University Press, 1992.
Schneider, David M., 'What is Kinship All About?' in Priscilla Reining (ed.), *Kinship Studies in the Morgan Centennial Year*, Washington, D.C., Anthropological Society of Washington, 1971, pp. 320-363.
Selanniemi, Tom, *Matka ikuiseen kesaan: Kulttuuriantropologinen nakokulma suomalaisten etelanmatkailuun* [*A Journey to Eternal Summer: An Anthropological Perspective on Finnish Sunlust Tourism*], Helsinki, Suomalaisen Kirjallisuuden Seura, 1994.
Soja, Edward W., *Postmodern Geographies: The Reassertion of Space in Critical Social Theory*, London and New York, Verso, 1989.
Strathern, Marilyn, *After Nature: English Kinship in the Late Twentieth Century*, Cambridge, Cambridge University Press, 1992.
Strathern, Marilyn, *Kinship at the Core: An Anthropology of Elmdon, a Village in North-West Essex*, Cambridge, Cambridge University Press, 1981.
Tunbridge, J.E. and G.J. Ashworth, *Dissonant Heritage: The Management of the Past as a Resource in Conflict*, Chichester and New York, J. Wiley, 1996.
Tylor, Edward B., *Primitive Culture*, London, J. Murray, 1871.
Urry, John, *Consuming Places*, London, Routledge, 1995.
Urry, John, *The Tourist Gaze: Leisure and Travel in Contemporary Societies*, London, Sage, 1990.
Veijola, Soile and Eeva Jokinen, 'the Body in Tourism,' *Theory, Culture and Society*, vol. 11, 1994, pp. 125-151.

Chapter 4

Openings To Each Other in the Technological Age

ROBERT MUGERAUER

Modernity versus Tradition

In this chapter I want to address the problems and possibilities that lie within the tension between modernity and traditional environments and peoples. At the core of this deep problem is modernity's negative definition of the 'other' – that is, the traditional – as lacking and disadvantaged according to modernist measures. This definition needs to be critiqued, and positive views of difference need to be developed in which 'self-identity' and 'the other' (including other environments and places) are made complementary. My method toward this end will consist of drawing from neglected strands of four historical-cultural sources: tradition itself, modernism, postmodern electronic telecommunications, and poststructuralist theory. The fundamental problem of how to describe, understand and nurture differences in a shared world can only be addressed by questioning 'our own' identity, 'our own' place, and 'our' relation to 'others'.

As the history of technology and culture shows, a remarkable appropriation took place in the period in Western history commonly called the 'Renaissance', 'Scientific Revolution', or 'Modern Age'. Through a complex series of events, the West took for itself the mantle of 'most scientific and technological', marking a transition not only in its self-image and claims on the rest of the world, but for the very 'measure' of itself and all others. This measure of cultural achievement and worth remains operative in today's contested phenomena of the Westernization or modernization of the planet through technology and capitalism.

That the appropriation was remarkable was not so much due to the rather sudden explosion of technological innovation, exploration, and capital development that took place, but to the oblivion or denial of the previous superiority of the 'other-than-Western' world. As a simple indication of the outlines of the story somehow forgotten by the West, one can point to the ancient development of astronomy and mathematics in Babylonia;[1] the astronomy and city-building of the Aztecs and Maya in the New World; the incredible science and technology of India (for example, the Hindu astronomical tradition) and China (too many factors – from printing and gunpowder – even to list); and the science and cultures of Islam and Judaism (which were not only vital in themselves, but which preserved the Greek heritage, re-introducing it to Europe via Spain). All of this sophisticated science and technology was developed while most of Europe's material culture remained relatively crude and undeveloped.

Thus, what happened in the West was not only that it simply developed powerful science and technology, but that it also forgot its position and relationship to other times and other cultures. Specifically, it developed scientific attitudes and technology that dismissed or subordinated spiritual-cultural values, making material culture and machines the measure of mankind.

As a poignant case, one only has to note the encounter of Australian-Tasmanian aborigines with Europeans. Able to mount successful long-distance colonial campaigns, the Dutch (from 1642), the French (from 1772), and the English (from 1802) found the Tasmanians leading a very simple way of life (which the English subsequently used as one justification for practices of systematic genocide). In cruel contrast with the new Western materialist values and technological measure of human worth, the Tasmanians had deliberately chosen a minimalist technology on the basis of a cultural-spiritual value complex. As a result of a major cultural shift some 3,500 years before, Tasmanians had deliberately stopped using many of their domestic and practical tools, any clothing, buildings, and even fire except for ceremonial purposes and for cooking meat – despite the severely cold climate – in order to spend 'more time at song, dance, and ceremony'.[2] In short, an admirable aboriginal spiritual culture and set of practices were flowering at the time of the arrival of European colonizers, for whom material culture increasingly was all that mattered.

The historical phenomena of Western modernization perhaps reached its climax in the nineteenth century, with the development of machines that allowed modernity to become identical with industrialization. Machines, especially coal- and steam-driven engines, maximized mineral extraction, transportation, manufacturing, and other forms of industrialization. This brought a transformation not only of the basis of material culture, but – self-reflexively – of the very forms and materials of the built environment: the new world of iron and steel, glass, electrification, and chemical production.[3]

Though it is not my place to discuss the relation of this secular mechanization to religious roots, the arrogance and imperialism that accompanied the spread of Western technology unfolded with a missionary zeal that oddly derived from the West's own spiritual traditions, even while it extricated itself from them and denied them. The assumption of authorization appeared not only in the drive across the North American continent, validated by belief in manifest destiny, but globally with a profane technological-evolutionary view. As Michael Adas has argued, 'Technological development was increasingly equated with the rise from barbarism to civilization, and machines were viewed as key agents for the spread of this new civilization . . .'[4]

Though architectural scholars know the well-rehearsed story of the development and spread of international architectural modernism, they tend to forget or overlook the way in which it was absorbed by technological modernism. Early-twentieth-century technological modernism, itself a fundamentally social-economic ideology, quickly subordinated its architectural subcategory, or variant, and the latter's radical social agenda. As a result, the cultural movement instead spun out a homogenous environment that displaced local, traditional ways of living and building. One of the initial drives of twentieth-century architectural and social modernity had been to displace outdated and harmful local regionalisms, which seemed no longer to have a place among the shifting of populations in the newly industrialized urban configurations that were well developed by the early part of the century. Old village patterns and stubbornly maintained tribal or group identities seemed to have little value among fellow brothers and sisters of the international industrial world. They were as irrelevant to the new age as the built forms and symbols of the obsolete Austro-Hungarian and Germanic identities must have been similarly irrelevant to the Holy Roman Empire. Yet these identities still legitimized the development of mechanized warfare among nation-states.

Given Western culture's sustained interest in the formal and aesthetic aspects of modernity, it is easy to forget or ignore today that subversive social agendas once drove much modern planning and design. Architects, planners, and designers of all sorts developed the clean forms and materials of modernity not only in response to the new industrial and social world around them, but as an explicit rejection of what they took to be atrophied traditional symbol and social systems. Their initial impulse also aimed to deny the reign of capitalism and the drive toward bourgeois comfort. According to James Holston: 'Modernist architecture claims to be an international movement that advances national development by building new kinds of cities which in turn transform daily life.'[5] Thus, modernist design embodied an epoch-making double-displacement of tradition: not only did it reject traditional built environments and life-worlds, but its advocates intended to supplant existing vernacular building traditions and local peoples by producing a new vernacular environment.

To comprehend the extent of its social agenda, one need only remember what a sweeping change it was to propose that architects should design housing for ordinary and working-class people – and that even beyond housing, they should take responsibility for designing workplaces, social-service spaces, and even entire cities. For instance, in *The Modulator*, Le Corbusier noted that 'the liberation had not yet taken place' which would produce the home through a genuine 'science of housing', which included not only the dwelling with programmes for the bachelor, the married couple, the family, and the nomad, but also extensions of the dwelling for communal services to lighten the housewife's burden and, outside the building, regulate relations to traffic and sports grounds, health clinics, schools, youth centers, and open spaces.[6]

Clearly, the modernist design project originally included validating, even supporting, new, non-traditional ways of living, especially for growing cities. The new urban forms would include a focus, for example, on the emerging life-worlds of single people (including working and professional women) and young couples, in addition to families. This demographic and emancipatory thrust was clearly evident in the Weissenhofsiedlung Experimental Housing Project built for the Deutscher Werkbund in Stuttgart in 1927. To cite but one of the participants using industrial construction materials and approaches, Mies van der Rohe argued in the Werkbund's publication, *Bau und Wohnung*, that what was needed was housing which 'makes more rational production possible while at the same time giving a great deal of freedom in the use of architectural space.'[7] It was no secret that the intention was to apply such new methods, forms and materials internationally, as was exhibited in a pair of postcards connected with the project which showed it both 'as is' and, via montage, as an Arabian Village (figures 4.1,4.2).

It is now clear that this ideology of technological modernization, which first announced itself in Western colonialism, passed to a new phase after World War II, when the dramatically increased power of the United States and its allies operated largely through 'effective' implementation policies. The attitudes and forms that had become operational in scientific-technological-industrial warfare were directly applied in new forms of the military-industrial complex – so aptly named by the warrior-statesman Dwight Eisenhower. More important, however, was the governing worldview of these international operations. As Ali Mazrui has convincingly shown, the process of global domination was guided by a 'social evolutionist teleology.'[8] In this effort, Michael Adas has added, 'Modernity is associated with rationality, empiricism, efficiency, and change; tradition connotes fatalism, veneration for custom and the sacred, indiscipline, and stagnation.'[9]

The forms and materials of modernization thus spread from central Europe across the world in a series of displacements of their original intentions. In the process, the initial impulses of figures such as Bruno Taut, Mies van der Rohe, and Le Corbusier were shifted to a simplified set of

Figure 4.1. Weissenhofsiedlung, 1927. (Postcard in author's collection.)

economically driven, management-oriented forms. As a result, forms and materials became identified as important not because of their liberatory potential for an industrialized population, but because they were cost-

Figure 4.2. Photomontage of Weissenhofsiedlung as Arabian Village, 1940. (Postcard in author's collection.)

effective and efficient formations that displayed a futuristic and status-affirming image of economic and social power. To be considered a player in the world's political arena, one needed to display success at, or commitment to, moving from tradition to technological modernity.

Major problems linger today because of this political-cultural development. Given a culture of forms and materials one might call 'internationalism' or 'international technological modernism', the problem of 'in what style shall we build' remains a subject of critical debate. At once, there are the tensed desires for and ideologies of tradition and modernity: of what seems desirable and attractive to ordinary people relatively innocent of theoretical debates, and to professionals and researcher-educators heavily invested in the latter dimensions. As Ibrahim Al-But'hie, Akel Kahera, and Aziz Hallaj have demonstrated, the issue of 'modern or tradition' has not been resolved.[10] Of course, this is to be expected, for as Henri Lefebvre has argued, even as new principles of organization reshape experiential and built environments, previous ones do not disappear. New patterns are inscribed in addition to, or on top of, older ones, which continue to operate in the background and at the margins where openings occur. Thus, it is imperative to examine simultaneously all built environments – both those emerging today and those that remain in effect from common heritages.[11]

Today environmental designers are well aware of the problem: it is that Modernist environments – because of forms, materials, and modes of construction – by their very success in deploying and symbolizing the international homogeneity of modernized development, tend to displace and ignore local identities and senses of place. This occurs *de facto* through the character of internationalized technologies. Thus, Euclidean-cartesian buildings communicate with other elements of the Euclidean-cartesian environment (infrastructure, transportation systems, etc.) and with internal heating, ventilation and air-conditioning (HVAC) systems that turn their backs both on bioclimatic conditions outside buildings and on traditional cultural responses to the natural environment embodied in local building traditions. In other words, the modernized technological environment is simultaneously imperialistic (on behalf, variously, of technological modernism, capitalism, or democracy), and oblivious of local identities and places. In its first movement, modernism expanded across the world, inserting itself everywhere; in its second, it has obliterated traditional environments. To cite only two empirical cases, Le Corbusier's housing complex in Algeria was inserted into one of the thickest sites of local 'otherness' (to the dominating French) as an instrument of political control; and his work in India, for all its stylistic influence and craftsmanship, turned away – and he turned admiring designers and engineers away – from engagement with the existing, historically sophisticated, local cultural-climatic environments.[12] As Adas has commented succinctly, such modernist practices have been assaults on the measures of human worth, for in place of traditional spiritual-cultural values and traditional environments, technological modernism has proposed to establish itself as the only real measure.

Such international development has obviously occurred without the realization of any corresponding liberating social agenda because modernist architects' and planners' ideologies – originally intended to develop social freedom from untenable nationalism and from bourgeois ideas of comfort and conformity – have been co-opted by government and private-sector power systems. Fortunately, it may still be possible today to retrieve this social agenda and the impulse of architects, planners, and designers to provide environments in which workers and ordinary people may positively produce and live. In this vein, critical theorists such as Jürgen Habermas and Henri Lefebvre have argued that such a modern possibility could remain a source for renewed public life. Though I cannot develop this theme here, I refer the reader to the substantial body of literature on the prospect for, and necessity of continuing, liberation through the development of post-Enlightenment critical-socialist modernism.[13]

From Modernity to Postmodernity

Much attention recently has been paid to the transition from modernism to postmodernism, and from the industrial age to the postindustrial. Thus, changes in technology, production and consumption have been identified as driving cultural and economic changes – changes that are transformations within capitalism. Capitalism appears to be entering a new and more powerful phase in which, while continuing to involve manufacturing and consumption of goods (now in new patterns of global dispersion and interconnection), it is simultaneously adding a new dimension or layer. Thus, the production of goods with use value first shifted to the production of goods with largely exchange value (as could be seen in the globalized demand for brand-name clothing, music, electronics, and so on). Now it has been shifted again to the production of goods with exchange values of information-entertainment technologies and services.

A result of this increased layering of modernism is that new means of generating and consuming wealth have appeared, ones which are related to traditional patterns in ways that are as yet not completely clear. On the one hand, information technologies such as computer and telecommunication systems have replaced older industrial-age technologies – infrastructure, telephone and telegraph, paper records, and so on – and on the other, they have created entirely new physical and cultural environments. The information and telecommunications sectors have developed everywhere, and nowhere, in forms that appear to be stabilizing not as postmodern but, instead, as 'high-tech', producing buildings with even more metal and glass on display than in the preceding phase of modernism. Their appearance in the forms and materials of modernism, postmodernism, late-modernism, or high-technology is historically understandable and also a sign of immaturity. Just as the first automobiles appeared as modified (animal-drawn) buggies, so the first systems buildings have appeared in various formations of modern glass and steel. The new forms and materials have yet to be generated, yet to be codified.

As I have argued elsewhere, however, what is radical and important about the logistical systems which provide the principle of organization of this new environment is that they are atopical and polymorphous.[14] They have neither any necessary locational demands or special placements nor inherent forms or materials. For example, given the internationalization of banking and credit systems, there is no need for local ‘safe-looking’ banks in which to place money – only convenient ATMs (automatic teller machines) and secure, user-friendly computerized credit-transfer systems. Note that the degree to which the new electronic systems establish their own logic or logistics is precisely the degree to which traditional types, forms, materials and patterns of social use or behaviour become irrelevant or even undesirable (figure 4.3). These placement-indifferent systems have appeared in their most developed forms in credit, energy, transportation, entertainment and administrative systems.

Congruent with the homogenizing development of systems environments, yet apparently moving in the opposite direction, is an attraction, almost fetishizing, of exotic, remote, authentic, and disappearing traditional environments. Travel once limited to the wealthy or leisure class has

Figure 4.3. Automatic teller machine: the place-indifferent measure of tourism. (Photo by author.)

become common to the paid labour force. Hence, international tourism has today become a major global phenomena with enormous economic impact. Tourism is one of the most developed and advanced technological-information-image systems. And the demand to see the foreign, exotic and authentic has created amazing and irresolvable tensions. Today's tourists seek traditional environments even as the capitalistic systems that produce the wealth (which allows them to desire and travel to such sites) destroy the traditional environments themselves. Moreover, in seeking traditional environments in large numbers, tourists destroy these environments, because they consume the cultural, natural and economic bases that generate and sustain them. Today tourists seek traditional environments, but insofar as they actually want them in terms of their images, fantasies and projections rather than the environments' own, they create enormous pressures to develop fake, caricatured or stereotyped environments in the midst of the places they visit (Cancun, the coast of Tunisia, Disney World, everywhere . . .). In short, today's tourists want other places to be interesting but safe, exotic but convenient, tasty but digestively friendly, full of character but antiseptically clean.[15]

Alternatively, generating and co-opting the exactly opposite alternative (as capitalism seems to have the genius to do), insofar as tourists want authentic traditional environments on the environments' own terms, they cause situations to be constructed that demand that peoples and places remain frozen as living museums, rather than partaking in the processes of normally occurring change and development. In this way, locals are led to dress-up in traditional costumes and display themselves for tourists. If a tradition is not healthy and strong enough to absorb such visitors and their expectations, the local people and places become fictions – if not lies (against which it is hard to maintain another, or deeper, identity and sense of place).[16] On the other hand, if a tradition is vibrant, its members will nevertheless be objectified while unavoidably enacting their tradition, in a process whereby their actions are compared with a version of tradition as preferred and specified by the tourism trade. Because of the dual possibilities of meanings that their activities imply, such subject peoples are no longer able simply to go about their ordinary daily lives within a continuing tradition.

These issues obviously are critical for people concerned with the vitality and future of traditional cultures, and for those who are fond of travelling and working in foreign and exotic places. As Caren Kaplan has put the question:

> So many of us desire to travel for fun or for education, even as many of us feel ambivalent about the mobile nature of employment and family organization. How do we sort out these perceptions of travel and come to understand the ways in which they are linked to a more postmodern

movement of destabilized nation-state, cultural and economic diasporas, and increasing disparities of wealth and power?[17]

Reconciling the individual experiences and structural inequalities of the visitors and the visited is a congruent responsibility, especially given the almost irresistible attraction to policy-makers of generating income by exploiting their own places and peoples:

[M]any have embraced tourism as a sustainable economic alternative to the unpredictable cycles and heavy environmental costs of extractive industry. Others see a darker side to tourism: the selling of place, history, and cultural identity in exchange for low-wage employment in an increasingly urbanized, economically divided, and corporate-dominated social environment.[18]

The importance of these concerns for researchers and practitioners is seen, for example, in the themes of recent conferences of the International Association for the Study of Traditional Environments. When the group met in Berkeley, CA, in 1996, it heard from figures such as Manuel Castells and Saskia Sassen about technology and the global city, and from Ali A. Mazrui and Suha Özkan about issues of contested identity. When it met in Cairo, Egypt, in 1998 the group vocally reported on and debated research on tourism.

Given that the desire for traditional environments and touristic practices are inextricably bound up with cultural production and reproduction of images and texts, any adequate postmodern examination of these important areas of investigation (the mix of technology, modernism, and no-longer-modernist rejuvenations of traditional-local identities), will also need to include attention to mass media. Forty years ago Daniel Learner contended that the process of modernization would depend largely on the effectiveness of mass communication and media, where 'newspapers, radio, and other forms of media would expose the increasing urbanized and literate populations of underdeveloped areas to a flood of information about the world beyond family and community,' thus orienting them toward change.[19]

Today, in addition to actually visiting other places as tourists, it is possible to visit them virtually through film, television, telecommunications, and the worldwide web. Though most images, represented environments, and experiences may be the occasion for mere voyeurism, hopefully some will run to the other end of the spectrum, providing at least the possibility of understanding participation that will provide ways for people to encounter each other. Obviously, film and television are largely modern, capital-intensive phenomena, dependent on capital systems for production and distribution. Since it is right to suspect or reject much of what mass media present as 'the' desirable, as 'the' norm, what matters most to such a project of liberation would be the opportunity, the possibility, and multiplication of self-articulated, chosen, and affirmed identities and differences.

In modern capitalist modes of dissemination, virtual visits via mass media and actual visits as part of mass tourism frequently function as instruments of stigmatization and stereotyping of the 'other'. But they may potentially also contain ways to present differences positively and consciously affirm self-identities with a combination of hope and good will. In making such visits, is it not possible to be open to others, to offer ourselves to their receptions and reactions?

There are some local, regional, and perhaps even national traditions of film, for instance, with which specific groups have begun to articulate and legitimate chosen identities and senses of place. I am not speaking of identities or any other dimension imposed or given 'unto them' by capital, governments, or an external elite, but only identities generated by and consented to by group members themselves. Other interesting shifts may also emerge through the wider availability of video and web pages on various free-nets which can be activated by individuals and groups in a self-determining manner.[20]

In a broad sense what I am saying is that it is important to note not only problems, but also possibilities. Principally, one must recognize there is no inherent form or material, no necessary placement or environment, in which the undeniably powerful, desirable, perhaps unavoidable, technological world-to-come will come. Instead, people have a chance to let it come, to open to it, in ways that affirm local modes of dwelling in bio-cultural regional traditions. If so, this would allow the new electro-technologies to be absorbed into traditional environments, ways of life, and value complexes.

As one example, my own recent work has explored the ways in which non-manufacturing high-technology activities (multimedia, graphics, code-writing, research and development) are colonizing existing environments in the United States.[21] At the same time that electronic-atopical technology is generating high-tech systems environments, it is also, almost invisibly, occupying domestic and industrial buildings from the nineteenth and twentieth centuries (figure 4.4). Part of the ethos of this high-technology work culture is that it is a 'cool', or 'hip' and free, form of work that allows non-formal dress and work habits with a high degree of amenities. The attitude common to work performed at home via telecommuting (as the much-discussed 'electronic cottage industry') applies equally to the workplace. The image of the latter has been made casual and imaginative with the perks of loosely structured (but long) hours of work, refrigerators full of drinks and snacks, couches for lounging, and such recreation equipment as ping pong and pool tables for relaxing.

Because the work is done on and via polymorphous electronic systems, connected to other systems via telecommunications, the work setting can just as easily be a converted house or warehouse as any place resembling a modified traditional office building. There currently seems a strong attraction to the industrial forms of warehouses, factories, and 'decommissioned' utility facilities and religious structures. On the one hand, this preference is economically and environmentally practical, since these buildings often were built to last (unlike many American structures), and because they recycle huge energy investments. They are also made with durable brick, wood and metal construction, with high levels of craftsmanship. Meanwhile, in terms of self-chosen image and symbolism, the strong materials and forms of these older industrial buildings connote a power and robustness which appeal to new entrepreneurs feeling their oats, just as the reuse of older homes connotes the comfortable unity of home and work and promotes continuity with the historical appearance and identity of local areas. These newly emerging groups, in the midst of formulating their identities and delineating their places of activity, exemplify an amazing new opportunity. For better or worse, right now, many such people are deciding what stock from their own traditional built environment to use, what heritage of identity and place to revitalize.[22]

But the larger point I wish to make with this example is that some technology subgroups in the United States (certainly not comprised only of

Figure 4.4. Townsend Building, San Francisco: once a coffee warehouse, now multimedia and legal offices. (Photo by author.)

US citizens) are choosing to define their shared identities and senses of place both in new ways they determine to be appropriate to who they are and how they live and within traditional environments that they have transformed to meet their needs. How other such subgroups around the world might choose to adapt their traditional cultural and natural environments to the atopical information technologies, I have no idea. That would be for them to work out. But one may already draw some clues or hints from previous environmental changes, and from the above-mentioned modes of film, video, and web pages that have been appropriated by individuals and local or regional groups.

Poststructuralist Theories of the 'Other'

As my treatment of modernity and postmodernity indicates, I believe it is fundamental to focus on the tensed relationships between traditional and non-traditional people and environments. Such responsibilities have also been the purview of those poststructuralist, postcolonial theorists who have positively elaborated an ethics of difference. Notable among these writers have been Emmanuel Levinas, Luce Irigaray, and Werner Marx – of whom I have space, unfortunately, to consider only the first.[23] Even with their entirely contemporary, *avant-garde* positions, interestingly they agree with some of the attitudes toward justice, law and compassion embodied in some of the West's oldest premodern traditions.[24] For example, an ancient formulation, reinvoked by Levinas, calls for responsibility to the 'widow, orphan, and stranger'.

In fact, it is not very difficult to have sympathy for and act to help the widow and orphan. In part, this is because they immediately evoke pity. Obviously, they need help. And, at one remove, it is clear all men and women are frail, and daily need the help of others. An ethics of compassion, then, does provide a basis to begin to overcome selfishness. But another reason it is somewhat easy to help the widow and orphan is that they are the widows and orphans of our relatives, of our neighbours and comrades, of our fellow citizens and co-believers. That is, they already are within 'our' group; they are here because they are 'one of us'.

Hence the importance of the 'other' in the third, most extreme, mode of appearance: as 'the stranger'. How much harder it is to respond helpfully to the stranger. Strangers all too often arouse our suspicion, distrust, fear, defensiveness, even hatred and anger because they are perceived and defined as different. Not 'one of us', they too often are experienced as the 'negatively other'.

The city is important not only because of the unprecedented urbanization that has occurred with modernization, but because the city is the scene – even paradigmatically the place – where all people encounter strangers and must find a way to interact sociably with them if they are not to be at war with each other and in violation of moral responsibilities. The city, where those with differences circulate in one place, is also where justice, law and charity are tested daily.[25]

At a deeper level, reflection shows how the distinction between the widow and orphan, who are 'of us', and the stranger, who is not, breaks down. This, however, requires rejecting the easy interpretation of the injunction 'to love our neighbours as ourselves', which might mistakenly be taken to mean to love only those who are like us, and not 'the others, the strangers'. A broader ethical or moral position is required in a city because the 'neighbour' in a city is not simply identical with 'those who are one of us, as over and against the alien others'. As Dietrich Bonhoeffer has argued, the neighbours we are enjoined to love as ourselves are those who simply appear nearby. Whomever is so placed, for whatever reason, makes a moral claim on us, is involved with us in mutual responsibility.[26]

If this is so, responsibility extends to everyone else, because in the shifting frames of place and time and in the course of individual and communal lives everyone has been a stranger in need, and all have come unto the homelands and households of others and taken or been granted new placement and identity. This conclusion is unarguable, as a diverse set of historical, cultural cases indicates. To begin, to contrast the settled peoples of Europe with the nomads coming in waves from the east offers an overly simple paradigm. Indeed, the difference between 'settled' and 'nomadic' perhaps is more one of the time frame or phase of a group's existence than of any unchanging characteristics. The same applies on a smaller scale, even within relatively settled peasant groups.

To take but one instance, dizzying shifts are evident even in cursory reflection on phenomena such as the movements of the Ural-Altaic racial groups. The Ural branch is represented by populations now known as Finns on the one hand, and Ugrians (especially Hungarians) on the other. Thus, both seemingly stable northern peoples and those in the middle of the movements back and forth between Europe and Asia in fact belong together as part of one very complicated historical-linguistic shift. In regard to the other branch (the Altaic group), the much-noticed movement of the 'Huns' resulted from their being pushed westward by a series of displacements originating in China, then rippling across Asia.[27]

In the case of the village, Oscar Handlin's empirical examination has shown how simplistic interpretations of peasant society 'exaggerate [its] stability and continuity.' Alpine village records show that the stranger is absorbed, that 'no family had been there from time immemorial.'

And that evidence conformed to the larger European pattern, which showed frequent thrusts to all the marches and a steady eastward drift of peasant population through the centuries.

Yet the impression of sameness was not merely the product of distortion through the observer's eye. Nor did it spring simply from the romantic fancies of folklorists. The villagers themselves believed that they had always been there, always worn the same costumes, made the same lace, sung the same music and danced the same figures. That belief was not evidence that they and their ancestors had actually done so, but it was evidence that the community possessed institutional devices for absorbing the effects of change and preserving stability over very long periods.[28]

A change of geography or time frame does not alter the story of the dynamic. The same tangled patterns of shifting roles of strangers and 'those at home' in a given place is as much the story of the Americas and Australia as Europe or Asia. Going forward temporally yields the same patterns. In the 1990s there continue to be vast transcontinental, even transglobal, migrations. The poor and the desperate, the well-off 'best and brightest', all circulate as they are able – by foot, by boat, and by jet. The flux of those already belonging and of strangers is everywhere in cultures and environments studied by scholars today.

Unrealized Potentials

As should be clear, to begin to move beyond current dilemmas and situations, traditional environments researchers are not without resources – resources that lie within the character of tradition itself, as well as within the original impulse of architectural modernism toward social-reform, in the new possibilities of freedom to be found in postmodern information technology, and in poststructuralist theories that positively indicate how to affirm differences. To emphasize the obvious, for all the attention to waves of modernism and postmodernism, a large number of people around the world still have the drive to keep traditions vital, renewing them appropriately to new situations or conditions – which, after all, is how traditions remain traditions for thousands of years. That traditional ways are still affirmed and lived should not be overlooked, as Mervat El-Shafie and Amr Abdel Kawi have shown in relation to Egyptian oases.[29] Nor should one overlook the dedication of professionals helping empower traditional groups to deal with tourism, as in William Bechhoefer's work on Amasya, Turkey.[30] Yet, as Caroline Swope's case study of the coastal Salish has indicated, one must also recognize the difficulty of understanding and maintaining traditions.[31] Of course, many researchers are also, or instead, undeniably a part of the modern world, and need to work within it lest they become romantics, nostalgically trying to live in or develop a world of which they never were nor ever can be a part. (Obviously this caution speaks to my own condition and that of others who, *de facto*, already are modernized.) As J.-F. Lyotard has pointed out, theoretically, because of its sceptical, critical attitude and belief in progress, the modern always has within it the drive and means to surpass itself, to become postmodern. To be modern means always to overthrow what comes before and to change it.[32] So, if tradition would be continually rejuvenated, it might find an ally in a part of modernism and postmodernism that would provide non-imperialistic opportunities to establish self-determining, differentiated or heterogeneous identities and senses of place. I believe this is possible.

Throughout this chapter, I have been speaking about (and from the perspectives of) not only tradition, modernism, postmodernism, and poststructuralism, but – reflexively – a sense of professional collegiality and friendship. Researchers and designers concerned with the built environment, through their international associations, share a common study of what to make of the world, and what to try to do next. This common ground in problems and in a range of approaches for research and practice explicitly recognizes differences of the other.[33] From everywhere researchers gather at places such as Berkeley, Paris, Tunis or Cairo to learn from each other and about the environments of other places, times and cultures.

Likely, nobody will arrive at definitive answers to the human questions about identity, place, and the 'other' which I have here outlined. But at least they may arrive at a sense for the questions' importance, open character, and timely-timelessness. Even more importantly, each of the major contending modes of constituting the world – traditional, modernist architectural, postmodern electronic-technological, and poststructuralist – contain not yet fully utilized, valid and promising theoretical and practical strategies with which positively to move forward as ethically responsible researchers and members of a global community. One need not continue with either methodologies or social practices that drive toward homogeneity at the cost of eliminating otherness or creating antagonists. Within a genuinely shared world one may affirm, even celebrate and rejuvenate, rich heterogeneities and fertile differences among traditions and their environments.

Notes

1. See O. Neugebauer, *The Exact Sciences in Antiquity*, New York, Harper and Brothers, 1962.

2. R. Lawlor, *Voices of the First Day*, Rochester, VT, Inner Traditions, 1991, pp. 77-84; J. Clark, *The Aboriginal People of Australia*, Tasmania, Tasmanian Museum and Art Gallery, 1986; and I. and T. Donaldson, *Seeing the First Australians*, Boston, George Allen & Unwin, 1985.

3. M. Adas, *Machines as the Measure of Men*, Ithaca, NY, Cornell University Press, 1989, e.g., p. 134; and J. Habermas, 'Modern and Postmodern Architecture,' *9 H*, no.4, London, 1983, pp. 9-14. On the obviously sexist language that connects 'men' and 'machine technology,' see C. Merchant, *The Death of Nature*, New York,

Routledge, 1989. Much analysis and repair also remains to be done by applying Luce Irigaray's strategies for dealing with 'sexual difference,' as indicated in note 23 below.

4. Adas, *Machines as the Measure of Men*, pp. 404-405.

5. J. Holston, *The Modernist City*, Chicago, University of Chicago Press, 1989, p. 10, cf.p. 6.

6. Le Corbusier, *The Modulator*, Cambridge, MA, MIT Press, 1968, pp. 110-111.

7. M. van der Rohe, *Bau und Wohnung*, Stuttgart, F. Wedekind & Co., 1927, p. 8.

8. A. Mazrui, 'From Darwin to Current Theories of Modernization,' *World Politics*, vol. 21, no. 1, 1968, pp. 69-83.

9. Adas, *Machines as the Measure of Men*, pp. 412-413.

10. I.M. Al-But'hie, 'Housing: Complexities of Tradition and Modernity. The Case of Riyadh,' in *Conservation, Rehabilitation, and Implementation*, Berkeley, IASTE, Traditional Dwellings and Settlements Working Paper Series, vol. 111, 1999; A.I. Kahera, 'Building, Aesthetics, and Technology,' in *Ecology, Tourism, and Traditional Settlements*, Berkeley, IASTE, Traditional Dwellings and Settlements Working Paper Series, vol. 108, 1999; O.A. Hallaj, 'Housing in the Age of Technology,' in *Invocations of Tradition in the Architecture of Tourist Development*, Berkeley, IASTE, Traditional Dwellings and Settlements Working Paper Series, vol. 107, 1999.

11. H. Lefebvre, *The Production of Space*, New York, Blackwell, 1991.

12. IASTE 2, Berkeley, 1990. Note the imperialistic connotations of technology in modernism discussed above may require a chilling rereading of Le Corbusier's motto: where 'the house as a machine for living' would provide the measure of progress at the scale of the dwelling, modernist forms and materials obviously would be more desirable than, and would displace, traditional ones.

13. For example, among recent work, see S. Moore, *Technology and Place*, Austin, University of Texas Press, forthcoming; and C.E. Irazabal, 'Architecture and the Production of Postcard Images: Tradition versus Critical Regionalism in Curitiba,' paper presented at the sixth IASTE conference, 'Manufacturing Heritage and Consuming Tradition,' Cairo, December, 1998. Among the 'classic' sources, see J. Habermas, *The Structural Transformation of the Public Sphere*, Cambridge, MA, MIT Press, 1992; 'Modern and Postmodern Architecture'; and Lefebvre, *The Production of Space*.

14. R. Mugerauer, *Interpretations on Behalf of Place*, Albany, SUNY Press, 1995; and Mugerauer, 'Electronic Communication and the Physical Community,' *ElectroCom 94*, Austin Software Council, 1994.

15. From the large literature on tourism, the critique is especially penetrating in C. Kaplan, *Questions of Travel: Postmodern Discourses of Displacement*, Durham, NC, Duke University Press, 1996; and S. Norris (ed.), *Discovered Country: Tourism and Survival in the American West*, Albuquerque, Stone Ladder Press, 1994.

16. This position is pointedly articulated in J. Kincaid, *A Small Place*, New York, 1988.

17. Kaplan, *Questions of Travel*, p.xi.

18. Norris, *Discovered Country*, back cover.

19. D. Learner, *The Passing of Traditional Society*, New York, 1958; and Adas, *Machines as the Measure of Men*, p.414. Note the counter arguments of contemporary continental hermeneutics: Hans-Georg Gadamer reinterpreted and argued for the positive basis of tradition in *Truth and Method*, New York, Continuum, 1989; and Martin Heidegger called for the reaffirmation of sense of place, identity, and local life-worlds in the face of placeless mass media and technology in *The Question Concerning Technology*, New York, Harper & Row, 1977.

20. See R. Mugerauer, 'Qualitative GIS: To Mediate, Not Dominate,' paper delivered at NCGIA's Varenius Project, Alisomar, California, 1998.

21. R. Mugerauer and L. Tatum, *High-Tech Downtown: Guidelines for Planning and Designing Building Conversions for Non-Manufacturing High-Technology Activities as a Key to Central City Rejuvenation*, Austin, Mike Hogg Endowment for Public Governance/University of Texas School of Architecture, 1998; and Mugerauer, 'Milieu Preferences Among High-Technology Companies: A Pilot Study,' Austin, Graduate Program in Community and Regional Planning Working Paper Series, 1997.

22. See Mugerauer and Tatum, *High-Tech Downtown*.

23. Space does not allow development of Irigaray's or Marx's work, both of whom powerfully develop the subject matter of this essay. Irigaray has argued, in a manner parallel to Levinas, that it is unexamined sexual difference that most effects our age metaphysically, epistemologically, ethically, and in terms of social action. For the ways that differences between the same and the other might be thought about via place, caress and sexuality, see her *An Ethics of Sexual Difference*, Ithaca, Cornell University Press, 1984. Werner Marx has developed the strategy that compassion may be the only remaining viable basis and measure for human conduct in an era where transcendence, order and rationality are being abandoned. See his *Towards A Phenomenological Ethics*, Albany, SUNY Press, 1992; and *Is There a Measure on Earth?*, Chicago, University of Chicago Press, 1987. In regard to the ideas of Emmanuel Levinas used here, see especially *Totality and Infinity*, Pittsburgh, Duquesne University Press, 1969; and *Otherwise than Being*, Dordrecht, Kluwer, 1991.

24. Given what appears to be a strain of nihilism in much of poststructuralist thought, those theorists who would support, for example, a liberatory leftist political agenda, have struggled to move beyond poststructuralist textual analyses to begin to formulate something like an ethics. There is considerable debate about which side of the nihilism/new ethics divide Derrida falls, as is the case with Baudrillard and Lyotard. In developing a 'safely leftist' position, many who in the past have argued against anything like objective meaning and value have taken up a post-Kierkegaardian tactic of reaffirming the unavoidability of obligations to others, not in the sense of a technical ethics – principled action grounded in a set of rational criteria – but in an attitude of responsibility and duty undertaken in 'fear and trembling' (the result is still what ordinary people would call an ethics). These theorists have regularly found the best routes lie through the unreproachable Jewish scholar Emmanual Levinas, whose works were cited in the note 23 above. For example, see J. Caputo, *Against Ethics*, Bloomington, Indiana University Press, 1993. The return of deconstructivists and poststructuralists to puzzle over issues of

obligation via biblical and parallel traditions obviously requires considerable reflection: at least they have in common the contention that purely rational conceptualizations have severe limitations in specifying interpersonal and social responsibilities.

25. M. Ignatieff, *The Needs of Strangers*, New York, Basic Books, 1960; J. Jacobs, *The Death and Life of Great American Cities*, New York, Doubleday, 1966; and C. Norberg-Schulz, *The Concept of Dwelling*, New York, Rizzoli, 1986. Of course, this does not deny that the problem of opening to the stranger is a serious social issue in rural settings, too (see the acknowledgement in Handlin's history of the village in this chapter), but only that rural settings often remain the enclaves of groups of who are substantially alike and resistant to 'the other' in a way that cannot happen in the heterogeneous population mix of the streets, public spaces, and commercial interchanges of the city.

26. D. Bonhoeffer, *Life Together*, New York, Harper & Row, 1954.

27. See E. Hildiner, *Warriors of the Steppe*, New York, Sarpedon, 1997; and J.B. Bury, *The Invasion of Europe by the Barbarians*, New York, W.W. Norton, 1967.

28. O. Handlin, *The Uprooted*, Boston, Little, Brown and Co., 1951, pp. 308-309.

29. M. El-Shafi, 'Siwa: Cultural Meaning and the Quest for Authenticity,' in *Preservation of Traditional Lifestyles and Built Form*,' Berkeley, IASTE, Traditional Dwellings and Settlements Working Paper Series, vol. 114, 1999; and A.A. Kawi, 'The Oasis of Farafra in the Eyes of Its Inhabitants,' in J.-P. Bourdier and N. AlSayyad (eds.), *Dwellings, Settlements, and Tradition*, New York, University Press of America, 1989.

30. W. Bechhoefer, 'Surviving Tourism: Report from Amasya,' paper presented at the Sixth IASTE Conference, 'Manufacturing Heritage and Consuming Tradition,' Cairo, December, 1998.

31. C. Swope, 'Raising the Stakes: Manufactured Heritage, Coast Salish Identity, and Casino Architecture,' in *Invocations of Ethnicity, Nationalism, and Religion in Heritage Strategies*, Berkeley, IASTE, Traditional Dwellings and Settlements Working Paper Series, vol. 106, 1999.

32. J.-F. Lyotard, *The Postmodern Condition*, Minneapolis, University of Minnesota Press, 1984.

33. This basically amounts to what, in *Truth and Method*, Hans-Georg Gadamer has called 'the fusion of horizons,' in which our genuine questions allow us to pass over to learn from other people and places. See the section on 'hermeneutics' in R. Mugerauer, *Interpreting Environments*, pp. xxvi-xxxii.

Bibliography

Adas, Michael, *Machines as the Measure of Men*, Ithaca, NY, Cornell University Press, 1989.

Al-But'hie, Ibrahim M., 'Housing: Complexities of Tradition and Modernity, the Case of Riyadh,' in *Conservation, Rehabilitation, and Implementation*, Berkeley, IASTE, Traditional Dwellings and Settlements Working Paper Series, vol. 111, 1999.

Bechhoefer, William, 'Surviving Tourism: Report from Amasya,' paper presented at the Sixth IASTE Conference, 'Manufacturing Heritage and Consuming Tradition,' Cairo, December 1998.

Bonhoeffer, Dietrich, *Life Together*, New York, Harper & Row, 1954.

Bury, J.B., *The Invasion of Europe by the Barbarians*, New York, W.W. Norton, 1967.

Caputo, John, *Against Ethics*, Bloomington, Indiana University Press, 1993.

Clark, Julia, *The Aboriginal People of Australia*, Tasmania, Tasmanian Museum and Art Gallery, 1986.

Donaldson, Ian and Tamsin, *Seeing the First Australians*, Boston, George Allen & Unwin, 1985.

El-Shafi, Mervat, 'Siwa: Cultural Meaning and the Quest for Authenticity,' in *Preservation of Traditional Lifestyles and Built Form*,' Berkeley, IASTE, Traditional Dwellings and Settlements Working Paper Series, vol. 114, 1999.

Gadamer, Hans-Georg, *Truth and Method*, New York, Continuum, 1989.

Habermas, Jürgen, 'Modern and Postmodern Architecture,' *9 H*, no.4, London, 1983, pp. 9-14.

Habermas, Jürgen, *The Structural Transformation of the Public Sphere*, Cambridge, MA, MIT Press, 1992.

Hallaj, Omar Abdulaziz, 'Housing in the Age of Technology,' in *Invocations of Tradition in the Architecture of Tourist Development*, Berkeley, IASTE, Traditional Dwellings and Settlements Working Paper Series, vol. 107, 1999.

Handlin, Oscar, *The Uprooted*, Boston, Little, Brown and Co., 1951.

Heidegger, Martin, *The Question Concerning Technology*, New York, Harper & Row, 1977.

Hildiner, Erik, *Warriors of the Steppe*, New York, Sarpedon, 1997.

Holston, James, *The Modernist City*, Chicago, University of Chicago Press, 1989.

Ignatieff, Michael, *The Needs of Strangers*, New York, Basic Books, 1960.

Irazabal, Clara E., 'Architecture and the Production of Postcard Images: Tradition versus Critical Regionalism in Curitiba,' paper presented at the Sixth IASTE Conference, 'Manufacturing Heritage and Consuming Tradition,' Cairo, December, 1998.

Irigaray, Luce, *An Ethics of Sexual Difference*, Ithaca, NY, Cornell University Press, 1984.

Jacobs, Jane, *The Death and Life of Great American Cities*, New York, Doubleday, 1966.

Kahera, Akel Ismail, 'Building, Aesthetics, and Technology,' in *Ecology, Tourism, and Traditional Settlements*, Berkeley, IASTE, Traditional Dwellings and Settlements Working Paper Series, vol. 108, 1999.

Kaplan, Caren, *Questions of Travel: Postmodern Discourses of Displacement*, Durham, NC, Duke University Press, 1996.

Kawi, Amr Abdel, 'The Oasis of Farafra in the Eyes of Its Inhabitants,' in Jean-Paul Bourdier and Nezar AlSayyad (eds.), *Dwellings, Settlements, and Tradition*, New York, University Press of America, 1989.

Kincaid, Jamacia, *A Small Place*, New York, Harper & Row, 1988.

Lawlor, Robert, *Voices of the First Day*, Rochester, VT, Inner Traditions, 1991.

Le Corbusier, *The Modulator*, Cambridge, MA, MIT Press, 1968.

Learner, Daniel, *The Passing of Traditional Society*, Doubleday, New York, 1958.

Lefebvre, Henri, *The Production of Space*, New York, Blackwell, 1991.

Levinas, Emmanuel, *Otherwise than Being*, Dordrecht, Kluwer, 1991.

Levinas, Emmanuel, *Totality and Infinity*, Pittsburgh, Duquesne University Press, 1969.

Lyotard, J.-F., *The Postmodern Condition*, Minneapolis, University of Minnesota Press, 1984.

Marx, Werner, *Is There a Measure on Earth?*, Chicago, University of Chicago Press, 1987.

Marx, Werner, *Towards a Phenomenological Ethics*, Albany, SUNY Press, 1992.

Mazrui, Ali, 'From Darwin to Current Theories of Modernization,' *World Politics*, vol. 21, no. 1, 1968, pp. 69-83.

Merchant, Carolyn, *The Death of Nature*, New York, Routledge, 1989.

Moore, Steven, *Technology and Place*, Seattle, Island Press, forthcoming.

Mugerauer, Robert and Lance Tatum, *High-Tech Downtown: Guidelines for Planning and Designing Building Conversions for Non-Manufacturing High-Technology Activities as a Key to Central City Rejuvenation*, Austin, Mike Hogg Endowment for Public Governance/University of Texas School of Architecture, 1998.

Mugerauer, Robert, 'Electronic Communication and the Physical Community,' *ElectroCom 94*, Austin Software Council, 1994.

Mugerauer, Robert, 'Milieu Preferences among High-Technology Companies: A Pilot Study,' Austin, Graduate Program in Community and Regional Planning Working Paper Series, 1997.

Mugerauer, Robert, 'Qualitative GIS: To Mediate, Not Dominate,' paper delivered at NCGIA's Varenius Project, Alisomar, California, 1998.

Mugerauer, Robert, *Interpretations on Behalf of Place*, Albany, SUNY Press, 1995.

Mugerauer, Robert, *Interpreting Environments*, Austin, University of Texas Press, 1997.

Neugebauer, Otto, *The Exact Sciences in Antiquity*, New York, Harper & Brothers, 1962.

Norberg-Schulz, Christian, *The Concept of Dwelling*, New York, Rizzoli, 1986.

Norris, Sean (ed.), *Discovered Country: Tourism and Survival in the American West*, Albuquerque, Stone Ladder Press, 1994.

Swope, Caroline, 'Raising the Stakes: Manufactured Heritage, Coast Salish Identity, and Casino Architecture,' in *Invocations of Ethnicity, Nationalism, and Religion in Heritage Strategies*, Berkeley, IASTE, Traditional Dwellings and Settlements Working Paper Series, vol. 106, 1999.

van der Rohe, M., *Bau und Wohnung*, Stuttgart, F. Wedekind & Co., 1927.

Chapter 5

Colonial Nostalgia and Cultures of Travel: Spaces of Constructed Visibility in Egypt

DEREK GREGORY

Egypt must soon be the favourite ground of the modern Nimrod, travel – who so tirelessly haunts antiquity . . . Thebes will be cleaned up and fenced in. Steamers will leave for the cataract, where donkeys will be in readiness to convey parties to Philae, at seven A.M. precisely, touching Esne and Edfoo. Upon the Libyan suburb will arise the Hôtel royal au Rameses le grand for the selectest fashion. There will be the Hôtel de Memnon for the romantic, the Hôtel aux Tombeaux for the reverend clergy, and the Pension Re-ni-no-fre upon the water-side for the invalids and sentimental – only these names will then be English; for France is a star eclipsed in the East.

(George William Curtis, 1856, *Nile Notes of a Howadji*)

Is getting to and from the registration desk to the elevators [at the Luxor Las Vegas] by boat along the river Nile any stranger than squeezing the Temple of Dendur into the Metropolitan Museum of Art in New York? Any stranger than traveling to Luxor, Egypt itself?

(Barbara Kirshenblatt-Gimblett, 1998, *Destination Culture*)

Traditions, Travel and Texts

In this chapter I want to disrupt some conventional appropriations of 'traditional environments'. 'Tradition' is at once an indispensable and an irredeemably compromised term. In one sense, my arguments can be read as merely another elaboration of the ways in which European modernity has 'invented' traditions – its own and those of other people. But I also depart from the usual terms of those discussions by connecting the invention of tradition to what Edward Said has identified as the citationary structure of Orientalism, in which successive writers cite and invoke one another, and thereby sustain a canonicized tradition that both invites and legitimates their claims to authenticity and truth.[1] I recover these connections between traditions and texts through the cultures of travel that were set in motion by European and North American tourists in Egypt between 1820 and 1920.

Eric Hobsbawm has suggested that 'invented traditions' are responses to novel situations – to the anxiety of the new – but this should not be limited to historical change: travelling through space is freighted with its own geographies of uncertainty in which travel writings are strategically implicated.[2] In the case that concerns me here, these writings helped to establish a tenacious continuity of disposition and practice whose chains reach beyond the nineteenth into the twenty-first century. More than this, these textualizations were embedded in the appropriation of Egypt as not so much an 'environment' as 'a space of constructed visibility' within which 'tradition' was seen in particular, partial, and highly powerful ways: where some traditions were illuminated, recuperated and privileged, while others were dimmed, marginalized or erased. Here, too, there is a vital continuity between cultures of travel in the past and in the present which becomes visible as a sort of 'colonial nostalgia'; and for this reason much of what follows is written under the sign of a postcolonialism that I need to clarify in advance.

Colonial Nostalgia and the Colonial Present

It has become commonplace to remark that postcolonialism revisits the colonial past in order to retrieve its impositions and exactions, its erasures and suppressions. Thus, Ali Behdad, in an essay that frames many of my own concerns, has offered an 'anamnesiac reading' of Orientalist cultures of travel in the age of colonial dissolution: a critical reading that 'unmasks what the object holds back and exposes the violence it represses in its consciousness.' If postcolonialism is thus 'on the side of memory', as he has suggested, then it declares its *parti pris* by staging a 'return of the repressed' to counter what he calls 'the nostalgic histories of colonialism'.[3] The inherent violence of the colonial past must not be forgotten; but what makes those histories so nostalgic – and so dangerous – is the *seductiveness* of colonial power. Hence, Leela Ghandi has argued that postcolonialism

> . . . can be seen as a theoretical resistance to the mystifying amnesia of the colonial aftermath. It is a disciplinary project devoted to the academic task of revisiting, remembering and, crucially, interrogating the colonial past. The process of returning to the colonial scene discloses a relationship of reciprocal antagonism and desire between coloniser and colonised. And it is in the unfolding of this troubled and troubling relationship that we might start to

discern the ambivalent prehistory of the postcolonial condition. If postcoloniality is to be reminded of its origins in colonial oppression, it must also be theoretically urged to recollect the compelling seductions of colonial power. The forgotten archive of the colonial encounter narrates multiple stories of contestation and its discomfiting other, complicity.[4]

Those seductions continue to exercise an extraordinary power at the start of the twenty-first century, which is why I prefer to speak not of the condition of 'postcoloniality' but instead of 'the colonial present', and why I wish to explore some of the ways in which the fatal attractions of colonial nostalgia are inscribed within contemporary cultures of travel.

To illustrate what I have in mind I offer two late-twentieth-century exhibits. In 1994 Gallimard published a guidebook to Egypt which was translated from French into English the following year and published in the United States by Knopf. Like several comparable texts it exquisitely aestheticizes and commodifies a particular visual economy of travel. The book opens with two double-page, silver-tone illustrations. The first juxtaposes 'Boats on the Nile' with 'Tourists returning from Karnak'; the second depicts 'Tourists picknicking in a temple'. All of these images were produced in the late nineteenth and early twentieth centuries. Significantly, whenever tourists are shown elsewhere in the text – as in a montage outside Shepheard's Hotel in Cairo, described in the caption as 'the epitome of European colonialism', or posing at the Pyramids – the illustrations are all taken from the last *fin de siècle*.[5]

My second exhibit is a tourist brochure produced for Thomas Cook's 1997-98 season on the Nile.[6] 'In the gracious days of Edwardian cruising', prospective clients are reminded,

Thomas Cook's palatial paddle-steamers dominated the Nile. Immaculately maintained and luxuriously furnished, they represented the era's best in comfort and technology, combining gleaming brasswork and deep-pile carpets with such things as electric light . . . And life aboard was described as 'the perfection of human existence.'

The present Nile fleet is advertised as 'a contemporary version of that grandeur'. Two vessels are singled out for special attention. The first is the MS *Eugénie*, 'named after the French Empress who opened the Suez Canal in 1869.' It was constructed in 1993 'in the style of the Belle Époque and bears a nostalgic resemblance to the paddle-wheelers of old.' In fact, everything about the ship, would-be travellers are assured, 'is designed to add to its turn-of-the-century allure.' In particular, 'the service, provided by 65 crew to 102 passengers, belongs to another age,' so lavish indeed that the voyage is promised to 'recall the grand opening of the Suez Canal, when Africa became an island and crowned heads of state sailed majestically through the new waterway – followed by Thomas Cook with a small party of adventurous tourists.' The second vessel is the MS *Prince Abbas*, where Thomas Cook has joined forces with the Nile Exploration Company to 'recreate the spirit of Victorian discovery tours, while incorporating cosseting comforts of modernity.'

On first sight of the delightful *Prince Abbas*, you are transported through time to years gone by, to the elegance and style of an Agatha Christie film set, to romantic Victoriana. Fashioned with the steam ships of old in mind, this comfortable craft has promenade decks and paddle-wheels to add to the aura of nostalgic authenticity.

'The door to every cabin and suite opens directly onto the covered decks,' just as they did in the past, 'but once inside you are met with all the modern conveniences of a first-class hotel.' From this privileged vantage point each cabin has 'a window onto the ancient world' where passengers can 'watch the timeless scenes,' travelling 'through 5,000 years, past almost biblical scenes – the billowing white lateens of feluccas, blue-gallabeahed men riding yellow camels, black-robed women bearing water pots on their heads.'

These exhibits are colonial nostalgia materialized and made visible, and they represent a visual thematic that needs to be taken with all possible seriousness. In particular, one needs to ask how this visual effect works and, simultaneously, what is hidden from view: what is *not* reflected in these silver-toned images and what is *not* seen in these 'timeless scenes'.

Spaces of Constructed Visibility

To respond to these questions I work with three sets of ideas. In the first place, Henri Lefebvre has suggested there is an intimate connection between what he has called 'the production of space' – a concept which now seems much less startling than when it was first proposed – and the systematic grid of power that inheres within modern scopic regimes. His ideas can illuminate some of the ways in which, in the course of the nineteenth and early twentieth centuries, 'Egypt' was under construction as a series of superimposed, overlapping and contradictory spaces through the investments and exactions of the ruling dynasty of Muhammad Ali and his successors, the appropriations of a wealthy land-owning class, and the various entanglements of European capitalisms with British, French and Ottoman imperialisms. Thus, 'modern Egypt' was being produced as a space of capital accumulation – a space of calculation and exploitation, of surveillance and supervision – in which an identity was forged between the abstractions of Space and the operations of Reason, an identity which appeared as an 'objectivity' that was seen as 'order itself'. There was a palpable intimacy between the production of this spatial order and the disciplining of human bodies: what Timothy Mitchell has called 'a common economy of order and discipline'.[7] There are of course significant connections between the production of this 'modern Egypt' and the production of a 'traditional Egypt', supposedly lying outside and yet alongside – and accessible from – the modern. But what interests me here is the formation of a visual economy that links political economy to cultural appropriation, and in particular the ways in which 'Egypt' was constructed as a particular sort of object.

In the second place, therefore, I draw on John Rajchman's highly suggestive reading of Foucault's spatial analytics to sketch the ways in which European and American cultures of travel were involved in staging Egypt as 'a space of constructed visibility': as a space within which 'Egypt' was made visible in particular ways for a particular audience.[8] The geography of this staging was dispersed through multiple sites, both inside and outside Egypt, and worked through a series of discourses and practices that installed a tensile apparatus of power, knowledge and geography. Clearly, 'traditional' Egypt was constructed in some measure through politico-economic and geopolitical formations that constituted it as both an obstacle to and an object of 'modernization' or 'development'.[9] But what interests me here is the way in which 'traditional' Egypt was also produced for travellers and tourists as a space that could be 'rationalized': as a space striated by routes and itineraries, triangulated by sights and views, and codified as a series of imaginative geographies through which its landscapes were made visible as a panoramic totality: 'timeless', 'authentic' and 'real'.

These productions were not the result of any transcendent logic or design. In the third place, therefore, I suggest that the micro-practices which were involved in the elaboration of these spaces can be brought into view through the actor-network theory developed by Bruno Latour and others. The construction of 'Egypt' in these ways was not the pure product of European dictation in which capitalism and colonialism inscribed their marks on an empty surface, filling a blank space awaiting its object. The investment schemes of international banks, the operating strategies of tour companies, and the emerging protocols of archaeology – to name only three of the powerful European agencies involved in the manufacture of 'Egypt' – were all immensely important in the formation of these partitioned spaces and in the mobilization and accumulation of the discourses through which they were made visible; but so too were the knowledges, skills and labours of countless local merchants, interpreter-guides, boat-owners, sailors and donkey-boys. This is not to oppose the 'power' of various capitalisms and colonialisms to the 'complicity' or 'resistance' of subaltern peoples; rather, actor-network theory allows for a *dispersed* and *distributed* understanding of agency by directing attention to the variable powers conferred upon all these actors by virtue of their enrolment in *heterogeneous networks*. This is to map a complex, foliated space of agency where agency, individual and collective, is always mediated by the spaces in which and through which it takes place. As 'Egypt' is constructed in these various ways, divided into the seemingly separate spaces of the 'modern' and the 'traditional', so some discourses and practices become privileged ('rational') whereas others become marginalized ('customary'). And yet the actor-networks through which these powers are conferred turn out to commingle the 'modern' and the 'traditional', the 'rational' and the 'customary' in such a way that the partitions between these spaces and the privileges accorded to them are constantly interrupted, confounded and dislocated.

The cultures of travel that I describe in this chapter involved attempts to construct what Jonathan Murdoch has called 'spaces of prescription', within which routes, sights and imaginative geographies could be regularized, standardized and made predictable; but as travellers and tourists made their way through these actor-networks they often found themselves within 'spaces of negotiation' that were fluid, individual and improvisational.[10] Their navigation was a slippery affair that usually required the recognition and even incorporation of 'local knowledges' that at once confirmed and capsized the valences of colonial discourse. The instability of these encounters was brought about in all sorts of ways, but the promiscuous entanglements of 'culture' and 'nature' within these hybrid actor-networks were of particular significance in punctuating the passage of tourists up and down the valley of the Nile. Sighting 'Egypt' within spaces of constructed visibility was thus always a precarious and conditional *achievement*.

In what follows I work with these ideas to recover an historical geography of the 'invented traditions' involved in the Nile voyage in the course of the nineteenth and early twentieth centuries.

Sails and Dreams: The Romantic Nile

Apart from those travellers who treated Egypt as a mere staging-post between Europe and Asia, most European and American visitors to Egypt in the middle years of the nineteenth century went up the Nile by *dahabeah*, and their collective experiences provided what came to be construed as one of the central organizing structures of Orientalist cultures of travel. A *dahabeah* was a large houseboat with triangular cross-sails. It had a cabin at the stern, divided into a saloon, individual sleeping quarters, closet, and bathroom. This space was reserved for the hiring party, who spent much of the day – and ate most of their meals – on the roof of the cabin under a large awning. The crew slept on the open deck around the forward mast; there would usually be 12-16 of them, including a captain (*reis*) and cook. Each hiring party would outfit their boat from the bazaars and stores in Cairo, under the eye of the *dragoman* (interpreter-guide) who had been engaged for the voyage; other provisions would be bought in towns and villages along the valley, while the gentlemen would shoot birds for the table en route, and small livestock and poultry would be towed behind in a small boat.[11]

Most of the canonical accounts of Nile travel were written about voyages on a *dahabeah*, so that each party was keenly aware that it was reproducing the experiences of its predecessors and, indeed, that its success in doing so – recorded in letters, diaries and journals – was in some sense a measure of its own experience. In consequence, this textualized chain also provided the yardstick against which later means of travelling on the Nile were to be gauged. The *dahabeah*, declared Thomas Knox, was 'the proper way to do the Nile trip.'[12] 'Proper' carries all sorts of connotations, but here it entangles privilege with authenticity. Three such privileges were of special importance.

In the first place, the experience of sailing up the Nile on a *dahabeah* was only open to people of independent means. It required the investment of considerable time, since it took most parties a week or more to reach Egypt from Europe, two weeks to 'do' Cairo and the Pyramids and to make arrangements for the Nile voyage, and then nine weeks or so to sail up to the First Cataract at Aswan and float back down to Cairo (and many elected to go as far as Abu-Simbel and on to the Second Cataract). It also required the expenditure of considerable sums of money.

In the second place, the *dahabeah* was a native craft that was made over into a 'White space': it had a native captain and crew, and its passage was made subject to local customs and convention, but travel writers and guidebooks issued elaborate advice to travellers on hiring, furnishing and commanding the vessel that would serve to establish their own sovereignty. Travellers who hired *dahabeahs* invariably prided themselves on being 'monarchs of all they surveyed', sailing under their own national flags, living in their own 'little worlds', following their own cultural codes and establishing a racialized order of power and precedence. These stage directions and scriptings transformed the *dahabeah* into a secure viewing platform from which Egypt was assumed to be available as a 'transparent space', fully open to the tourist gaze.[13]

In the third place, the *dahabeah* was supposed to give a privileged access to Egypt as an 'anachronistic space'. This was made possible in part by the imaginative disassociation it effected between the frantic world of European modernity and the tranquil Orient. Within this extravagantly imagined geography the *dahabeah* became what Laporte called 'the child of antiquity, wafted by the breath of heaven.' To sail on the Nile thus opened a passage directly into the past, and travel writers of the period constantly impressed on their readers the power of reverie. The imagery of a dream licensed the quintessential Orientalist fantasy of entering into the 'slumbering Orient' and becoming physically immersed in another world. According to the young Florence Nightingale in 1849:

> You feel, as you lie on the divan, and float slowly along, and the shores pass you gently by, as if you were being carried along some unknown river to some unknown shore, leaving for ever all you had known – a mysterious feeling creeps over you, as if it were the passage to some other world . . . You lose all feeling of distance [and] all feeling of identity too, and everything becomes supernatural.[14]

For Nightingale that 'other world' was the world of the ancient past and of the Bible, and she eschewed contact with the *fellahin* in the villages of the Nile Valley. But many other travellers drew parallels between the bas-relief scenes they saw on the temples and tombs and the other 'living Egypt' outside. In the end, they privileged the view from the *dahabeah* precisely because it allowed for a proximity, even an intimacy between them and local people.

But it was not all plain sailing; and while the *dahabeah* voyage was elaborately scripted its containing assumptions and conventions could nonetheless be challenged and called to account. There were three main ways in which the privileges of security and sovereignty that accreted around these invented traditions could (in principle) be dislocated: the breakdown of civil order, the eruptions of 'culture-nature', and the interventions of local people.

For much of the nineteenth century most travellers thought of the Nile Valley (at least as far as the First Cataract) as a space of civil order, and as the voyage on what Emily Beaufort called the 'now fashionable and crowded Nile' became routinized so the anxieties of Europeans and Americans about their personal safety were banished to the margins of their mental landscapes. They frequently met up with other touring parties and enjoyed 'the Society of the River' and its security of numbers, but they were also afforded considerable physical protection by the civil and military authorities. Local villages were required to supply men to watch over each *dahabeah* when it moored for the night, for example, and while travellers were divided over the effectiveness of the measure – the watchmen frequently slept soundly until morning – most of them thought the precautions unnecessary. They were in no doubt that the authorities would take very seriously indeed any attacks on 'Franks' and their boats.[15] Much of this could be attributed to the emphasis placed by Muhammad Ali on law and order: the Nile was a vital economic artery, and once he assumed power he lost no time in making the river secure. Travelling through Egypt in 1835-36, American John Lloyd Stephens was shocked by Muhammad Ali's cruelty and oppression, but he accepted that his despotic rule had 'made Egypt, from the Mediterranean to the Cataracts, as safe for the traveler as the streets of New York.' Similarly, Harriet Martineau thought it 'the least that European travellers can do in acknowledgement of the security and facilities which the Pasha's government affords to testify to that security and those facilities.' By the 1840s it was widely accepted that brigandage and piracy on the river had been more or less extinguished by 'stationing well-armed sloops on the Nile and by destroying some of the most notorious of their haunts,' and many travellers saw the ruins of the village of Beni Hassan as a monument to the Pasha's success in imposing order on an otherwise 'anarchic' landscape.[16]

Of course, such a sensibility was disrupted from time to time by political and military adventures, such as the local rebellions against the central authorities in the 1860s, the terrible British bombardment of Alexandria and the subsequent occupation of Egypt in 1882, and the Sudan campaigns in the closing decades of the nineteenth century. Yet little of this seems to have compromised the extraordinary self-confidence and assertiveness of most travellers. On the contrary, while these episodes disrupted their plans and threatened the envelope of security within which they were accustomed to moving, the restoration of order through the exercise of military power did much to confirm in their minds the supremacy of the colonizing flag under which they sailed and its conjunction of 'might' and 'right'.

Equally striking was the capacity of travellers to detach themselves from the landscape of violence through which they passed. As the 1865 rebellions were being put down, a party of Englishmen including the Reverend Smith and his clerical father were sobered by the sight of vultures wheeling in the sky above the corpses and the smoking ruins of villages which had been razed to the ground by the Pasha's troops. But forty pages later in his account of the trip Smith was able to conclude that he could not imagine any more 'agreeable or even luxurious mode of passing through any country than that afforded by the Nile boat,' where travellers could 'dream away their time on the deck under the awning, drinking in the balmy air of Egypt, and interested and amused by the shifting scenes on the river banks.'[17]

However, cultures of travel were also marked by a concern for the travelling body – for its vulnerability and integrity – that reached beyond the danger of interpersonal violence. And in many cases these supremely self-confident performances seem to have been more profoundly de-scripted by the instabilities and eruptions of culture-nature. Disease was one of the most threatening of these mutant hybrids. Cholera and plague were even more alarming prospects in Egypt than they were at home, and most travellers fled at their approach. Few were brave (or foolhardy) enough to follow A.W. Kinglake's example of remaining in Cairo in 1834-35 when 'the plague was so master of the city, and stared so plain in every street and every alley.' Kinglake claimed he stayed because his experience was 'sharpened by the sting of the fear of death.'[18] Books listed many other diseases and ailments, and advised travellers on the medicines they should take with them. As Charles Warner wryly observed, it made a 'cheerful prelude to [the voyage] to read that you will need blue-pills, calomel, rhubarb, Dover's powder, James's powder, carbolic acid, laudanum, quinine, sulphuric acid, sulphate of zinc, nitrate of silver, ipecacuanha and blistering plaster.' It is difficult to know how often travellers used any of these remedies: Warner and his friends 'never experienced a day's illness, and brought them all back,' but he agreed that 'the knowledge that we had them was a great comfort.'[19]

The vulnerability of travellers was also shaped by the physical landscapes through which they moved. Some writers thought it impossible to bring Egypt's landscapes, at once 'cultural' and 'natural', within the compass of their own language-games because they were read as a visible testimony to the 'queerness' of the Orient. To Florence Nightingale, for example, 'the whole Nile [was] so unnatural, if one may use the expression, so unlike nature' that it lay outside the competence of any European language.[20]

And, in fact, many European and American writers held this 'nature' at a distance by imagining it as a landscape painting, at once controlled and contained, though few of them saw it through such jaundiced eyes. This visual effect – nature as a still life – was reinforced by the emphasis constantly placed on the play of colour and light, and on the minimalist, skeletal and geometric forms of the landscape, which, to Traill, was 'the slightest of impressionist sketches, dashed off as it were in half a dozen strokes of Nature's most careless brush.'[21] If the colonial picturesque was, as Sara Suleri has suggested more generally, a gesture of self-protection, in Egypt its poetics played into an Orientalist fantasy of a passage through a placid dreamworld:

> Thus we glided on through the scenery of a dream – without effort, peaceably, silently. Silently, for nature in its happiest moods has a silence of its own, articulate and musical. There is a silence made up of all the stray notes in the broad landscape – the song of birds, the murmurous hum of summer insects, the distant lowing of oxen, the rippling of the stream, a kind of invisible harmony . . .[22]

But from time to time and from place to place that 'harmony' was shattered, and 'nature' exceeded its containing frame, and these excesses and interruptions called into question the conventional ways in which travellers understood both 'culture' and 'nature'. The stable geometries of the red and the black, for example, the parallel lines of baked desert and fertile valley, could be redrawn by violent windstorms that could rage for days.

The Nile itself was forever changing its course from one season to the next; banks were undermined, new shoals appeared, and some stretches were notorious for sudden squalls and shifting currents that could capsize a boat. For travellers intending to sail into Nubia there was also the First Cataract to negotiate. Some hired another boat for the journey from Aswan to the Second Cataract, but for most the ascent through the turbulent waters assumed the status of a rite of passage that soon became one of the central traditions of the Nile voyage. This 'opera of the Cataract', as T.G. Appleton called it, was not only a dangerous performance, as the *dahabeah* was whirled around, hurled against the rocks, and slowly dragged up river against the force of the waters by an army of local men and boys. It was also a terrifying glimpse of 'a world turned upside down'.[23]

The way in which travellers saw this inconstant, inverted and even 'unnatural' nature cannot be disentangled from the way in which they understood – and misunderstood – local culture. Most of them privileged their experience of sailing on other rivers over the local knowledge of their boat crews. According to J.W. Clayton, these sailors knew as much about sailing as they would about the building of Nelson's Column or the novels of Henry James.[24] Shipwrecks were not uncommon, and travellers who escaped (not all did) frequently attributed the disaster to the ignorance and incapacity of the crew, and their survival to their own foresight and fortitude. Thus, Julian Arnold blamed the wreck of his family's *dahabeah* on the 'Arab indecision' of their *reis*. Running into a squall, the 'unusual commotion' brought Arnold's father out on deck, 'who, realising the need of some decisive move, told the bewildered [*reis*] to furl the trinkeet sail or, if not possible, to let everything fly.' Either way it was too late; the boat foundered. The family and most of the crew scrambled onto the submerged hull, where the sailors 'continued to howl and lament in a most ridiculous fashion.' 'They seemed to have no manhood left,' Arnold wrote, 'appearing not to retain the slightest reliance on their own personal exertion.'[25]

These responses flowed from a complex reservoir of assumptions about local people. Although their actions were frequently dismissed as the products of stupidity and superstition, many writers also insisted that local people had become such skilled participants in the expanding networks of travel and tourism that there was a cunning and even a cupidity behind them too. 'Taking strangers up the Nile seems to be the great business of Egypt,' Warner cautioned his readers, and for those strangers – 'innocents abroad', as Mark Twain famously called them – 'all the intricacies and tricks of it are slowly learned.' When Arnold described tourists and local people making moves on 'the chess-board of *dahabeah* travelling', he was configuring tourism as an elaborate game, a battle of wits, which was supposed to be part of the fun. 'The traveler who thinks the Egyptians are not nimble-witted and clever,' Warner advised, 'is likely to pay for his knowledge to the contrary.'[26] Local people were skilled in the arts of resistance and had long deployed them against the military and civil authorities, and so it is scarcely surprising that they should deploy similar tactics against the exactions of wealthy travellers and tourists.[27]

For the tourist there were all sorts of ways of keeping these 'games' within bounds. Books provided advice on how to draw up a formal contract setting out the requirements of the hiring party and the responsibilities of the *reis* and his crew.[28] Although travellers presumably took some comfort from this and frequently declared themselves 'monarchs of all they surveyed', their authority did not go unchallenged. Guidebooks and travel writings usually recommended fairness and firmness in dealing with the boat crew, but many of them recognized that a persistent refusal to comply with the instructions of the travellers might require physical punishment. This was invariably represented as being 'traditional'. 'The discipline of the stick,' W.H. Bartlett wrote in one of the classics of Nile travel, is 'perfectly well understood in Egypt.' Such an extreme measure 'revolts at first one's English prejudices' (or so he said), and he declined to allow his *dragoman* to beat their 'lazy' and 'worthless' *reis* on his behalf. But he admitted that he soon came to regret his 'ill-timed interference with established usage,' because 'reluctance to harsh measures passes for facility and weakness.' Others disagreed. The more temperate Wilkinson was 'far from advising that constant use of the stick which is sometimes resorted to most unnecessarily,' and Bayle St. John insisted that 'the Arab sailor has none of the qualities of a man fit to be beaten, except that he does not return the blow and forgives it in an hour.'[29]

If sentiments like these were coloured by paternalism, they did at least establish parameters which many travellers were reluctant to transgress. Others elected to apply to a local magistrate or governor to bastinado the offender – which involved beating the soles of his bare feet with whips – but once the sentence had been passed they often intervened to have the punishment cut short or even set aside. 'We decided between ourselves to let the punishment commence,' Bayard Taylor recalled, 'lest the matter should not be considered sufficiently serious, and then show our mercy by pardoning the culprits.' Their decision, so Taylor assumed at any rate, 'was received with great favour; the two culprits came forward and kissed our hands.'[30] Ida, Countess Hahn-Hahn, did not care for such squeamishness at all:

> If the *reis* and steersman were to receive a thorough bastinado, we should get along as well as possible, but they have learnt the abhorrence of Europeans to such measures, and therefore take advantage of our forebearance. You see, my dear brother, that the voyage of the Nile has its shady side.[31]

The Countess would undoubtedly have regarded the experience of Adam Kennard as positively sinister. Kennard and his companion sailed up the Nile in 1855 and were unusually abusive towards their crew, who 'mutinied' one by one. Their *dragoman* was the first to leave, abandoning them far above the Second Cataract: 'Our sole turnpike to all Moslem joys left us,' Kennard complained, 'five hundred miles up the country, surrounded with every species of Arab knavery and ferocity, and unable to speak a word of the language.' They subsequently had a violent argument with the *reis* over his refusal to proceed at night; pistols were fired and the whole crew deserted apart from one sailor and the cook. The hapless pair then had no choice but to help haul the boat back to Aswan: 'Harnessing ourselves into the rope, [we] commenced . . . the laborious undertaking of tracking a Nile boat of some twenty tons' burthen for seven miles against a strong current and a contrary wind.'[32] This was a stark reversal of the usual roles, since it was supposed to be the crew who tracked boats up the Nile, while the hiring party lounged on the deck exchanging remarks about the 'laziness' of the Arabs.

The purpose of stories like this was to promote the privileges of the travellers, of course, and accounts of the ways in which such trials were overcome served to establish both the authenticity of the experience and the authority of the narrator. In a sense, the very existence of obstacles and dislocations was testimony to the continued presence of the 'traditional', whereas the fact that they were surmounted was testimony to the power of the 'modern': and both of those terms were unambiguously racialized. In most cases we have no direct access to what the largely unlettered labourers thought about such matters. And yet accounts like these *can* bring into view the colonizing system of power, exploitation and exaction that lay behind the 'traditions' of the Nile voyage. In reading these texts against the grain, reversing the privileges that their narrators arrogated to themselves and exposing an underbelly of violence, it becomes evident that the 'space of constructed visibility' within which these cultures of travel were staged was never stable nor one-dimensional. The ways in which the established assumptions, expectations and practices of the Nile voyage were brought to crisis reveal that space to have been a field of chronic tension, displacement and compromise.

And that space was soon to be displaced in quite other ways.

'A Perpetual Cloud of Smoke'

A more or less regular steamboat service between Alexandria and Cairo had been established by the early 1840s to speed the overland transit of passengers to and from India. Although this was not universally popular, few travellers wanted to linger in Alexandria, and most of them were keen to reach Cairo as soon as possible.[33] They saw the fabled 'city of the Arabian Nights' as the real threshold of the Orient, and while it was possible to hire a *dahabeah* in Alexandria, it was Cairo that became the usual base camp for the ascent of the Nile.

Attempts to establish a steamboat service above Cairo met with little success until the 1870s. Several small, English-built steamers had been operated in the early 1860s on behalf of the Khedive by the Azizeh Company, but they were entirely at the ruler's beck and call, and all in all it seems clear that tourist traffic was never an important consideration. Despite this problem, such was the hegemony of the *dahabeah* that it seems doubtful that many travellers would have taken advantage of the steamers.

When Thomas Cook took his first party of tourists to Egypt early in 1869, he hired two of these old steamers – the *Benha* and *Beniswaif* – from the company which had now gone bankrupt to transport his thirty-odd clients from Cairo up to Aswan and back. The venture soon revealed the shortcomings of the existing arrangements. The boats were infested with fleas; they constantly ran aground on sand-banks; and when one of them broke a paddle-wheel, so one passenger recorded in her diary, the captain of the other 'wanted a written order from Mr Cook to proceed to Thebes without us [but] Mr Cook would not put pen to paper as he had little power over the Viceroy's regulations.' There were many other minor incidents, eventually provoking the exasperated diarist to exclaim: 'So much for Nile travelling where nothing can be believed or depended upon.'[34] Later in the year Cook returned with another party to witness the opening of the Suez Canal, and while he despaired at the way in which the Canal had attracted 'greedy and reckless speculators and a race of avaricious adventurers, accompanied by a race of harpies of the vilest composition, who pander to the worst passions of corrupt humanity,' he still described his passage through the Canal as 'one of the red-letter days of my tourist life.'[35] The following year his son John Mason Cook hired the large new steamer *Beherah*, which could accommodate forty-four passengers, and which had hitherto been reserved for guests of the Viceroy, and – as the company was later to boast – 'thus personally conducted to the First Cataract and back the largest party of English and American tourists that had to that date ascended the river as one party.' He returned convinced that 'the traffic of the Nile might be considerably developed.'[36]

To do so, however, required the introduction of a regulated system so that – contrary to the frustrations visited upon Cook's first party – *everything* 'could be believed and depended on.' As I want to show, the production of such a system involved both appropriating the privileges of security and sovereignty that accreted around the 'traditional' Nile voyage and guarding them against the usual sources of dislocation and disruption. Neither of these was a foregone conclusion and, as I also want to show, the production of the space of constructed visibility within which – and through which – Thomas Cook & Son operated was thus a conditional *achievement*.

In the early 1870s the company obtained the exclusive agency for the passenger service of the Khedive's steamers, and soon set about promoting the advantages of steam over sail. The arguments were driven home in the columns of Cook's Excursionist and Tourist Advertiser throughout the decade. Progress on a *dahabeah* was so slow, it was claimed, that passengers rapidly tired of the 'monotonous' scenery; dependent on the vagaries of the wind, they could find themselves becalmed 'for a whole week', and then the only way forward was for the crew to track, 'a very painful sight and really galley-work.' Then there were the inevitable social tensions that arose among a (too) small party 'living day and night in close proximity [where] continual frictions are unavoidable.' Worse, the trip was positively dangerous. 'If the truth were told of the disasters, delays and deaths in connection with *dahabeahs* that went up the river last season, Egyptian visitors would be very cautious how they committed themselves to the dangerous and dilatory service.' The steamer was not only quicker and more sociable, in Cook's estimation it was even more rational. On a *dahabeah* the custom was to sail upriver as fast as possible, taking advantage of the sporadic winds from the north, and reserve the inspection of the tombs and temples for the slow journey down river with the current, a convention which Amelia Edwards – no friend of Cook – complained was tantamount to reading the 'Great Book' backwards. But the steamer made its major stops on the journey upriver so that, as the company put it, 'the order of sight-seeing is perfectly in harmony with the importance of the monuments.'[37]

Seasoned travellers were not only unconvinced by all this, they were absolutely horrified at Cook's programme. Their twin criticisms were really only different ways of putting the same, profoundly class-structured objection.

First, the steamer was vilified as the ensign of European modernity. By bringing the 'modern' so abruptly into the heart of the 'traditional', the very presence of the steamer was a forceful reminder of the precariousness of 'the timeless Orient' – of the predicament of belatedness – and noisily intruded the 'modern' into scenes where most travellers expressly did not wish to see or hear it. 'When you visit a country of the past,' Laporte advised his readers, 'do not be persuaded to despise the institutions of the past.'

But it was not only the machinery of the steamboat that provoked angry denunciations; it was also the 'mechanical' culture of organized tourism itself and its sense of unvarying regimentation. Although Samuel Manning travelled on a steamer himself, he allowed that

> The delicious sense of repose, the Oriental *Kief*, the Italian *dolce far niente*, which constitutes so large a part of the enjoyment of the Nile trip, is impossible on board a steamer. [T]he rate of progress . . . is yet far too rapid to let us abandon ourselves to the lotus-eating indolence which can expect to find anything in Egypt so refreshing to the wearied frame and over-wrought brain of the traveller in search of health. Then, too, it is impossible to linger where we please. We must hurry on. Two hours may be enough for the tombs

of Beni Hassan, three hours for the temple at Esneh, four days for Luxor and Karnak; but it is distressing to feel that we cannot stop if we like. Haunted by the fear of being too late, we complete our survey, watch in hand, to be sure of catching the steamer before she leaves her moorings in the river.[38]

Second, since the steamer was both faster and cheaper than the traditional *dahabeah*, it opened Egypt to tourists of lesser means. By the early 1870s Cook was taking parties of his 'personally conducted' – what his critics called 'Cookies' or 'Cookites' – up the Nile every fortnight: a venture that was often criticized for its social pretensions.

The most outspoken of Cook's *fin-de-siècle* critics was the French romantic novelist Pierre Loti, whose description of a visit to the temple at Abydos was, in its way, exemplary of both nationalist *hauteur* and class condescension:

But what is this noise in the sanctuary? It seems to be full of people. There, sure enough, beyond a second row of columns, is quite a little crowd talking loudly in English. I fancy that I can hear the clinking of glasses and the tapping of knives and forks . . . Behold a table set for some thirty guests, and the guests themselves – of both sexes – merry and lighthearted, belong to that special type of humanity which patronises Thomas Cook & Son (Egypt Ltd.) They wear cork helmets, and the classic green spectacles; drink whisky and soda and eat voraciously sandwiches and other viands out of greasy paper which now litters the floor. And the women! Heavens! what scarecrows they are! And this kind of thing, so the black-robed Bedouin guards inform us, is repeated every day so long as the season lasts. A luncheon in the temple of Osiris is part of the programme of pleasure trips. Each day at noon a new band arrives, on heedless and unfortunate donkeys.[39]

Farther up the Nile, Loti recoiled at the arrival of yet another party:

But all at once there is a noise of machinery, and whistlings, and in the air, which was just now so pure, rises noxious columns of black smoke. The modern steamers are coming, and throw into disorder the flotillas of the past: colliers that leave great eddies in their wake, or perhaps a wearisome lot of those three-decked tourist boats, which make a great noise as they plough through the water, and are laden for the most part with ugly women, snobs and imbeciles . . .[40]

If sentiments like these preserve – in acid rather than aspic – the privileges of the little world of the *dahabeah*, nonetheless by the turn of the century the larger world had already turned.

Full Steam Ahead

By the end of the nineteenth century Egypt had become an extremely fashionable destination for well-to-do European and American travellers. Cairo, wrote Reynolds-Ball in 1901, 'is indeed emphatically a society place, and of late years especially, as an aristocratic winter-resort it ranks with Cannes or Monte Carlo.' For its winter residents – the *hivernants* – at the fashionable hotels like the Continental, Shepheard's and the Gezireh Palace, 'from January to April there is one unceasing round of balls, dinner parties, picnics, gymkhanas and other social functions.'[41] Much the same was true at Luxor and at Aswan, where new, modern hotels were also the centres of the tourist's social circle: 'During the day there are sports of all sorts, and in the evenings concerts, balls and bridge-parties. From time to time gymkhanas, donkey and camel races, paper-chases &c. are held, when ladies, gentlemen, children, young and old, take part with extraordinary enthusiasm.'[42]

Not surprisingly, the Nile voyage had been transformed too. It was still possible to make the trip by *dahabeah*, but the boats were now less expensive and could even be hired through Thomas Cook. This was still an expensive proposition, of course, but once Thomas Cook had entered the market the company lost no time in trumpeting its belated discovery of the 'many' advantages of the *dahabeah*. In particular, the traditional Nile voyage conferred both 'absolute privacy' and 'perfect independence' upon its clients. In a bold understatement, Cook conceded that the company had 'merited, to a certain extent, the accusation that we have ridiculed the expensive luxury of the *dahabeah*,' but in mitigation added that 'during the past nine years we have, at the request of private families, arranged for the voyage to the First or Second Cataract by *dahabeah*, and during the seasons of 1879-80 and 1880-81 we had the pleasure of organizing and carrying out no less than *nine special private parties*.' The social aspirations of that emphasis were unmistakable: so much so, indeed, that the company reversed its recoil from the tyranny of the wind, the monotony of the landscape and the horrors of tracking. 'The crews are practiced in "tracking" and real delays are of rare occurrence,' it was now announced, 'but as the Nile is full of variety, and places of interest abound, enforced detention is seldom wearisome.'[43] Still, the number of available boats dwindled and Cook slowly edged out many local operators. By 1888 Edwards already noted that 'the increase of steamer traffic has considerably altered the conditions of Nile travelling' and 'fewer *dahabeahs* are consequently employed.'[44]

The tourist steamer had in fact become highly fashionable, as Cook moved resolutely up-market in what John Mason Cook described as a 'voyage from vulgarity'. In 1879 he had decided 'to reduce the numbers of our parties and keep them more manageable than they have hitherto been.'

Parties were now to be 'select and private', conducted by 'gentlemen of experience and culture', and catering for members 'uniformly of the most intelligent, refined and cultured class.' In 1880 Cook signed a ten-year agreement with the Khedive, 'under which we take a more perfect control of the steamboat traffic on the Nile than we have hitherto done'; the company now assumed 'full control of the fittings and appointments of the steamers as well as the working of them.' All seven ships were refitted and the number of cabins reduced 'so that with every berth occupied the steamers will not be inconveniently crowded, as they have hitherto been.'[45]

The company's plans were interrupted by the Urabi revolt in 1881, however, and by the British occupation in the summer of 1882. In late October John Mason Cook travelled out with three companions for a lightning tour to assess the situation. In Alexandria they breakfasted 'in the only hotel left standing', and found the destruction in the rest of the city 'fearful'. Undeterred, they hurried on to Cairo where they were immensely relieved at the 'great fun' still to be had in the bazaars of the old city and at the Pyramids.[46] On their return, Cook moved quickly to reassure his would-be passengers that order had been restored and that the envelope of security was intact. But the immediate consequences were serious. In December 1882 an American archaeologist working in the Nile valley recorded that 'the Cook steamer had nine passengers and but four *dahabeeyehs* have gone up; there are scarcely any travelers; the occupation of the *dragoman* and the *anteekeh*-seller is well-nigh gone.' The interruption turned out to be protracted. Not only were regular steamboat services suspended during the military operations, but soon afterwards a cholera epidemic hit the tourist traffic; then in 1884 the company was contracted to provide transport for Gordon's ill-fated mission to Khartoum, and later in the same year to ship the relief column of 18,000 British and Egyptian troops from Alexandria to the Second Cataract.[47]

The company's complicities in the colonial project were not confined to the materialities of these military campaigns; its involvement also had a powerful symbolic charge. The Gordon relief expedition was fraught with anxieties and punctuated by arguments over payment, and a committee of inquiry concluded that the company had made 'a very considerable profit' out of the army's operations and that any further financial demands would be 'exorbitant'.[48] Most historians agree that John Mason Cook's motive was patriotic, and he was plainly bruised by allegations of this sort. More significant, however, was the way in which the company capitalized on its adventures. Cook was now seen as a symbol not only of British presence but of British power in and even British sovereignty over Egypt. An imaginative chronicle of the company's success in transporting the relief column to Wadi Halfa described John Mason Cook as 'King of Upper and Lower Egypt' and 'Pharaoah of the boats of the north and south', filling the heart of Queen Victoria by bringing her soldiers to Wadi Halfa: all of which was recorded in hieroglyphics 'found at Assouan'. The exorbitant language and the textual form amounted to a comic conceit, but it was also a profoundly serious one which identified Cook with – and inscribed Cook within – the space of 'traditional' Egypt.[49]

This appropriative gesture spiralled beyond the company's military services to underwrite Cook's *system*, a combination of political economy and disciplinary power by means of which the company established and enlarged its own space of prescription. It was this conjunction that had made Cook's military service possible. When the company declared that 'the Government must either buy us out or they must give us the work,' this was the unvarnished truth: but it was a truth that had been successfully manufactured by the company itself.[50] After the Anglo-Egyptian campaigns this regime of truth – this apparatus of power-knowledge – was put to work to produce a space within which 'traditional' Egypt could be made visible to a travelling audience in ways that were so routinized and so rationalized that they seemed perfectly natural and naturally transparent.

The suture this installed between the logistics of military operations and the logistics of modern tourism was recognized by a British journalist:

> In Egypt he who puts himself in the hands of Cook can go anywhere and do anything. Whether it be the transport of an army or the regulations for the use of a steamer's bath-room, you will find every point thought of and every point thought through.[51]

By this means 'Egypt' and its 'exhibition' were made to appear as both incontrovertible and interchangeable. That identity was underscored by the same writer who described the Khedive as merely 'the nominal Governor' of Egypt; 'its real governor' was Thomas Cook. He gave that combination of political economy and disciplinary power an explicitly colonial gloss:

> Mr Cook is a blessing to Egypt – perhaps the only one of England's recent blessings which nobody disputes. It is not only the vast amount of money he brings into the country, nor the vast number of people he directly employs. Besides that, you will find natives all up the Nile who practically live on him. Those donkeys are subsidised by Cook; that little plot of lettuce is being grown for Cook, and so are the fowls; those boats tied up on the bank were built by the sheikh of the Cataracts for the tourist service with money advanced by Cook.
>
> Therefore, when the Governor is pleased to travel up and down his Nile, you may see the natives coming up to him in long lines, salaaming and kissing his hand. When he appears they assemble and chant a song with refrain 'Good-mees-ta-Cook'. Once he took Lord Cromer up the Nile, and they went to visit a desert sheikh somewhere at the back of Luxor. The old man had no idea that the British had been possessing Egypt all these years – barely knew that the late Khedive was dead,

'Haven't you ever heard of me?' asked Lord Cromer. No; the sheikh had never heard of Lord Cromer.

'Have you heard of Mr Cook?'

'Oh, yes; Cook Pasha – everybody knows Cook Pasha.'[52]

Whether the vignette was fanciful or not, the symbolic management of Cook's extraordinary and seemingly ubiquitous presence was unmistakable. As Douglas Sladen remarked, surveying the company's tourist steamers berthed at Cairo, 'Cook is the uncrowned King of Egypt, and this is the navy with which he won his battle of the Nile.'[53]

It was quite a navy. Throughout 1885 the company had waited with growing impatience for the old steamers to be released from their military contracts. By the end of the year it was obliged to announce that 'we have not received back from the [Egyptian] Government the Tourist Steamers in a fit state to justify us advertising the commencement of our regular steamboat service before the first week of January 1886.' Cook accepted that once the steamers had been returned they would need extensive repairs and refitting before being used for commercial passengers; but it soon became clear that they could never again be used for a 'first-class passenger service'. Instead, several of them were to be 'remodelled and fitted up' to provide a cheap express service (which would convey both tourists and 'native passengers'), while the company would build its own Nile fleet for the exclusive use of its foreign clients. Contracts were placed for four new vessels, and in 1886 Cook launched the *Prince Abbas*, the first of the company's fleet of 'floating palaces'.[54]

Cook was determined to attract a clientele to match. Mabel Caillard, who travelled up the Nile with John Mason Cook on several occasions, recalled that he desperately wanted 'to charm into his net the big fish who so far had disdained to be classed in the category of Cook's travellers.'[55] In 1888 the Reverend Charles Bell, travelling on Cook's *Rameses*, declared that 'We were very fortunate in our fellow-passengers on the Nile, some of them highly cultured and of wide reading, in whose society time went not only lightly but profitably. There were scientific men, and men of travel; ladies who sketched admirably, and had a true feeling for art.' More than twenty years later it was even easier for Philip Marden to insist that:

For the great majority the tourist steamer must always remain the popular choice – and, be it said, the perfectly satisfactory choice. I have small patience with the supercilious disdain which superior persons see fit to bestow on the common herd who are forced to take their Egypt under the chaperonage of either of the well-appointed steamer companies that now exploit the Nile. For most of us there is no other way – and for the reasonable traveler there is no need of a better.[56]

The fact of the matter was that Cook had acquired a certain cachet. The company still had its critics, and traditionalists continued to rail against the speed at which it passed tourists through Egypt on what one writer described as 'the Rameses Limited Express'.[57] Even so, Cook's tourist steamers no longer automatically invoked images of 'mechanical' tourism, and by 1906 George Ade among many others did not hesitate to describe the three weeks devoted to their ascent and descent of the Nile as 'a loafing voyage'. Neither were Cook's clients the object of a universal upper-class disdain, and in 1907 Norma Lorimer, like countless others, advised that 'where money is a matter of no consideration, Cook's boats are, of course, the most pleasant and luxurious method of travelling.'[58] Effortlessly joining Ade's leisure to Lorimer's luxury, Douglas Sladen (the snob's snob) wrote of his several trips up the Nile on board a Cook's tourist steamer:

Breakfast is a country-house meal . . . In the height of the season, when the tourist steamers are full, on the days when there are no excursions, the particular young man sometimes breaks out into silks suits and wonderful socks, or at any rate rare and irreproachable flannels, just as the girl who has come to conquer Cairo society rings the gamut of summer extravagances. They have the moral courage for at least two different costumes between breakfast and dinner.

Conjuring up the world of the country-house weekend, the garden party and the golf club, of Hurlingham and Royal Ascot, Sladen re-inscribed the same distinctions as Loti in a parallel social register. 'Cook's trips up the Nile cost a great deal of money,' he added astringently, and 'the English people who go on them belong mostly to the class of the unemployed rich.' By the turn of the century Cook's 'floating palaces' lived up to their name, and the company's published programme included a list of the 'names of some of the royal and distinguished persons who have travelled under the arrangements of Thomas Cook and Sons' which included royalty from all over Europe and beyond.[59]

It is thus not surprising that many of the privileges formerly associated exclusively with the *dahabeah* were successfully appropriated by those who travelled on the steamers. For G.W. Steevens in 1898:

[T]he keynote of Nile life is peace; it is an existence placid, regular, reposeful. There is just enough variety in it to keep your mind awake, and just enough sameness to keep it off the stretch. There is just enough excursioning ashore to persuade you that you are not lazy, and just enough lazing aboard to assure you that you are enjoying rest. You pick up letters on the way, enough to remind you that you are of the world, and to convince you blessedly that you are not in it. A vision of half-barbarous life passes before you all day, and you survey it in the intervals of French cooking. You are not to worry, not to plan, not to arrange about anything; you are just to sit easy and be happy . . . Rural Egypt at Kodak range – and you sitting in a long chair to look at it.[60]

Recalling a trip in 1913, Rudyard Kipling wrote:

[T]he land of Egypt marched solemnly beside us on either hand. The river being low, we saw it from the boat as one long plinth, twelve to twenty feet high of brownish, purplish mud, visibly upheld every hundred yards or so by glistening copper caryatides in the shape of naked men baling water up to the crops above. Behind that bright emerald line ran the fawn- or tiger-coloured background of desert, and a pale blue sky closed all. There was Egypt even as the Pharaohs, their engineers and architects had seen it . . . When the banks grew lower, one looked across as much as two miles of green-stuff packed like a toy Noah's Ark with people, camels, sheep, goats, oxen, buffaloes, and an occasional horse.[61]

This sense of enframing – of 'traditional' Egypt as a succession of framed pictures or a moving panorama – was an exceedingly powerful organizing trope, but its power derived not only from enduring cultural conventions but also from their mobilization within a new actor-network. The services that Cook had provided for the Anglo-Egyptian campaigns helped to identify the company with the stability and security – the civil order – that had been achieved through the exercise of political and military power. Indeed, by the end of 1885 the company affirmed that 'the present state of Egypt, combined with the security always imparted by the presence of English soldiers, conduces to the conviction that the country is once more open to tourist travel.'[62] The envelope of security had been not only restored but strengthened. And yet if the company's activities were to give a new lease of life to the cultures of travel that had sustained the traditional *dahabeah* voyage, but which were now to be set in motion under the sign of a distinctively *modern* tourism, then the other sources of dislocation that had hitherto plagued the Nile voyage had to be contained: the eruptions of culture-nature and the interventions of local people.

Although Egypt was still seen as an unhealthy place for its native population, haunted by endemic diseases like opthalmia, it was increasingly advertised as 'the first of health resorts' for those who visited from colder climes, where 'the charms of its lovely climate, the restorative effects of reposeful air, give relief and new life to the invalid and the toil-worn.' This was a sort of colonization – certainly a cultural appropriation – of its climate. Prospective tourists were now informed that 'the risks run by visitors of contracting disease' were 'probably neither more in kind nor greater in degree than at home.' The exception was Cairo, where 'the dangers are increased mainly because the old town is insanitary and its inhabitants careless and uncleanly; but even so they may be rendered insignificant by exercising ordinary precaution.' Even there, by 1906 Cook's *Handbook* noted with pleasure 'considerable improvement in Cairo as a health resort,' not only in the continued development of 'the European quarter' of modern hotels but also in 'the native quarters of the city' where 'the Egyptian is no longer allowed to live amid dust and dirt.' On the Nile, where a *cordon sanitaire* could be put into effect more readily – where, as a relieved Elizabeth Butler put it, 'a few yards of water afford you complete immunity from that nearer contact which travel by road necessitates; and in the East, as you know, this is just as well'[63] – fears of illness were consistently downplayed.

'Culture-nature' was also domesticated through the production of other modalities that imposed new orderings – new modes of regulation – on the admission of 'nature' into 'culture'. In the past most travellers had been delighted by the food on board their *dahabeah*, and writers regaled their readers with accounts of well-stocked pantries and groaning tables. Although it was the responsibility of the *dragoman* to superintend the purchase of supplies and of the cook to prepare the meals, however, this had often required considerable planning and supervision on the part of travellers. Guidebooks included detailed lists of provisions to be brought from England and provisions to be bought in Alexandria and Cairo; anything else had to be purchased from the villages and towns strung out along the valley. But the spaces of prescription within which Cook and the other travel companies operated now allowed tourists to occupy different positions within the new actor-networks, and responsibility for provisions passed to the company and its agents. Clients who hired a *dahabeah* from Cook were guaranteed an excellent table. Clients could also receive 'from the steamers constant supplies of fresh fruit, vegetables, poultry, &c., &c., and specially fresh beef which cannot be had on board an ordinary *dahabeah*, unless seven or eight *dahabeahs* happen to meet together, enabling them to divide a whole bullock amongst them.'[64]

In fact there was a close articulation between tourist traffic on the river and the new hotels on its banks. The Karnak Hotel at Luxor had a kitchen garden of more than four acres which supplied 'the finest fresh vegetables'; cattle, sheep and poultry were fattened for the hotel table; and the hotel's dairy provided fresh butter, milk and cream 'to meet the demand of [Cook's] steamers and *dahabeahs*.'[65] The Cataract Hotel at Aswan, opened by Cook in 1899, boasted a vast dining room designed to resemble the interior of a mosque, which de Guerville pronounced 'delicious':

> It is easy to understand the enormous difficulties which have to be overcome to conduct really well such a place, and to offer daily, at six hundred miles from Cairo, a varied and excellent menu worthy of any of the big Paris restaurants, to hundreds of guests with appetites whetted by an open-air life. It is really extraordinary to find on the frontiers of Nubia, at a reasonable

price, all the comforts and luxuries to which we are accustomed, and for which we are willing to pay a large price, at Ostend, Baden Baden, Nice or Monte Carlo.[66]

These technical interventions coupled 'culture' and 'nature' in new ways, and I suspect that many of them passed unremarked by tourists who had indeed grown accustomed to them. But from the closing decades of the nineteenth century these changes in agricultural production and marketing systems were connected to the construction and extension of barrages and irrigation schemes that dramatically increased the cultivable area, and few tourists missed the significance of projects on such a grand scale. By 1891 the French-built barrages north of Cairo had been restored; a new barrage was completed at Asyut in 1898; and in the same year work was started on the Aswan Dam at the head of the First Cataract, which was completed in 1902. When John Ward visited the construction site at Aswan he reversed the Orientalist image of 'traditional' Egypt as a dreamworld to conjure up the magical transformations wrought by British modernity: 'At first one thinks it must be a dream, that Harland and Wolff's Works, Crewe Station, Wolverhampton iron forges, the Aberdeen granite quarries may have got mixed, and all have been dropped down together.'[67] Many British and American travel writers represented these vast public works as spectacular colonizations of an unruly and incontinent nature. In a vivid illustration of the sexualized tropes that so often framed these schemes Frederic Penfield, the US Consul-General, praised the British architects of the Aswan Dam for 'compelling the great river to pay tribute to agriculture rather than wasting its virtue in the Mediterranean Sea, knowing that the pregnant soil could in a few years be made to defray the cost of any reservoir that human agency could construct.'[68]

And yet – even as it sought to regulate the admission of a virile 'nature' into a modernizing 'culture' – the construction of the Aswan Dam allowed this colonized 'nature' to spill over into other 'cultures'. Local villages were submerged by the rising waters of the lake, and the Roman temples on the 'sacred island' of Philae were flooded for part of the year. Soon after the dam opened one observer reported that

Steamers that once lay twenty-five feet below the post-office at Shellal can now steam over the roof of that building, if it were in existence. Tourists who have had a very stiff and dusty climb to reach the kiosk on Philae can now enter between the columns in a rowboat. Philae to Kalabsheh is now devoid of cultivated land, the water reaching to the foot of the rocky hills; the villages which once stood on along the banks have disappeared, as have also the inhabitants, with the exception of a few who have built their huts on the mountain-side. To reach the temple of Taffeh, the steamer must pass under overhanging date-palms and over the house-tops of a once existing village. *It is gratifying to note that the water has in no place reached the temples* . . . At some of the villages the natives beg for bakshish, on a plea that the water has covered their land, and therefore deprived them of a living.[69]

As my added emphasis indicates, it was the threat to the temples on Philae that concerned most writers: not least because the construction of the dam heightened the tension between the 'traditional' and the 'modern' and in doing so visibly dramatized the predicament of belatedness. This was the object of Loti's lament in *Mort de Philae*, where he described how the Aswan Dam had destroyed the First Cataract and sentenced Philae to death.

The dam was raised a further five metres between 1907 and 1912, doubling its storage capacity, and by 1914 Sidney Low was reprimanding the 'aesthetic sentimentalism' of the dam's critics. 'We cannot sacrifice the interests of millions of Egyptians, living and to come, in order that a few genuine students and a considerable number of idle tourists may gaze at some interesting, though not supremely important, examples of Ptolemaic art.' Low evidently did not think much of tourists, least of all 'the clients of Cook'. At Aswan, he wrote, 'one finds oneself whirled tumultuously into the full stream of Egyptian pleasure-seekers.' For the most part these modern tourists did not take the antiquities 'too seriously': 'They visit the monuments in parties and in high spirits.' In Low's eyes, the pity of it all was that 'of modern Egypt – the real, living Egypt – they know even less than they do of that ancient Egypt which still lies half buried under the dust.' The only Egypt that tourists really appreciated, so he said, was 'the Egypt of Messrs Cook, the Egypt of the hotels and the palace-steamers, the Egypt of the *dragoman* and the donkey-boy.' And it was 'modern Egypt' – colonial Egypt, more emphatically *Britain's* Egypt – that captivated Low, and for this reason he believed that the tourist gaze at Aswan ought to be diverted from the ancient temples:

[T]he lover of the aesthetic has his compensation in the charm of an imposing and significant contrast. The temples rise like islands out of the broad sheet of water – the huge artificial lake into which this reach of the Nile has been converted by the dam. The stone colonnades, looking more Greek than Egyptian in their lightness and grace, are beautiful in their way; but there is a beauty of another kind – the beauty of stern majesty and purposeful strength – in the mighty bar of granite that lies athwart the river and curbs its pace or holds the tremendous energy of its impact in suspense.[70]

In fact, that 'stern majesty and purposeful strength' – the virility of British colonial power – had already captured Cook's attention, and the company had added the dam to its itineraries. Two hours were devoted to 'exploring [the] enchanting island in every part,' and passengers then rejoined the steamer for the short trip to the dam. 'After viewing this stupendous work for storing the waters of the Nile for irrigation purposes, and lunching in the newly erected Chalet overlooking the Cataract, donkeys will be in waiting for the return journey.'[71]

These vast engineering projects also affected navigation through their various impacts on winds and channels. Some writers claimed that the 'increased canalization and the added area of vegetation [had contributed] to a diminution of the frequency and force of the *khamsin*,' and though this was not without controversy Cook's *Handbook* agreed that in some places these changes had helped 'to make the winters less dry and the summers less hot.'[72] If, on some reaches of the river, *dahabeahs* were now less likely to run foul of violent windstorms, however, they could still find themselves becalmed; but in such circumstances Cook's clients could call upon the company's powerful steam launches for a tow. And if the modern barrages regulated the flow of the Nile, tourist steamers could still run aground at some times of the year; but this was now frequently treated as a lark. 'I fancy this occasional contact with the African continent under our feet will become so common an occurrence,' Marden continued, 'that we shall soon pay little heed to it.'[73] Indeed, colonial modernity's power over the physical landscape had become so impressive that Lorimer was surprised to find that she had to transfer from one steamer to another at the First Cataract:

> I imagined, I suppose, if I ever thought about it at all, that there was some splendid lock system on the Nile which would wipe out all the difficulties and enable Cook's tourists to continue their journey undisturbed. But not at all! For once in the course of our luxurious trip under his guidance we have been put to a few hours' inconvenience.[74]

In fact there was an elaborate system of four locks to the west of the dam, but since most tourists still elected to travel only as far as Aswan it made sense for Cook to run separate steamer services on the Upper and Lower Nile. The curtain had fallen on 'the opera of the Cataract' for the traditional *dahabeah* voyage too: 'The American sun-seeker or English milord, making the voyage to Wadi Halfa by his own *dahabiyeh*, will no longer have his craft hauled up the cataract by a hundred shrieking Arabs and Berberins, for most likely it will be taken up the rapids and through the locks by electricity generated by the rushing Nile itself.'[75] If 'nature' continued to be seen as a spectacle to be witnessed from the comfort of the deck, therefore, it was a 'nature' which – for its modern spectators at least – was now contained by a modern techno-cultural frame.[76]

The actor-networks of modern tourism also considerably diminished the capacity of local people to intervene in the plans of tourists. There was still a raw violence to some of Cook's ancillary operations, perhaps most obviously coaling the steamers:

> A dirty disagreeable operation it was, filling the atmosphere with clouds of fine black dust, and making a mess of everything about the decks. The work is performed by a number of the poor peasants, chiefly women and children, who are forced to do this without any remuneration from the authorities. The unfortunate *fellaheen*, who were half-naked and miserable, carried the coals on their heads in small baskets, and they ran up and down the bank in Indian file, urged on by one or two men with whips, who superintended the work . . . When it is over, they all congregate on the bank and look with interest upon the steamer and its passengers who, on such occasions, appear on deck, and occasionally fling copper coins and oranges among the dusky throng.[77]

And there were the usual warnings about crafty merchants in the bazaars, about sellers of antiquities who turned out to be manufacturers of the same 'antiquities', coupled with constant complaints about the incessant clamour for *baksheesh*. But these excesses and cautions also served to bring into view Cook's own system of disciplinary power. According to John Ward:

> The very name of Cook becomes in Egypt a magic talisman securing all who trust in it immunity from fraud and protection from rudeness, incivility [and] petty annoyances of any kind . . . Messrs Cook take an ignorant Arab or an ebon-tinted Nubian from his native village, put him through some mysterious training known only to themselves, and in a short time he is fit for use, is labelled 'Cook' in large letters and lo! he at once becomes a patient, efficient and trustworthy servant of all bearers of their tickets.[78]

Steevens employed a different metaphor, but its burden was exactly the same. 'Cook's representative is the first person you meet in Egypt, and you go on meeting him,' he explained.

> He sees you in; he sees you through; he sees you out. You see the back of a native – turban, long blue gown, red girdle, bare brown legs; 'How truly Oriental!' you say. Then he turns round, and you see 'Cook's Porter' emblazoned across his breast. 'You travel Cook, sir,' he grins; 'all right'. And it is all right: Cook carries you, like a nursing father, from one end of Egypt to the other.[79]

That sense of security, of paternalism and protection, was installed from the moment of arrival. Early in the twentieth century one French writer described 'pandemonium' – 'a Tower of Babel' – as his ship dropped anchor at Alexandria and crowds of porters swarmed on board to claim the luggage. Out of this chaos, as if by magic, would appear Cook's sturdy porters. And once on the river the same sense of ordered ease – of order as ease – prevailed. Brushing aside the traditional images of local people,

Cook's advertisements proclaimed that its *dahabeahs* would be under the command of 'an experienced and trustworthy *reis*' and that its parties would be accompanied by 'a skilled and competent *dragoman*.' The presumption was not that the old stereotypes had been wrong; it was that a significant change had been brought about by the power and perfection of Cook's own system. 'Thanks to the extent and permanence of our organization,' the company explained, 'we have at our disposal the pick of native employees.'[80] It was the same on the tourist steamers. The officers were all British, and from the beginning the company announced its intention 'to appoint an improved class of waiters and servants' on its new ships.[81]

It seemed to work. 'The servants on Cook's boats spoil you for any other servants,' Sladen marvelled. 'They hang about you like shadows in soft white robes, wondering what you could want next.'[82] This was more than a matter of service, however, because the 'white robes' were part of the 'reality-effect': they confirmed the persistence, the substantial presence, of the Orient as a *tableau vivant* to which tourists on the steamers enjoyed a privileged access. In much the same way, Sladen was delighted to find that the chief *dragoman* on board the steamer was 'Mohammed, the *doyen* of Cook's,' not only because he conjured up what Sladen called 'the good old days' – though he was now, of course, answerable to the company for his actions – but also because he was 'even more endeared to the tourist by the picturesqueness of himself.'[83] 'Picturesqueness' was an absolute requirement; it was essential for the actors to look the part. In the same vein, Lorimer echoed the sentiments of many of her companions when she declared she would be perfectly satisfied 'if there was nothing else to watch all day but the antics of our Soudanese crew':

> There is the ostrich-feather broom boy who watches for a speck of dust to brush away, and the brass boys who lift up rugs and mats to find some hidden treasure in the way of knobs to polish, indeed there is a boy with a grinning smile and flashing teeth for every mortal occupation you can imagine. I often wonder if there is a special crew kept to do nothing but say their prayers, for there is always a group of black-skinned Soudanese in white drawers on their knees in the bows of the boat. Perhaps Thos. Cook and Son recognise how valuable they are for 'off days' on the Nile, for tourists to kodak.[84]

The 'kodak' was the apotheosis of the appropriation of Egypt as a late Victorian and Edwardian exhibition in which scopic pleasure was not compromised by physical vulnerability.[85] For clients who hired one of its *dahabeahs* 'all trouble and annoyance incidental to disbursements in strange money and among a rather grasping people are altogether avoided,'

Cook's brochure promised, and 'every charge, every probable outlay is included in the hire of the ship.' It would no doubt be wrong to exaggerate the degree to which Thomas Cook and other travel companies succeeded in establishing their own spaces of prescription. There must have been all sorts of ways in which their local employees – and the countless others with whom tourists came into casual contact – were able to make their own accommodations and negotiations with the actor-networks of modern tourism but which are largely absent from the accounts provided by writers who travelled on the first-class tourist steamers. But the reason for that absence is revealing. For what is so striking about Cook's *fin-de-siècle* operation is its success in contracting the space of trans-culturation and minimizing such fluid, sensuous and spontaneous interactions between tourists and local people. The modern culture of travel was identified with a visual economy that was, in some substantial sense, de-corporealized; the bodies of local people were reduced to silhouettes or silent spectacles, glimpsed in the distance working the *shaduf* or witnessed on board mopping the deck.

For those who wished to do so, it was possible to close that distance, which is how Cook advertised the remodelled express post-boats it operated on behalf of the government:

> Native passengers are conveyed on the lower deck, and as this necessitates the calling at many of the small riverside towns and villages, glimpses are afforded of quaint native life not always compassed by the tourist programme. The lower deck of a post boat is quite removed from the travellers above, and presents a varied and motley crowd of passengers – natives of all classes, parties of troops, rich proprietors and the fellaheen, sometimes a *zenana* had been improvised with close drawn curtains to shroud the Mahomedan ladies from all observation.[86]

Cook was not alone in these observations, which offered travellers the prospect of a sort of 'natural history' of other cultures, allowing them to gaze over 'native life' unchallenged and perhaps even undetected. Some writers recommended the post-boat over the tourist steamer for these very reasons:

> Native passengers travel in the lower deck of the mail boats; they provide themselves with their own food; consequently, at various places where the steamer stops, there are the most lively scenes between the native passengers and their compatriots on shore, bargaining for sugar cane and other articles of food; then there are farewells on departure, and greetings on return, which are full of little touches of tragedy or comedy.[87]

'From these scenes of native life,' she continued, 'the gilded seclusion of the tourist steamer is removed.' But it was precisely that 'gilded seclusion' – what Kipling described as sitting 'on copiously chaired and carpeted decks, carefully isolated from everything that had anything to do with Egypt'[88] – and the sense of a panorama unfolding effortlessly before one's eyes – 'rural Egypt at Kodak range' – that had become the attraction for most tourists. The work of production – and perhaps even the possibility of performance – had been made faint; what was powerfully present was projection. Indeed, seventy years before Thomas Cook was to conjure up 'an Agatha Christie film set' to advertise the company's Nile cruises, Grenfell declared that 'Egypt had become the first materialized movie film on record, where the audience does the moving.'[89] Egypt had been reduced to its exhibition.

Colonial Nostalgia and Cultures of Travel

And so let me finally return to the two exhibits from our own *fin de siècle*: the guidebook and the tourist brochure. What can be made of them? John Frow has suggested, I think persuasively, that there may be an irredeemable nostalgia contained within most modern forms of tourism.[90] But it should be clear why I place the emphasis on a distinctively *colonial* nostalgia. In doing so, however, I depart from the usual meaning of the phrase. When Renato Rosaldo originally described 'imperialist nostalgia', he had in mind what he took to be a characteristic of cultures of colonialism and imperialism – namely, a desire for the very cultures that had been destroyed by their encroachments and inscriptions.

> Curiously enough, agents of colonialism . . . often display nostalgia for the colonized culture as it was 'traditionally' (that is, when they first encountered it). The peculiarity of their yearning, of course, is that agents of colonialism long for the very forms of life they intentionally altered or destroyed. Therefore, my concern resides with a particular kind of nostalgia, often found under imperialism, where people mourn the passing of what they themselves have transformed.[91]

Yet it seems to me that the two exhibits with which I began are signs of colonial nostalgia re-inscribed as a conspicuously *therapeutic* gesture that fuses two moments into a single constellation.

The first is a nostalgia *for colonialism itself*, a desire to re-create and recover the world of late Victorian and Edwardian colonialism as a culture of extraordinary confidence and conspicuous opulence: in a word – Thomas Cook's word – 'majesty'. This is to redeem a culture of travel that was, long before the British military occupation of Egypt in 1882, profoundly colonizing in its gestures, practices and appropriations. But it is to redeem it in such a way that the inherent violence of its dispositions is lost from view. Colonialism is recuperated – and rehabilitated – at a distance, so to speak, through the production of what Paul Smith calls an 'inflated truth' whose manifestations fill the visual frame and crowd out the other figures. Thus the Edwardian tourists at lunch in the tomb leave in the shadows those who cooked for them, waited on them, and guided them through Egypt; the opening of the Suez Canal is invoked without summoning the forced labour that was expended in its construction or recalling its disastrous effect on the Egyptian economy.[92]

I am not surprised that these other histories should have been erased, needless to say, although fleeting traces of them reappear in other guidebooks to modern Egypt. But I want to insist on the complex ways in which notations of class are folded into this history of the colonial present. Orientalism and colonialism more generally are marked by 'race', gender and sexuality in highly significant ways – and colonial history and postcolonial criticism have done much to bring these to our attention – but in most of this discussion the *class* positions, privileges and practices carried within its cultural formations seem to have gone largely unmarked and unremarked. Yet this is central to understanding the power – and the seductiveness – of both colonialism and nostalgia.

The second moment is more oblique. It involves the recovery not only of *fin-de-siècle* claims to power but also its claims to knowledge. For these guidebooks and brochures imply that we are not 'too late' – that we can steal a march on the anguish of 'belated Orientalism' – and that we can do so, ironically enough, by imaginatively transporting ourselves back to its definitive period. By accepting this invitation, responding to its visual interpolations, and situating ourselves as late Victorian and Edwardian tourists, we are assured (and, I think, reassured) that it is possible to regain an intimacy with a 'timeless' Egypt and – by virtue of the privileges that are attached to this subject-position – to guarantee the authenticity of this 'Orient'. The formation of this subject-position, and the constellation of power, privilege and nostalgia that is accreted around it, is not independent of the exhibition of 'traditional' Egypt as a space of constructed visibility. Indeed, throughout the nineteenth and into the twentieth centuries it became a commonplace for travellers and tourists to describe Egypt through a series of visual metaphors: as a kaleidoscope, a diorama, a panorama, a phantasmagoria, a magic lantern show, a movie film. We might now revisit that past, go behind the elaborate stage-machinery and finally realize that we have been the architects of the illusion all along.

Notes

An earlier version of this paper was presented at the Sixth Conference of the International Association for the Study of Traditional Environments,

'Manufacturing Heritage/Consuming Tradition,' Cairo, December 15, 1998. This work is part of a larger unpublished manuscript tentatively titled *Dancing on the Pyramids: Orientalism and Cultures of Travel.*

1. E. Said, *Orientalism*, London, Penguin, 1979, revised 1995, pp. 176-177.

2. E. Hobsbawm, 'Inventing Traditions,' in E. Hobsbawm and T. Ranger (eds.), *The Invention of Tradition*, Cambridge, Cambridge University Press, 1983, pp. 1-14.

3. A. Behdad, *Belated Travelers: Orientalism in the Age of Colonial Dissolution*, Durham and London, Duke University Press, 1996, pp. 6-8, my emphasis.

4. L. Ghandi, *Postcolonial Theory: A Critical Introduction*, Edinburgh, Edinburgh University Press, 1998, pp. 4-5. There is a considerable – and fractious – debate over the meaning of the 'post' in postcolonialism, but Ghandi's arguments seem to me to be compelling both politically and intellectually. See also D. Gregory, *The Colonial Present*, Oxford, U.K., and Cambridge, MA, Blackwell, forthcoming.

5. *Egypt*, New York, Knopf, 1995.

6. Thomas Cook Holidays, *Egypt, Jordan and Israel,* January-December, 1997, pp. 12-18.

7. H. Lefebvre, *The Production of Space*, Oxford, U.K., and Cambridge, MA, Blackwell, 1991 translation, first published in French in 1974; T. Mitchell, *Colonising Egypt*, Cambridge, Cambridge University Press, 1988. Mitchell deploys Foucault rather than Lefebvre, and there are significant differences and oppositions between these two French theorists, but I cite Mitchell here because I think his work can be used to illuminate the equally important parallels between them.

8. J. Rajchman, 'Foucault's Art of Seeing,' in Rajchman, *Philosophical Events: Essays of the '80s*, New York, Columbia University Press, 1991, pp. 68-102.

9. T. Mitchell, 'The Object of Development: America's Egypt,' in J. Crush (ed.), *Power of Development*, London, Routledge, 1995, pp. 129-157.

10. J. Murdoch, 'The Spaces of Actor-Network Theory,' *Geoforum*, vol. 29, 1998, pp. 357-374.

11. The English transliteration varies from author to author; *dahabeah* is the most common form, but there were several variants, and I have not attempted to standardize them in the quotations that appear throughout the essay. The party hiring a *dahabeah* would usually be between three and eight people. Smaller parties would sometimes hire the smaller (and faster) *cange*: see, for example, W.H. Bartlett, *The Nile Boat, or Glimpses of the Land of Egypt*, London, Virtue, 1849; and L. Pascal, *La Cange: Voyage en Égypte*, Paris, Hachette, 1861.

12. T. Knox, *The Oriental World*, Hartford, CN, A.D. Worthington, 1877, p. 529.

13. D. Gregory, 'Scripting Egypt: Orientalism and the Cultures of Travel,' in J. Duncan and Gregory (eds.), *Writes of Passage: Reading Travel Writing*, London and New York, Routledge, 1999, pp. 120-127.

14. F. Nightingale, *Letters from Egypt: A Journey on the Nile 1849-50*, A. Sattin (ed.), New York, Weidenfeld and Nicolson, p. 48. This passage – and the relationship between Nightingale's 'dream-images' and her obsessive 'dreaming' – is more complex than I can explicate here: see D. Gregory, 'Between the Book and the Lamp: Imaginative Geographies of Egypt, 1849-50,' *Transactions of the Institute of British Geographers*, vol. 20, 1995, pp. 29-57.

15. E. Beaufort, *Egyptian Sepulchres and Syrian Shrines*, London, Longman, Green, 1861, vol. I, p. vii. 'Today the inhabitants of a village in a district where a traveller has been robbed are required to pay compensation to the victim; all in consequence have an interest in maintaining good order and keeping a vigilant watch': A.C. Bey, *Aperçu Générale sur l'Égypte*, Bruxelles, Meline, Cans, 1840, vol. II, p. 97.

16. J.L. Roberts, *Incidents of Travel in Egypt, Arabia Petraea, and the Holy Land*, San Francisco, Chronicle Books, 1991, originally published in 1837, p. 25; H. Martineau, *Eastern life, Present and Past*, Philadelphia: Lea and Blanchard, 1848, pp. 159-160; and Countess I. Hahn-Hahn, *Letters of a German Countess*, London, Colburn, 1845, originally published in German, 1844, III, p. 182. See also A.L. Al-Sayyid Massot, *Egypt in the Reign of Muhammad Ali*, Cambridge, Cambridge University Press, 1984, pp. 100-101. There were periodic outbreaks of peasant violence against the central authorities, but these were localized and seldom prolonged: outside pockets of protest in Upper Egypt and the Delta, Massot claims, 'the country was remarkable for its tranquillity,' pp. 135-136.

17. A.C. Smith, *The Attractions of the Nile and its Banks*, London, John Murray, 1868, vol. II, pp. 116-131, 173.

18. A.W. Kinglake, *Eothen*, Oxford, Clarendon Press, 1928 ed., originally published in 1844, p. 186. Although most travellers recoiled from that particular *frisson*, many of them realized that the landscapes through which they passed were saturated with the signs of death. Some of the most vivid testimony can be found in Flaubert's writings describing his *voyage en Orient* in 1849-50, where Egypt becomes not only a memorial to the dead – a landscape of temples and tombs – but a land where death is still ever-present: his journal and his letters home are pock-marked with fragments of half-buried mummies, the cries of the dying, the muzzles of hunting dogs purple with clotted blood, and the carcasses of camels and donkeys: see Gregory, 'Between the Book and Lamp,' p. 42.

19. C.D. Warner, *Mummies and Moslems*, Tornot, Belford, 1876, p. 104. In fact, Warner was exasperated by the preparations: 'I bought chunks of drugs, bottles of poisons, bundles of foul smells and bitter tastes. And then they told me I needed balances to weigh them in. This was too much. I was willing to take along an apothecary's shop on this pleasure excursion; I was not willing to become an apothecary,' p. 103.

20. Nightingale, *Letters*, pp. 45, 47.

21. Beaufort, *Egyptian Sepulchres*, p. 23; A. Edwards, *A Thousand Miles up the Nile*, London, 1877, 2nd ed., 1888, p. 72; and H.D. Traill, *From Cairo to the Soudan Frontier*, London, John Lane, 1896, p. 201. All this said, it should be noted that many travellers railed at the inadequacy of their own watercolours and sketches.

22. H. Hopley, *Under Egyptian Palms, or Three Bachelors' Journeyings on the Nile*, London, Chapman and Hall, 1869, pp. 89-90. Cf. S. Suleri, *The Rhetoric of English India*, Chicago, University of Chicago Press, 1992, p. 75; Suleri also accentuates the ways in which desire and sexuality surface through the picturesque.

23. T.G. Appleton, *A Nile Journal*, Boston, Roberts, 1876, p. 131; the diabolical imagery is from Florence Nightingale's description of the island of Elephantine at the northern end of the cataract in her *Letters*, p. 86; Warner's reflections on the 'convulsion of nature' are from his *Mummies*, pp. 217-232.

24. J.W. Clayton, *Letters from the Nile*, London, Bosworth, 1854, p. 53.

25. J. Arnold, *Palms and Temples: Four Months' Voyage upon the Nile*, London, Tinsley Brothers, 1882, pp. 63-75; see also his father's account in E. Arnold, *Wandering Words*, London, Longmans Green, 1894, pp. 49-70.

26. Warner, *Mummies*, pp. 52-53, 92; M. Twain (Samuel Clements), *The Innocents Abroad*, New York, American Publishing Co., 1869; Arnold, *Palms and Temples*, p. 157.

27. See E. Toledano, *State and Society in Mid-Nineteenth-Century Egypt*, Cambridge, Cambridge University Press, 1990.

28. See Gregory, 'Scripting'.

29. W.H. Bartlett, *Nile Boat*, pp. 109-110; J.G. Wilkinson, *Modern Egypt and Thebes*, London, John Murray, 1843, vol. I, p. 216; and B. St. John, *Village Life in Egypt*, London, Chapman Hall, 1852, p. 223.

30. B. Taylor, *A Journey to Central Africa, or Life and Landscapes from Egypt to the Negro Kingdoms of the White Nile*, New York, Putnam, 1854, pp. 63-64. There was probably a formulaic quality to these exchanges on both sides, though Taylor seems to have convinced himself that the outcome was as he said: he later wrote that 'as we make no unreasonable demands we are always cheerfully obeyed,' and that 'the most complete harmony exists between the rulers and the ruled' on board the *dahabeah* (p. 87). Other travellers consoled themselves that the punishment was far from severe. 'It was over in less time than it takes to write it,' Smith recorded, 'and we were told it was a light punishment and did not hurt him much. It was a comfort to have the intelligence, for I was suffering beyond description': J.V.C. Smith, *Pilgrimage to Egypt*, Boston, Gould and Lincoln, 1852, p. 219. Other travellers were markedly more sensitive to the suffering of the unfortunate victim. Stephens witnessed the bastinado being applied after a dispute between two Egyptians: 'When I heard the scourge whizzing through the air, and, when the first blow fell upon the naked feet, saw the convulsive movements of the body, and heard the first loud piercing shriek, I could stand it no longer': see J.L. Stephens, *Incidents of Travel in Egypt, Arabia Petraea, and the Holy Land*, San Francisco, Chronicle Books, 1991, originally published in 1837, pp. 147-149.

31. Hahn-Hahn, *Letters*, pp. 155-156.

32. A.S. Kennard, *Eastern Experiences Collected during a Winter's Tour in Egypt and the Holy Land*, London, Longman, Brown, Green and Longman, 1855, pp. 129-131, 178-181, 189-193.

33. E. Warburton, *The Crescent and the Cross, or Romance and Realities of Eastern Travel*, London, Colburn, 1855, original ed. 1844, pp. 17, 29.

34. P. Brendon, *Thomas Cook: 150 Years of Tourism*, London, Secker and Warburg, 1981, pp. 125-127; *Miss Riggs's Diary*, Thomas Cook Archives, London, February 18, March 3, 1869.

35. Brendon, *Thomas Cook*, p. 131.

36. *Cook's Excursionist and Tourist Advertiser*, October 11, 1886, p. 3; Brendon, *Thomas Cook*, p. 130. Thomas Cook was neither the only nor even the first European tour operator in Egypt. Henry Gaze was there before him and there was a considerable rivalry between the two companies until Gaze's sons declared

bankruptcy in 1903: see Brendon, *Thomas Cook*, p. 185; and Knox, *Oriental World*, p. 531.

37. Edwards, *Thousand Miles*, p. 70:

> It is the rule of the Nile to hurry up the river as fast as possible, leaving the ruins to be seen as the boat comes back with the current; but this, like many another canon, is by no means of universal application . . . [For] those who desire not only to see the monuments, but to follow, however superficially, the course of Egyptian history as it is handed down through Egyptian art, it is above all things necessary to start early and see many things by the way.
>
> For the history of ancient Egypt goes against the stream. The earliest monuments lie between Cairo and Siout, while the latest temples to the old gods are chiefly found in Nubia. Those travellers, therefore, who hurry blindly forward with or without a wind, now sailing, now tracking, now punting, passing this place by night, and that by day, and never resting till they have gained the farthest point of their journey, begin at the wrong end and see all their sights in precisely inverse order.

My summary version of Cook's arguments is from R. Etzensberger, *Up the Nile by Steam*, London, Thomas Cook, 1877, pp. 7-10. The company used its publication to take a pot-shot at the authority of 'guide-books and [published] diaries of tourists' which sung the praises of the *dahabeah*; they were the product of 'a single short visit,' whereas *Up the Nile by Steam* was 'the result of personal observation during eight seasons' (p. 6).

38. Rev. S. Manning, *The Land of the Pharaohs: Egypt and Sinai Illustrated by Pen and Pencil*, London, Religious Tract Society, 1875, pp. 66-67.

39. This was no exaggeration: countless books describe picnicking among the tombs and temples. See, for example, W.C. Maughan, *The Alps of Arabia: Travels in Egypt, Sinai, Arabia and the Holy Land*, London, Henry King, 1875, pp. 48-49.

40. P. Loti, *Egypt*, London, Werner, 1909, translated from the French, pp. 135-136,160. The title in the original French – *Mort de Philae* – used the construction of the Aswan Dam and its threat to the island of Philae as a symbol of what Loti took to be the downfall of Egypt – of a particular, romantic and quintessentially Orientalist 'Egypt' – brought about by the British occupation.

41. E. Reynolds-Ball, *Cairo: The City of the Caliphs*, London, Fisher Unwin, 1901, p. 130; cf. T. Mostyn, *Egypt's Belle Époque: Cairo 1869-1952*, London, Quartet, 1989.

42. A.B. de Guerville, *New Egypt*, London: Heinemann, 1906, p. 221.

43. *Cook's Excursionist and Tourist Advertiser*, November 1, 1881, p. 3; *Programme of Thomas Cook's international tickets to Egypt*, London, Thomas Cook, 1898-99, Thomas Cook Archives, London, pp. 5-7. By 1889 Cook had built three first-class *dahabeahs* of its own, and soon added three more.

44. Edwards, *Thousand Miles*, p. 195 (note added to 2nd ed.).

45. *Cook's Excursionist and Tourist Advertiser*, November 29, 1880, p. 3; Brendon, *Thomas Cook*, pp. 137,183.

46. W. Bemrose, *Recollections of Egypt and Palestine*, 1882, manuscript, Thomas Cook Archives; Bemrose was a personal friend of John Mason Cook and accompanied him on his tour 'to visit Alexandria, Cairo, the Pyramids, the Nile and the late battlefields'.

47. Brendon, *Thomas Cook*, pp.189-200. The company's own fleet was not big enough to undertake such an operation on its own; twenty-seven steamers and 650 sailing ships were used to transport the relief column of troops, supplies and coals as far as Wadi Halfa, and John Mason Cook hired 'a little army' of 5,000 local men and boys to work on the mission: *Cook's Excursionist and Tourist Advertiser*, February 2, 1885, p. 3.

48. Brendon, *Thomas Cook*, pp. 190-192, 197-199.

49. 'Cook and Son in Egypt (according to hieroglyphics found at Assouan),' Thomas Cook Archives. The conceit was not confined to the company. According to *Vanity Fair*, March 9, 1889:

The nominal ruler [of Egypt] is Twefik; but Tewfik takes his orders from Baring; and Baring, I suspect, has to take his orders from Cook. The latter Sovereign becomes more and more potent as we get further up the Nile and here at Luxor, where a special hotel has arisen under the light of his countenance, he figures quite as a modern Ammon-Ra. It seems likely too that his might and majesty will increase.

50. *Cook's Excursionist and Tourist Advertiser*, February 2, 1885, p. 3.

51. G.W. Steevens, *Egypt in 1898*, London, Blackwood, 1898, p. 269.

52. *Ibid.*, pp. 68-69, 270-271.

53. D. Sladen, *Queer Things about Egypt*, London, Hurst and Blackett, 1910, p. 388. The epithet was a cliché even amongst non-British writers: see, for example, G. Montbard, *The Land of the Sphinx*, London, Hutchinson, 1894, p. ix; de Guerville, *New Egypt*, p. 3.

54. *Cook's Excursionist and Tourist Advertiser*, February 2, 1885, p. 3; November 2, 1885, p. 3; October 1, 1886, p. 3; February 1, 1887, p. 4; and September 12, 1887, p. 3. The new ships were planned to provide state-rooms for thirty-two first-class passengers, a dining saloon on the upper deck and a private saloon for ladies, and two or more bathrooms whose use was now 'included in the fare'. The first-class return fare from Cairo to the First Cataract was set at £50. The *Prince Abbas* was a magnificent ship, but old prejudices lingered in some quarters. Wilbour declared that 'it is really fine and looks comfortable'; so much so that he asked Mrs Goadison, Ruskin's next-door neighbour near Lake Windermere, why she did not go in it or on one of Cook's *dahabeahs*. She said, 'I could never face Mr Ruskin again if I were to go in a Cook boat.' C.E. Wilbour, *Travels in Egypt, December 1880 to May 1891*, Brooklyn, NY, Brooklyn Museum, 1936, p. 411.

55. M. Caillard, *A lifetime in Egypt 1876-1935*, London, Grant Richards, 1935, p. 55.

56. P.S. Marden, *Egyptian Days*, Boston, Houghton Mifflin, 1912, p. 11.

57. L. Bacon, *Our Houseboat on the Nile*, Boston, Houghton, Mifflin and Co., 1902, pp. 155-159. Bacon satirized the journey from start to finish, from the 'record-breaking transatlantic greyhounds', through the express mail steamer to Alexandria and the train to Cairo, and then the steam-*dahabeah* to Luxor: 'So they return to Cairo [by train], having indeed broken the record for speed but having failed to come into touch with anything along the way.'

58. G. Ade, *In Pastures New*, New York, McClure and Phillips, 1906, p. 197; N. Lorimer, *By the Waters of Egypt*, London, Methuen, 1909, p. 419. There were

other companies operating steamboats on the Nile, including the rival Anglo-American line and the Express Nile Company, but Cook had captured the lion's share of the market by the turn of the century.

59. *Programme*, pp. 73-87.

60. Steevens, *Egypt*, pp. 214-215.

61. R. Kipling, *Letters of Travel (1892-1913)*, London, Macmillan, 1920, pp. 246-247.

62. *Cook's Excursionist and Tourist Advertiser*, December 11, 1885, p. 3.

63. E. Butler, *From Sketch-book and Diary*, London, Adam and Charles Black, 1909, p. 55.

64. *Cook's Exursionist and Tourist Advertiser*, November 1, 1881, p. 4.

65. *Cook's Traveller's Gazette*, December 10, 1903, p. 23.

66. de Guerville, *New Egypt*, p. 221; cf. Loti, *Egypt*, pp. 279-280.

67. J. Ward, *Pyramids and Progress*, London, Eyre and Spottiswoode, 1900, p. 247.

68. F.C. Penfield, *Present-day Egypt*, New York, Century, 1903, p. 151. That this was a colonial project – and a colonial triumph – was made clear over and over again. Thus, for example, the construction of these barrages was 'the greatest of all the tasks which Englishmen have accomplished in Egypt': S. Low, *Egypt in Transition*, London, Smith, Elder, 1914, p. 127.

69. Penfield, *Present-day Egypt*, p. 184; emphasis added.

70. Low, *Egypt in Transition*, pp. 133, 140, 144. Appropriately, the introduction to Low's book was written by Lord Cromer who, as Britain's Consul-General, had approved the construction of the Dam and had been instrumental in the arrangements for its financing.

71. *Cook's Arrangements for Egypt and the Nile, 1908-9*, Thomas Cook Archives, pp. 35-37. The 'Chalet' – a sort of rest-house – had been built by Cook for the use of its clients.

72. Penfield, *Present-day Egypt*, pp. 362-363; and E.A.W. Budge, *Cook's Handbook for Egypt and the Sudan*, London, Thomas Cook, 1906, p. 39.

73. Others were less fortunate and necessarily more attentive. Kipling, *Letters*, March 11, 1913, described being stuck on a sandbank for 28 hours:

The thermometer stood at 55° (this on the edge of the Tropics) but the wind it never stood still for one wicked minute. It blew like Hades and cut like a knife and our miserable steamer swung round and round (but never got off the sand-bank) and other steamers (four others) came up and got stuck too and a little express steamer who was taking cheap tourists on a cheap trip had to stop and throw us a rope and help to haul us off bodily.

74. Lorimer, *By the Waters*, p. 248.

75. Penfield, *Present-day Egypt*, p. 163.

76. At least some visitors recognized that this modern techno-cultural frame was an imperfect construction:

Behold how nature mocketh at man's vain efforts to improve her workmanship! This [Aswan] dam which was to furnish an inexhaustible supply of water, and abolish famine in Egypt, is now cursed with all the fervour of Oriental hyperbole by the very *fellaheen* who expected untold benefit from its construction.

Water there is in plenty, but the crops grow less and less, the cotton poorer and poorer, its fibre ever more brittle. The scientists must once more go to school and learn to undo the harm their famous barrier has wrought. Impounding the water has likewise precipitated the rich sediment, which, brought from the upper reaches of the Nile, had fertilized and enriched the lower valley, and had produced the fabulous crops of former uncontrolled high Niles.

See B.M. Carson, *From Cairo to the Cataract*, Boston, Page, 1909, pp. 219-220.

77. Maughan, *Alps of Arabia*, p. 64. Tourists were no strangers to sights like this. In the closing decades of the nineteenth century many of the temples and tombs were cleared using forced labour, but most spectators seem to have had little difficulty in externalizing the scene. At Luxor, for example, the Tirards saw 'five hundred children . . . carrying away the dust in baskets, and emptying them down the steep bank of the river; they did their work singing, scampering and dancing, as if it were a huge game, notwithstanding that they were pursued now and then by the whip of the overseer': H. and N. Tirard, *Sketches from a Nile Steamer*, London, Kegan Paul, 1891, p. 69.

78. Ward, *Pyramids and Progress*, p. 173. He attributed this system to a staff of 'British gentlemen . . . under whose intelligent control at their offices at Cairo and branches at the large towns these natives have been licked into shape' (pp. 173-174).

79. Steevens, *Egypt*, pp. 68-69.

80. *Programme*, pp. 5-6.

81. *Cook's Excursionist and Tourist Advertiser*, October 1, 1886, p. 3.

82. Sladen, *Queer Things*, p. 391.

83. D. Sladen, *Egypt and the English*, London, Hurst and Blackett, 1908, p. 427.

84. Lorimer, *By the Waters*, p. 169.

85. See D. Gregory, 'Emperors of the Gaze: Photographic Practices and the Captivation of Space in Egypt 1839-1914,' forthcoming.

86. *Programme*, pp. 12-13.

87. Mrs. H. Fawcett, in W.J. Loftie (ed.), *Orient-Pacific Line Guide*, London, Sampson Low, 1901, pp. 95-96.

88. Kipling, *Letters*, p. 241.

89. Sir W.T. Grenfell, *Labrador Looks at the Orient: Notes on Travel in the Near and Far East*, Boston and New York, Houghton Mifflin, 1928, p. 20. For obvious reasons the brochure never identifies the 'Agatha Christie film set' as that of *Death on the Nile.*

90. J. Frow, 'Tourism and the Semiotics of Nostalgia,' *October*, vol. 57, 1991, pp. 123-157.

91. R. Rosaldo, 'Imperialist Nostalgia,' in Rosaldo, *Culture and Truth: The Remaking of Social Analysis*, London, Routledge, 1993.

92. P. Smith, 'Visiting the Banana Republic,' in A. Ross (ed.), *Universal Abandon? The Politics of Postmodernism*, Minneapolis, University of Minnesota Press, 1989, pp. 128-148.

Bibliography

Anon, *Cook's Arrangements for Egypt and the Nile, 1908-9*, Thomas Cook Archives.

Anon, *Cook's Excursionist and Tourist Advertiser*, various issues.

Anon, *Cook's Traveller's Gazette*, various dates.

Anon, *Egypt*, New York, Knopf, 1995.

Anon, *Miss Riggs's Diary*, Thomas Cook Archives, London.

Anon, *Programme of Thomas Cook's International Tickets to Egypt*, London, Thomas Cook, 1898-99, Thomas Cook Archives, London.

Ade, George, *In Pastures New*, New York, McClure and Phillips, 1906.

Appleton, T.G., *A Nile Journal*, Boston, Roberts, 1876.

Arnold, Edwin, *Wandering Words*, London, Longmans Green, 1894.

Arnold, Julian, *Palms and Temples: Four Months' Voyage upon the Nile*, London, Tinsley Brothers, 1882.

Bacon, Lee, *Our Houseboat on the Nile*, Boston, Houghton, Mifflin and Co., 1902.

Bartlett, W.H., *The Nile Boat, or Glimpses of the Land of Egypt*, London, Virtue, 1849.

Beaufort, Emily, *Egyptian Sepulchres and Syrian Shrines*, London, Longman, Green, 1861, vol. I.

Behdad, Ali, *Belated Travelers: Orientalism in the Age of Colonial Dissolution*, Durham and London, Duke University Press, 1996.

Bemrose, William, *Recollections of Egypt and Palestine*, 1882, manuscript, Thomas Cook Archives.

Bey, A. Clot, *Aperçu générale sur l'Égypte*, Bruxelles, Meline, Cans, 1840.

Brendon, Piers, *Thomas Cook: 150 Years of Tourism*, London, Secker and Warburg, 1981.

Budge, E.A. Wallis, *Cook's Handbook for Egypt and the Sudan*, London, Thomas Cook, 1906.

Butler, Elizabeth, *From Sketch-book and Diary*, London, Adam and Charles Black, 1909.

Caillard, Mabel, *A Lifetime in Egypt 1876-1935*, London, Grant Richards, 1935.

Carson, Blanche Mabury, *From Cairo to the Cataract*, Boston, Page, 1909.

Clayton, J.W., *Letters from the Nile*, London, Bosworth, 1854.

de Guerville, A.B., *New Egypt*, London, Heinemann, 1906.

Edwards, Amelia, *A Thousand Miles up the Nile*, London, 1877; second edition, 1888.

Etzensberger, R., *Up the Nile by Steam*, London, Thomas Cook, 1877.

Fawcett, Mrs Henry, in W.J. Loftie (ed.), *Orient-Pacific Line Guide*, London, Sampson Low, 1901.

Frow, John, 'Tourism and the Semiotics of Nostalgia,' *October*, vol. 57, 1991.

Ghandi, Leela, *Postcolonial Theory: A Critical Introduction*, Edinburgh, Edinburgh University Press, 1998.

Gregory, Derek, 'Between the Book and the Lamp: Imaginative Geographies of Egypt, 1849-50,' *Transactions of the Institute of British Geographers*, vol. 20, 1995.

Gregory, Derek, 'Emperors of the Gaze: Photographic Practices and the Captivation of Space in Egypt 1839-1914,' forthcoming.

Gregory, Derek, 'Scripting Egypt: Orientalism and the Cultures of Travel,' in James Duncan and Derek Gregory (eds.), *Writes of Passage: Reading Travel Writing*, London and New York, Routledge, 1999.

Gregory, Derek, *The Colonial Present*, Oxford, U.K., and Cambridge, MA, Blackwell, forthcoming.
Grenfell, Sir Wilfrid Thomason, *Labrador Looks at the Orient: Notes on Travel in the Near and Far East*, Boston and New York, Houghton Mifflin, 1928.
Hahn-Hahn, Ida, Countess, *Letters of a German Countess*, London, Colburn, 1845; originally published in German in 1844.
Hobsbawm, Eric, 'Inventing Traditions,' in Hobsbawm and Terence Ranger (eds.), *The Invention of Tradition*, Cambridge, Cambridge University Press, 1983.
Hopley, Howard, *Under Egyptian Palms, or Three Bachelors' Journeyings on the Nile*, London, Chapman and Hall, 1869.
Kennard, Adam Steinmetz, *Eastern Experiences Collected during a Winter's Tour in Egypt and the Holy Land*, London, Longman, Brown, Green and Longman, 1855.
Kinglake, A.W., *Eothen*, Oxford, Clarendon Press, 1928 ed.; originally published in 1844.
Kipling, Rudyard, *Letters of Travel (1892-1913)*, London, Macmillan, 1920.
Knox, Thomas, *The Oriental World*, Hartford, CN, A.D. Worthington, 1877.
Lefebvre, Henri, *The Production of Space*, Oxford, U.K., and Cambridge, MA, Blackwell, 1991 translation.; first published in French in 1974.
Lorimer, Norma, *By the Waters of Egypt*, London, Methuen, 1909.
Loti, Pierre, *Egypt*, London, Werner, 1909, (translated from the French).
Low, Sidney, *Egypt in Transition*, London, Smith, Elder, 1914.
Manning, Rev. Samuel, *The Land of the Pharaohs: Egypt and Sinai Illustrated by Pen and Pencil*, London, Religious Tract Society, 1875.
Marden, Philip Sanford, *Egyptian Days*, Boston, Houghton Mifflin, 1912.
Martineau, Harriet, *Eastern Life, Present and Past*, Philadelphia, Lea and Blanchard, 1848.
Massot, Afaf Lutfi Al-Sayyid, *Egypt in the Reign of Muhammad Ali*, Cambridge, Cambridge University Press, 1984.
Maughan, William Charles, *The Alps of Arabia: Travels in Egypt, Sinai, Arabia and the Holy Land*, London, Henry King, 1875.
Mitchell, Timothy, 'The Object of Development: America's Egypt,' in Jonathan Crush (ed.), *Power of Development*, London, Routledge, 1995.
Mitchell, Timothy, *Colonising Egypt*, Cambridge, Cambridge University Press, 1988.
Montbard, Georges, *The Land of the Sphinx*, London, Hutchinson, 1894.
Mostyn, Trevor, *Egypt's Belle Époque: Cairo 1869-1952*, London, Quartet, 1989.
Murdoch, Jonathan, 'The Spaces of Actor-Network Theory,' *Geoforum*, vol. 29, 1998.
Nightingale, Florence, *Letters from Egypt: A Journey on the Nile 1849-50*, Anthony Sattin (ed.), New York, Weidenfeld and Nicolson.
Pascal, Louis, *La Cange: Voyage en Égypte*, Paris, Hachette, 1861.
Penfield, F.C., *Present-day Egypt*, New York, Century, 1903.
Rajchman, John, 'Foucault's Art of Seeing,' in Rajchman, *Philosophical Events: Essays of the '80s*, New York, Columbia University Press, 1991.
Reynolds-Ball, Eustace, *Cairo: The City of the Caliphs*, London, Fisher Unwin, 1901.
Rosaldo, Renato, 'Imperialist Nostalgia,' in Rosaldo, *Culture and Truth: The Remaking of Social Analysis*, London, Routledge, 1993.
Roberts, J.L., *Incidents of Travel in Egypt, Arabia Petraea, and the Holy Land*, San Francisco, Chronicle Books, 1991, originally published in 1837.

Said, Edward, *Orientalism*, London, Penguin, 1979; revised 1995.
Sladen, Douglas, *Egypt and the English*, London, Hurst and Blackett, 1908.
Sladen, Douglas, *Queer Things about Egypt*, London, Hurst and Blackett, 1910.
Smith, A.C., *The Attractions of the Nile and its Banks*, London, John Murray, 1868, vol. II.
Smith, J.V.C., *Pilgrimage to Egypt*, Boston, Gould and Lincoln, 1852.
Smith, Paul, 'Visiting the Banana Republic,' in Andrew Ross (ed.), *Universal Abandon? The Politics of Postmodernism*, Minneapolis, University of Minnesota Press, 1989.
St. John, Bayle, *Village Life in Egypt*, London, Chapman Hall, 1852.
Steevens, G.W., *Egypt in 1898*, London, Blackwood, 1898.
Stephens, John Lloyd, *Incidents of Travel in Egypt, Arabia Petraea, and the Holy Land*, San Francisco, Chronicle Books, 1991; originally published in 1837.
Suleri, Sara, *The Rhetoric of English India*, Chicago, University of Chicago Press, 1992.
Taylor, Bayard, *A Journey to Central Africa, or Life and Landscapes from Egypt to the Negro Kingdoms of the White Nile*, New York, Putnam, 1854.
Thomas Cook Holidays, *Egypt, Jordan and Israel*, January-December 1997.
Tirard, H. and N., *Sketches from a Nile Steamer*, London, Kegan Paul, 1891.
Toledano, Ehud, *State and Society in Mid-Nineteenth-Century Egypt*, Cambridge, Cambridge University Press, 1990.
Traill, H.D., *From Cairo to the Soudan Frontier*, London, John Lane, 1896.
Twain, Mark (Samuel Clements), *The Innocents Abroad*, New York, American Publishing Co., 1869.
Warburton, Eliot, *The Crescent and the Cross, or Romance and Realities of Eastern Travel*, London, Colburn, 1855; original edition 1844.
Ward, John, *Pyramids and Progress*, London, Eyre and Spottiswoode, 1900.
Warner, C.D., *Mummies and Moslems*, Tornot, Belford, 1876.
Wilbour, Charles Edwin, *Travels in Egypt, December 1880 to May 1891*, Brooklyn, NY, Brooklyn Museum, 1936.
Wilkinson, J.G., *Modern Egypt and Thebes*, London, John Murray, 1843.

Chapter 6

Everyday Attractions: Tourism and the Generation of Instant Heritage in Nineteenth-Century San Francisco

J. PHILIP GRUEN

Immediately before us lights were twinkling from ships' masts, and long straight lines of lamps climbed up a conical hill, and small steamers shot across our path like meteors, with their cabins brilliant with light. There was a sound of bells, the shrill whistle of the engine, the rattling of many carriages, the gleam of a red light, and we were gliding along the side of a pier covered with a sea of anxious faces. Hardly had we stopped than the owners of these faces were upon us. They boarded us like pirates, and then arose a Babel of cries, among which I could distinguish as follows: 'Who wants to go to the Cosmopolitan?' 'Who wants a carriage?' 'Grand Hotel, sir?' 'This way for the What Cheer Coach!' 'Carriage, sir?' 'Take you up for a dollar, sir!' 'Want a hand-cart for your luggage, sir?'[1]

London tourist John Player-Frowd hardly expected the scene at the San Francisco pier when he first arrived in 1872. Player-Frowd, like many residents of Great Britain, had been exposed to conflicting accounts regarding San Francisco since the Gold Rush of the late 1840s. These accounts portrayed the city either as a paradise on earth, where riches were plentiful and civility reigned, or as a den of iniquity, inhabited by the scum of the earth. What he found, as did many San Francisco tourists between 1870 and 1890, was that the city featured a little of both – but not an overabundance of either. The pier activity he described was somewhere in between: here was a noisy and seemingly chaotic environment where hotel runners scrambled to sell chauffeured transportation, often by means of luxurious carriage, to the city's more urbane lodgings.

But it was not simply the contrast between iniquity and gentility that tourists found compelling. Tourists were interested in the activity itself, and they often wrote about it in their memoirs. With hindsight, the existence of such a thriving tourist-related industry at the pier in the 1870s may seem unusual to contemporary scholars, who have generally not considered San Francisco's role as a tourist destination in the nineteenth century. That role may appear even more startling to those who have assumed that early tourist ventures to the American West were principally concerned with viewing the wide-open spaces of the natural landscape. But accounts of urban activity in San Francisco, usually laudatory, were the rule rather than the exception, as tourists came to familiarize themselves with this new metropolis on the edge of the American continent. And by the 1870s, San Francisco had established itself as the premier tourist city in the United States west of the Mississippi River, and – along with the Yellowstone and Yosemite areas – one of the West's principal attractions as well.

Along with the Gold Rush, the Vigilance Committee, and the Comstock Lode, tourism and tourists formed an important, yet often overlooked, part of San Francisco's nineteenth-century heritage. Beginning in the early 1870s, guidebooks and other official publications lured tourists to various sites, including the city's buildings, parks, and Chinese quarter, carefully manufacturing and packaging them and helping to generate a sense of instant heritage. The literature catered for wealthy tourists, most of whom hailed from the major American cities of the Eastern Seaboard (such as Philadelphia, Boston, and New York), but many of whom also came from major European cities (particularly London and Paris). All through the nineteenth century tourists from these cities had been heading to 'exotic', less-developed locales in non-Western countries – particularly British travellers to the colonies and territories of India, Egypt and Australia. But San Francisco's rapid development in a spectacular natural setting, its connections to the Pacific Rim, and its sheer distance from the rest of the heavily occupied world offered a potentially exotic experience all its own. In fact, tourists began to visit the city as early as 1865, increased in numbers after the completion of the trans-continental railroad in 1869, and were common to the city's landscape by the early 1890s. The presence of a large tourist population in San Francisco, along with a built environment which its boosters marketed as resembling the best of Europe and the East Coast, helped the city shed its more popular early reputation as a rowdy, dysfunctional frontier town. Acknowledging the importance of the tourist presence in San Francisco as early as the 1870s also contradicts the view that tourism did not emerge as an important activity in the city until the Panama-Pacific International Exposition of 1915 – or in the United States more generally until the 1890s, when tourist services became 'rationalized', the railroads reduced travel rates, and tourism was 'transformed' into an 'industry'.[2]

Acknowledging the existence of tourists in early San Francisco by examining guidebooks and other official publications, however, does not automatically indicate *how* tourists may have experienced the city. While some travellers may have experienced San Francisco just as the guidebooks intended (as a series of genteel, free-standing monuments), for many travellers it was the city's energy, crowds, and otherwise unregulated, non-manufactured and everyday reality that were most appealing (Figure 6.1). This interest in the everyday is occasionally revealed through less official publications such as traveller's books and the prints that illustrated them, articles and imagery in newspapers and national magazines, and unpublished diaries and letters. The more staid portrayals of the city presented in the official publications did not prepare tourists for the city's everyday life. Tourists, however, explored and consumed San Francisco in ways that both included *and* resisted the features of instant heritage promoted in the guidebooks. This combination of a sense of gentility (often experienced by visits to carefully promoted sites) and a feeling for the overall high level and density of urban activity meant 'San Francisco' to many tourists, helping them form a more complete picture and break down the assumption that the city represented a radically different kind of urban

Figure 6.1. Market Street, San Francisco, looking west toward the Ferry Building – photograph by I.W. Taber, about 1885. Cable railways, horses, carriages, pedestrians, power lines, and signage all jostle for position in this nineteenth-century Taber photograph of San Francisco's busiest street, one of several views contained within a Taber photographic souvenir album. The view shows a somewhat less-than-genteel city, but nevertheless one that would have been familiar to tourists. The Palace Hotel is the third prominent building from the right. (Photo courtesy of the Bancroft Library, University of California, Berkeley.)

experience than they had encountered elsewhere. It was ultimately this larger urban experience – not simply the associations tourists made with individual sites – that inspired tourists to consider San Francisco among America's and Europe's more 'civilized' destinations.

Such consideration of the significance of the tourist landscape to the early San Francisco experience suggests that there was more to the city's early years than the unevenness associated with its Gold Rush period and the alleged chaos and urban disorder that accompanied and followed it. While tourists were only a part of the transformation of the city from its Gold Rush beginnings, they nevertheless contributed to the cultural landscape, and should be considered in any serious study of nineteenth-century San Francisco. This chapter attempts to fill at least part of that gap.

Tourists in Nineteenth-Century San Francisco

The story of nineteenth-century San Francisco is popularly summarized by a few events that mark its rise from a provincial settlement to a bustling metropolis. Following the 1848 discovery of gold, the story goes, the once-tiny settlement on the shores of San Francisco Bay became an 'instant city' characterized by waves of immigration, overcrowding and crime. The 1850s were a time of high-stakes gambling and heavy drinking. Thugs wandered the streets, looting and harassing at will. Prostitution was rampant. To provide order, groups of citizens formed committees of 'vigilance', taking the law into their own hands and doling out justice as they saw fit.

By the 1860s and 1870s, the story continues, San Francisco shook its image as a rowdy city on the edge of the frontier by concentrating its efforts on shipping, manufacturing, real estate development, housing construction, and public works. Jobs provided by new industries, railroad construction, and the economic frenzy that accompanied the 1859 discovery of silver in Nevada's Virginia City stimulated further growth, as money flowed freely through the West Coast's major port of entry. From approximately 1,000 residents in 1848, San Francisco's population grew to 233,956 by 1880, and 300,000 by 1891. By the early 1890s San Francisco had become a place of extraordinary production: an active, almost frenetic city – the first metropolis on the western shores of the United States.

If one wholly subscribes to this sweeping narrative, it seems that there was little room during San Francisco's early years for the casual visitor or pleasure-seeking tourist. Indeed, the rapid growth of nineteenth-century San Francisco – and the chaos and urban disorder that occasionally accompanied it – has absorbed much of the scholarly attention that has been paid to this period of the city's history. To date, most writers have ignored the effect of tourists on this process of urbanization, and few have considered that tourists may have taken an active interest in this growth.[3]

San Francisco is not alone among American cities in this regard. Tourists have been overlooked in histories of the early urban American West, as they have been more generally in histories of early urban America.[4] This is puzzling, considering that by the 1890s New York City – not Niagara Falls, the White Mountains, Yellowstone, or Yosemite – was the most widely visited place in America, and that nearly every early 'Western' railroad trip included urban encounters along the line (or just off it). Furthermore, many cities were more than stopovers; they were sold as attractions in and of themselves, and tourists explored them.[5]

With the completion of the Union and Central Pacific Railroad lines – the first of the trans-continental railroads – most tourists who ventured west to explore the 'frontier' and to visit its natural wonders eventually completed their railroad journey in San Francisco (or, more specifically, across San Francisco Bay in Oakland). Even if their trip had been inspired by a desire to explore the natural landscape of the West (which many domestic travellers believed marked a distinct 'American' heritage unlike anything in Europe), few tourists hopped off the railroad in Oakland and took the first excursion train along the coast or into California's interior. Nearly all tourists boarded ferries for the short trip across the bay to the city of San Francisco, where they remained for at least a couple of days.

Despite these realities of travel, the historiography of tourist encounters with the nineteenth-century American West (and twentieth, for that matter) has had a decidedly anti-urban bias. Western tourism at that time has meant visits to 'natural wonders': The Rocky Mountains and Pike's Peak, Colorado's Garden of the Gods, Yellowstone, Yosemite, the Calaveras Big Trees, and the geysers near Calistoga. Generally, Western tourism has been considered a history of travelling to these natural sites – first by railroad, then by automobile, and finally by recreational vehicle and packaged bus tour. Tourists in early San Francisco appear in secondary sources (when they do appear) because some travellers stayed briefly in the city before transferring from an overnight Pullman car to an excursion train heading for Yosemite or the Hotel del Monte in Monterey. Thus, when Earl Pomeroy set out in the late 1950s to write the first history of early Western American tourism (and still one of the few), his focus was on the wide-open spaces of the West, not its urban centres.[6] Most scholars who have written about tourism in the American West since then have used Pomeroy as a foundation, providing more detailed analyses of how tourists visited the gems of natural scenery described in his book.

It is, of course, justified to understand the tourist in the American West as an explorer of nature, for these scenic natural wonders were tirelessly promoted by railroad companies, guidebooks, journalists, writers and photographers in the nineteenth century, and they were visited by numerous tourists in the nineteenth century. Thomas Cook and Sons, the British travel company, offered package tours to view the natural 'wonders' of the American West as early as 1870. Even the ‘See America First’ campaign, initiated in 1906, was designed to infuse American nationalism and generate domestic revenue by exploring the ‘scenic wonders’ of the country – particularly those in the West.[7] Furthermore, those few scholars who have bothered to explore urban tourism in nineteenth-century America have generally seen it as a pursuit of the natural, in part because many urban dwellers, on their days off, may have desired to ‘escape’ the density, poverty and corruption of the cities for tranquil suburbs or picturesque urban parks. All too often, scholars of American tourism have assumed that nineteenth-century travellers would have sought out similar places of refuge during their visits to cities elsewhere. This view of life in the nineteenth-century city stems in part from a range of accounts that dwell upon the urban ‘ills’ of rapidly industrializing cities and the attempts of reformers to make them more liveable and genteel.

Understanding tourists as distinct from cities also ignores the efforts of business and civic leaders to promote their downtowns as visitor attractions during a time that, in San Francisco, was frequently associated with lawlessness and unchecked growth. The view also fails to recognize that many people were fascinated with industrial and big-business America, and that city skylines, with their office buildings, factories and smoke, may have come to mark the notion of ‘city’ and ‘opportunity’ to many travellers, just as it had to many immigrants coming to America’s Eastern Seaboard during the nineteenth century.[8] And while tourists certainly did enjoy visiting the suburbs, understanding tourism solely as an escape from cities disregards the possibility that many travellers enjoyed participating in the urban energy of the places they visited.

While the ‘opening up’ of the vast spaces of the ‘frontier’ to Anglo-American control and, later, tourist exploration represents part of the history of the American West, much of the region’s nineteenth-century history was one of urbanization. At the time the ratio of urban to rural residents in the American West was far greater than in the East (and even the Midwest), as people clustered together in cities from Denver to Portland. San Francisco, of course, was the most urban of these western cities, and this was part of its visitor appeal. The city provided the added attraction of being the principal nineteenth-century port of entry to the Far East, giving many travellers their first chance to glimpse Asian cultures, particularly the Chinese. With the Chinese routinely vilified in the contemporary literature for their ‘uncivilized’ and ‘heathen’ ways, a visit to San Francisco also provided some tourists with an opportunity to see what they perceived as a piece of the ‘undeveloped’ world on native soil.

That all this was framed by a scenic natural landscape marked by rolling hills and sparkling water could only have heightened the overall experience: here was a city that combined the ‘natural’ and the ‘artificial’ in ways that tourists could not have experienced elsewhere. Europeans and Americans heading across the land for the first time must have imagined San Francisco entirely as a spectacle of difference – an urban 'other'. And yet what travellers experienced once they arrived – and what they seemed to be most interested in – were not only the differences or the exoticness of it all, but the fact that a fully American city, with grand monuments within a vibrant urban landscape, could thrive at such a distance from the rest of the urbanized world.

But who were the nineteenth-century San Francisco tourists? This question is not so easily answered, for tourists belonged to no single class, race, gender or geography. One early traveller from England, for example, complained about the 'admixture' of strangers and sexes on one of Thomas Cook's 'package tour' trains.[9] Class differences were fairly acute in the 1870s and 1880s, when the high costs of rail travel made it prohibitive for any but members of the wealthy, leisured classes to visit San Francisco.[10] But by the 1890s, with more railroads, more tour companies, and a national economic depression, increasing competition and lowered rates made it possible for members of the more middling classes to tour California as well. Furthermore, many non-wealthy travellers came to San Francisco before 1890 hoping to find work or seek a fortune in the gold country, and passed their interim time in the city. These travellers often expressed fascination with many of the same things – from packed streetcars to leisurely drives in the park – encountered and enjoyed by wealthy vacationers staying at first-class hotels. Upon arrival they, too, may have found the guidebooks and other official literature promoting the city and its sites useful.

Surviving traveller's accounts were also written by women as often as they were by men, even though women were excluded from participating in many aspects of the alleged 'public' sphere in the nineteenth-century city (and, in certain cases, even from aspects of the tourist landscape – particularly those associated with night-time activities). But enough of the tourist landscape was enjoyed by both sexes, and tourism – while hardly shifting commonly held conceptions about women's roles in public space – nevertheless transcended some of the gender barriers that separated men from women in civic life. Both men and women expressed fascination with the city's urbanity and its 'civility' and were impressed with its weather, its setting, and its overall appearance.[11]

The problem of defining the nineteenth-century San Francisco tourist is compounded by the fact that while most visitors came from the East Coast or from Europe, some also came for the day or the evening by ferry, stagecoach or railroad from San Francisco's outlying suburbs such as Oakland, San Rafael, or San Mateo. Still others emerged from within San Francisco itself, taking the afternoon or evening to engage in activities that could also be associated with those of tourists such as trips *to* the suburbs or night-time visits to restaurants and the theatre. Many travellers were also lured to San Francisco by railroad companies and other corporations among whose principal motives were to sell lots to potential settlers or investors along the route; and these travellers 'toured' the areas while they considered resettling or investing. Others may have ventured to San Francisco as pleasure-seekers initially, and later decided to settle there permanently. It is possible that tourists in the nineteenth-century urban West more generally have been ignored because of these very difficulties of definition.

Some of the difficulty about discussing tourism at all – anywhere – is that there is rarely a consensus as to who exactly *are* tourists, and whether those tourists should be defined by their motives or their experiences. Much of the scholarly work concerning tourism thus far has placed a greater weight on the former in defining tourists; what happens once tourists finally arrive at the intended destination seems of little consequence compared with what they expect to find, from what conditions or situations they are trying to 'escape', how 'leisured' one must be in order to be considered a tourist, and how a tourist 'industry' – with its guidebooks, package tours, and other amenities – is set up to manipulate and manufacture the tourist experiences in particular ways.[12] Furthermore, most analyses of tourism have assumed that tourists are motivated to 'see' things in the places they visit, whether other people or physical objects. Thus, there has been an emphasis in the literature regarding the tourist 'gaze' and a privileging of the visual over the other senses.[13]

While tourist motivation and the 'gaze' are important and play a role in the tourist experience, they rarely paint a convincing, all-inclusive picture of tourists, and they rarely account for the *range* of tourist experiences. Instead, they create a rigid methodological framework that fails to consider the possibility that tourists – as consumers – have choices in what they consume, or that tourists may actually travel to discover, or engage with, the everyday practices of others, and not always because these practices *differ* from the tourists' understanding of the 'everyday'. It also assumes that tourists are not only funnelled toward certain sites, as manufactured and directed by official literature, but that they blindly follow that literature, and therefore that this official perspective is key to understanding the process of tourism. Studies of tourism have traditionally adopted this tactic, ignoring the view from the ground – that is, of the tourists themselves. Did tourists not have a say in their travel experiences? Tourism, like everyday life, is a complex process that defies simple categorization.

Rather than creating boundaries (which are forever penetrable), this chapter therefore attempts to offer a broader view of tourists. In San Francisco such people included pleasure-seekers, potential settlers, and other explorers whose stay (or visit) was sufficiently memorable for them to put it into words in the form of books, diaries or letters, and whose presence was significant enough that a whole range of literature and imagery – from guidebooks to lithography – catered for them. By examining some of their own words in addition to that of the official literature, this chapter suggests that experiences are as important as motivation and promotion in understanding tourists and tourism.

Generating Instant Heritage

Different types of official media intended to attract visitors to the city played a key role in the generation of instant heritage in early San Francisco. The official perspective was best represented in guidebooks, but on occasion, it was also evident in newspaper and magazine articles and their accompanying illustrations. Such publications commonly directed tourists in particular ways, encouraging them to partake in those aspects of San Francisco that – with the exception of the Chinese quarter – lent the city an air of gentility. The publications were usually produced by private companies, whose interest in distributing guidebooks and pictorial imagery lay in attracting wealthy travellers to invest in company-owned commercial properties.[14] View-photographer Isaiah West Taber, for example, published guidebooks and 'souvenir' booklets in the 1880s and 1890s asking San Francisco guests to visit his establishment while they toured the city – whether or not they intended to purchase any of the items for sale there.[15] Other locally published guidebooks (often called 'strangers' guides') resembled broadsheet newspapers and provided visitors or newly arrived residents with the 'essentials' for orienting themselves to the city. These sources usually included transportation routes, streetcar fares, and the locations of city and county offices, along with the city's 'attractions'. Together, these official documents worked to promote the city as a refined and cultured place to visit.

Other common forms of official documentation were photographic panoramas and colour lithographs portraying bird's-eye views of the city. While the mid-to-late-nineteenth-century photographic panoramas remained luxury souvenirs even after the rise of tourism in San Francisco, the bird's-eye views were commonly used to promote American cities, particularly in the West, and could be reproduced quickly without incurring a great cost either to producer or consumer. Despite the differences in cost, the overall goal of representation in these two different types of images was similar: while both often depicted the city as a dense landscape of commerce and activity, they also cleaned up this otherwise messy process. Carleton Watkins and Eadweard Muybridge's photographic panoramas of San Francisco from Nob Hill in the 1860s and 1870s, for example, depicted a city that was somehow too perfectly organized. Similarly, bird's-eye views of San Francisco commonly set its landscape of commerce against a sparkling blue bay, or portrayed it at sunset, usually –

perhaps intentionally – from too high up to reveal any of the particulars of everyday life (with the exception of numerous ships crowding the wharves). To be sure, whether they originated from the photographer's eye, the artist's brush, or the lithographer's stone, these were largely romantic, *constructed* views as seen from particular vantage points.[16] Only tourists who scaled the city's hills could glimpse it in any fashion remotely resembling them. Even when tourists did obtain such a view firsthand, it only comprised a sliver of their overall urban experience.

While the bird's-eye views and the photographic panoramas offered potential tourists the view from above, guidebooks transported the space of 'constructed visibility' to the ground, where the city became an organized document of individual sites.[17] To help attract visitors, the guidebooks promoted the city's physical appearance, recommending that tourists visit discrete sites both in and out of it. Among other sites (or 'sights'), these publications invariably promoted the Chinese quarter (Chinatown), hotels and theatres inside the city boundaries, and Golden Gate Park and the Cliff House in what were San Francisco's nineteenth-century suburbs. The proliferation of official documentation in the nineteenth century could be seen as a widespread civic effort to establish a 'permanent' sense of the city – at least in the imagination – in the face of constant change. To some extent, this marking of a permanent site was a way of generating a heritage for a city that, in the 1870s and 1880s, still lacked a lasting tradition.[18]

In an effort to stem the rapid tide of change, official publications boasted that new – and presumably permanent – private and public buildings rivalled or, at times, surpassed the 'best examples' in London, Paris, or New York. On the one hand, this represented an effort to infuse the new metropolis with a sense of civility that would seem familiar to visitors from elsewhere. On the other, it claimed for San Francisco a distinct architectural heritage that was uniquely 'San Francisco'. The process of civic beautification through new architecture and the promotion of San Francisco as a 'Paris of the West' by civic leaders and entrepreneurs not only attempted to represent the city's advance into the world of high culture, but also pointed up significant differences between 'civilized' San Francisco and the 'humble' dwellings, 'exotic' practices, and 'peculiar' people of the Chinese quarter.

The official literature frequently illuminated this distinction, treating the Chinese quarter as an oddity and a tourist attraction by the 1850s, thus allowing publishers to highlight what might have been the most significant cultural difference between urban San Francisco and other major American cities at the time. Despite the alleged 'risks' involved in exploring the Chinese quarter, these publications often reserved the lengthiest amount of text for its description and encouraged intrepid travellers to experience it. At the same time, the description of the quarter as a 'mysterious' and 'strange' place served to exacerbate the separation in thought and mind of the Chinese from 'Americans' (White, and generally male, San Franciscans), just as the Chinese were legally separated by land, forced as they were into a nine-square-block area near the downtown business district. While laws physically separated Chinese from Whites, the denigrating written portrayals kept them psychologically at bay, allowing travellers to peer voyeuristically into what was depicted as a den of moral contagion before returning safely back to the hotel by horse-drawn carriage.

While secondary literature on nineteenth-century tourism in America has generally argued that tourists desired to escape from the city, with its impersonal office buildings, soot and crowds, into the 'picturesque' refuge of parks, cemeteries, and leisurely drives, the guidebooks and other publications often promoted the typically urban aspects along with the picturesque.[19] Visitors also made little distinction, revelling equally in the urban 'danger' of the Chinese quarter and the city's busy pulse, and meandering by horse-drawn carriage through the supposed tranquillity of Golden Gate Park.

But not all travellers experienced San Francisco as a series of separate sites, nor did they necessarily view the Chinese quarter as a peculiar environment to be feared as well as admired. In many cases, tourists were not content to remain within the frame of constructed visibility spelled out by the guidebooks and depicted in the idealized pictorial imagery. Instead, tourists bled off the edges, encountering the city's interstitial spaces, ordinary built environments, and everyday life, just as they toured the pristine monuments, ate at fine restaurants, and relished the city's more genteel aspects. A more comprehensive picture of tourist San Francisco begins to emerge if one examines national magazines and the many books written by travellers from the East Coast and Western Europe, and often published and marketed in those parts of the country and world. Such publications were a type of official literature, but their intents and loyalties were less explicit; it is often difficult, for example, to determine whether they intended to promote travel for the purposes of boosting civic pride, to generate business for local commercial establishments, or simply to recount adventures for a general reading public.[20]

The widespread dissemination and consumption of magazines like *Harper's Weekly*, *Scribner's Monthly*, and the *Overland Monthly* often introduced the public to San Francisco and played an important role in promoting the city to potential tourists. Rather than a singular view of San Francisco, showing stand-alone monuments in tranquil settings, such magazines combined these views – in written and pictorial form – with scenes of everyday activity in the Chinese quarter, in the streets, or at the waterfront. Unlike much of the official literature, not all of these views, or the articles that accompanied them, portrayed the city in a positive light. Together, they provided an arguably more representative picture of city life than that spelled out in the guidebooks or shown in the bird's-eye views.

Like the official literature, the travel books often included descriptions of San Francisco along with other locations in California and the West, and the authors usually detailed particular sites while suggesting that San Francisco should be included in any trip to the Pacific coast. Unlike the guidebooks, however, the writers of these accounts commonly described the city from their own particular points of view, thus providing portraits of the city based on experiences already experienced, rather than experiences imposed from elsewhere. Often, this meant *discussing* the very rapid change and growth of the city, even when it was not exactly picturesque. London visitor Harry Jones, for example, wrote that San Francisco was 'thriving' and spreading 'like fire'; and to convey this growth he mentioned the 'moving of whole houses down the streets from one part of the city to another' (an image of this process accompanied his text).[21] Those illustrations which accompanied George A. Sala's massive tome on his travels across America similarly depicted a San Francisco of motion and activity, where people, horses, carriages and ships put the stately monuments into action.[22]

This phenomenological view of the city is supported by the unedited letters and diaries written by travellers during this period, which provide perhaps the closest approximation of raw, uncut responses to San Francisco. Like the travel literature, these documents both supported and subverted the official view disseminated by the guidebooks, suggesting that travellers not only experienced the city as a series of separate monuments privileging the eye, but on different sensual levels, including those of sound – and, on occasion, even smell.[23] This multi-layered sensory experience further activated the otherwise staid monuments within the crowds that, for tourists, constituted San Francisco as a living, breathing place, full of life, energy and diversity.

Picturesque San Francisco

So where did tourists go? If one believes much of the literature about tourism in nineteenth-century America, one might assume that tourists had little interest in the city and only desired to flee San Francisco for the picturesque attractions in its adjacent suburbs. To be sure, the suburban attractions were heavily advertised in the promotional literature, and they were also visited by tourists. But while the suburban experience was also included, it was the urban experience which mattered to most nineteenth-century travellers in San Francisco. Carriages (functioning as taxis) were stationed outside most of the luxury hotels, with attendants ready to drive visitors along winding roads to Golden Gate Park, the Presidio, the Cliff House, Woodward's Gardens, Twin Peaks, or Mission Dolores – places commonly listed among the city's 'getaways', 'excursions', and 'suburban points of interest'.[24] The tourist could experience these placid roads and 'escape' the city on foot, by carriage, or by streetcar. Newspapers at this time also wrote about local San Franciscans making trips to the suburbs, suggesting that the 'suburban points of interest', at least initially, were geared more toward locals than visitors.[25] Even in the 1850s advertisements and accounts of boat or ferry excursions to more remote areas such as Contra Costa, Mare Island, the Farallon Islands, Alameda, or the Napa Valley were frequently published in local newspapers.[26]

While some tourists certainly did head into the suburbs, the assumption that nineteenth-century tourists should be *defined* by those who sought pleasure by escaping the city for the tranquillity of the suburbs is misleading. Such a view is largely rooted in a generalized understanding of nineteenth-century urban centres as places of crime, overcrowding, and moral decay – places that needed to be 'repaired' and infused with space, greenery, and 'lungs'.[27] While San Francisco boosters were not immune to the appeal of the rapidly disappearing natural landscape, and pushed, among other things, for the creation of Golden Gate Park in the early 1870s, their efforts did not turn San Francisco into a suburban retreat. While Golden Gate Park was intended to provide scenery, moral uplift, and to 'keep the poor and the young from the temptations scattered around them', it was actually located a considerable distance from the city's most urbanized areas – at least originally.[28] While citizens engaged in the occasional 'greening' of downtown with little urban parks, neither these, the exotic plantings around the entrance court of the Palace Hotel, the creeping vines and array of flowers decorating many of the city's private residences, the placid serenity of the Laurel Hill and Lone Mountain cemeteries, nor the landscape paintings by Albert Bierstadt gracing the Lick House dining room walls were enough to constitute a ruralized city.[29]

If anything, the variety of literature concerning tourism and travel indicated that there were two nineteenth-century San Franciscos: a sparsely-populated suburban one with Golden Gate Park, leisurely drives, Mission Dolores, beaches, and the Cliff House; and an urban one consisting of a financial district, a Chinese quarter, shops, grand hotels, tightly packed houses, office buildings, and industries along the waterfront. Tourists visited both of these San Franciscos, but guidebooks rarely separated the city into 'suburban' and 'urban' areas. John S. Hittell, compiler of the massive *Bancroft's Pacific Guide Book* and other guidebooks, for example, considered San Francisco among California's 'Pleasure Resort Districts' – the city was a resort that combined urban and suburban attractions into an overall landscape of leisure:

> Among [pleasure resort districts], the first place belongs to San Francisco, with her leading business streets; her cable railroads; her ocean beach; her Seal Rocks; her spacious bay, studded with islands; her active stock market; her cemeteries; her park; her Chinatown; and her hundred hills, some of them crowned with palaces.[30]

Exotic San Francisco

Among the writers promoting San Francisco in the nineteenth century, New York-based journalist Charles Nordhoff was one of the most prolific. From the 1860s through the 1880s, Nordhoff wrote numerous articles in *Harper's Magazine* and *Scribner's Monthly*, the collection of which resulted in a number of books about tourism and travel.[31] The city of San Francisco was one of Nordhoff's favourite topics – a city, he wrote, that was 'one of the pleasantest and most novel of all the sights of California.'[32]

For Nordhoff, what was most 'novel' about San Francisco were the 'strange sights' of the Chinese quarter. He said the quarter could occupy a tourist's 'leisure' for several days, depending upon visitor curiosity.[33] Like most writers and guidebooks discussing the quarter in the nineteenth century, Nordhoff combined an interest in the area with a nativist condescension and religious fervour that portrayed the Chinese people and their non-Christian ways in an alien, almost barbarous light.

Nordhoff essentially divided the Chinese quarter into two principal sectors for tourists: one 'safe' environment that included the main streets by day and the 'extraordinary' Chinese theatre by night; the other a dangerous night-time environment of alleyways, gambling, and opium dens advisable only for men accompanied by a police officer.[34] Unlike the rest of San Francisco, Nordhoff's promotion of the Chinese quarter by day or night was predicated on a broad cultural fascination with the unusual, the strange, and the extraordinary – a fascination generated in part by the 'non-fictional urban sensationalism' common to a genre of reform literature that emerged in England and America in the mid-nineteenth century. Much of this literature highlighted the cultural landscape of the working and non-working poor; and in America, it sensationalized these conditions for the intended purpose of delivering the people – usually by means of religious conversion – from poverty, intemperance, crime and ignorance.[35]

Some of this reform work carried over into tourism as well; thus, by the late nineteenth century, slum tours had become popular in certain areas of New York City. Furthermore, disparaging portrayals of the Chinese – along with laws that refused them citizenship and basic rights afforded Western and Northern European immigrants – helped solidify American national identity as particularly White and of European descent.

With the exception of the Chinese theatre, Nordhoff emphasized that the Chinese quarter was not a place of amusement. Rather, it was a 'blot' on the city, a place where 'vile' and 'heathen' practices took place (such as the smoking of opium and the worship of pagan gods), and a slum that he compared on more than one occasion to New York's Five Points. Nordhoff described the quarter as so aesthetically distasteful that he recommended its wholesale demolition and replacement with new accommodations. If this could be accomplished, 'civilization and Christianity and free government on the Pacific coast would make a great gain.'[36]

Nordhoff's book also included a chapter entitled 'John', which detailed the 'heathen' ways of a man Nordhoff saw as typical of Chinese male residents of the quarter.[37] This section had a manifold purpose: at the same time that it was rooted in Christian notions of moral propriety and the belief that a better society could be forged if everyone became a good Christian, its portrait of the allegedly 'heathen' ways of the Chinese also secured 'civilization' for White San Francisco (and, by association, White America), and simultaneously enticed tourists to visit the quarter by accentuating its differences and overall peculiarities.

Nordhoff's description of the Chinese quarter and of the Chinese more generally was not particularly unusual in comparison with those of other writers, or to those in guidebooks discussing the area in the late nineteenth century; it was simply among the more glaring. Accounts of the quarter in the official literature routinely castigated the Chinese, while travel writers described their built environment in language typical of the Orientalist discourse of the day. One representative article in *Scribner's Monthly* described the geography of the Chinese quarter as a 'system of alleys and passages, labyrinthian in their sinuousities, into which the sunlight never enters; where it is dark and dismal, even at noonday.'[38]

James Mason Hutchings, whose California accounts Nordhoff had read prior to his own initial visit, and who was one of the first writers to help forge the California mystique as a land of pleasure and natural beauty, published insolent descriptions of the Chinese and their quarter in his *California Magazine*. Like Nordhoff, one Hutchings writer noted the slum-like conditions of the quarter and the 'humble', 'filthy', and 'unpleasant' dwellings occupied by the Chinese. He described their manners as 'singular, and in some respects, amusing.'[39] Willard Glazier, an American soldier-turned-urban-traveller and popular writer in the nineteenth century, described the Chinese quarter's tenement houses as 'crowded and filthy beyond description, and the breeding places of disease and crime.'[40] Frank Green, a British traveller who visited the quarter in the 1870s, wrote of the 'abominable stenches' emanating from it, 'to which hog slaughtering at Chicago', by comparison, was but 'a trifle'.[41] One guidebook to San Francisco published by the Central Pacific Railroad (much of which was built using Chinese labour) included a fictionalized account of a tourist's night-time stroll through the quarter:

> Strange faces and odd figures throng the street. Men with half-shaven heads and outlandish dress and manners, and women with painted faces and queer eyebrows, and semi-barbarian attire, flit by us. The walls of the buildings have curious hieroglyphics – and hark back to the voices – what jargon is that? We are foreigners at home! This is Asia. This is Canton.[42]

References to the Chinese quarter as a transplanted version of mainland China were also common, even if the geography was incorrect. Thus, it was equally likely to find the area compared to Canton as to Peking, although very few Chinese who came to San Francisco in the nineteenth century came from northern China, and most had never experienced the rigors of city life. Guidebooks, strangers' guides, and travellers' accounts also referred to the Chinese interchangeably as 'Celestials' or 'Mongolians'.[43]

Geographical accuracy was considerably less important to this literature than was an apparent desire to highlight the dangerous and exciting 'mysteries' of the area in order to appeal to travellers and to manufacture the Chinese experience as an attraction for visitor consumption. The promotion of this community as an exotic yet lurid tourist attraction served to accentuate differences between Whites and Chinese, helping Whites to see the Chinese as fundamentally different from themselves – an attitude that legitimized White superiority and, among other things, prevented efforts by the Chinese to obtain citizenship or assimilate into American culture. That this densely-packed community within an already congested city was one of the most heavily promoted and widely visited sites in all of the American West also helps debunk the notion that early Western tourism was marked by the desire of a leisure-bound populace to escape their own densely-packed urban cores and head for the open spaces of the natural landscape. Yet, at the same time, the existence of the Chinese quarter threatened to topple the notions of gentility the city's leaders desired to promote. It may have been partly for this reason that the differences between the Chinese quarter and the rest of San Francisco were accentuated in the official literature: so that the quarter could appear as foreign, undeveloped and 'uncivilized' within a larger, cultured and refined American city.

But not all travellers experienced the Chinese quarter as if they were visiting a separate city within San Francisco. Diaries and travellers' accounts, in fact, suggest that the tourist experience of the quarter may have been different from the guidebook version. However condescending and xenophobic their writing may have been, many tourists pointed out the efficiency and organization of the quarter, describing its well-dressed, well-behaved, and rarely – if ever – intoxicated residents. They were also interested in the everyday life of the community as a space of work, often noting the industriousness of the Chinese – characteristics to which they hoped American workers could aspire (figure 6.2). One diarist, writing in 1872, saw the Chinese quarter as a place of everyday work and efficiency, and took an interest in this everydayness:

The Chinese Quarter there is a great sight. There are in California at least 100,000 Chinese, and more are coming daily – The Chinaman is an admiral [sic] man, the best cooks, washmen, nurses, domestic servants, gardeners, workmen, labourers & artisans in California are Chinamen. In their persons they are marvelously clean & they have a patience which is wonderful.[44]

Others were astute enough to realize that even the Chinese quarter was not a cohesive entity that could be summed up with a totalizing view. Miriam Florence Leslie (who referred to herself as 'Mrs. Frank Leslie') noted apparent class differences among the Chinese, pointing out both the 'fashionable shops' along Kearney Street and their prosperous merchants with 'silken caps and fine cloth clothes', and the poor residents of the smaller streets with their 'dingy' houses.[45] Harry Jones, too, noted the mercantile 'progress' of the Chinese, explaining that a visitor could not only see the 'humble laundry' of Ho Ki, but also the more impressive offices of 'Ho Sing, Wo Ching & Co'.[46]

Many others recognized that denigrating portrayals of Chinese were not only inaccurate, but helped support the 'lowest class' of citizens who were

Figure 6.2. Provision market in alley in Chinatown, San Francisco – photograph by I.W. Taber, about 1880. Taber owned and operated one of San Francisco's most prolific nineteenth-century photographic studios, marketing many of his views to tourists in the form of souvenir books and individual photos. Traveler Herbert C. Leeds included this view, depicting a scene of everyday life in the Chinese quarter, in a book of photographs documenting his trip across the country. (Photo courtesy of the Bancroft Library, University of California, Berkeley.)

helping drive state and national legislation keeping the Chinese separate from White society. While some authors wrote positively of the Chinese and their quarters because they saw the Chinese population as necessary to keep wages low (and thus production and profits high for White America), others were more genuinely sympathetic to the plight of San Francisco's Chinese population, which included constant harassment from local 'hoodlums'. Charles Loring Brace, writing a travel book from New York about his journeys to the 'new' West, for example, noted that the laws against the Chinese were 'oppressive', and that the 'truth' regarding the Chinese and their situation 'is the best tribute a traveler can pay to the sense of justice of the more civilized Californians who detest these abuses equally with ourselves'.[47]

Some travellers noted, too, that the 'heathen' Chinese posed less of a 'problem' to San Francisco's 'civility' than the inhabitants of the Barbary Coast, which occupied a few blocks along Pacific Street just east of the Chinese quarter. Here were the 'vilest class of poor whites' living in an area notorious for brawling, drinking, gambling and prostitution. While the Barbary Coast provided the image of the 'Wild West' in its early days, guidebooks – if they mentioned it at all – rarely recommended visiting it. Tourists, however, occasionally did, and some pointed out that its existence revealed problems considerably more pertinent than those associated with the so-called Chinese 'problem'.[48] In noting the work habits and impeccable hygiene of the Chinese, London traveller Henry Hussian Vivian explained that the Chinese quarter was a 'garden of roses compared to many Continental towns, Berlin among the number, not to mention Cologne, Italian towns par excellence, and some French.'[49]

Finally, Leslie also seems to have understood that the Chinese may have considered Whites to be odd and outlandish in their own cultural practices, including tourism. She noted that while Whites 'arrogantly try and civilize and Christianize [the Chinese] by our own standard, [the Chinese] complacently seat themselves upon the heights of their own civilization, their own religion, and consider us as outside barbarians whom it is not worth their while to convince of error or ignorance.'[50] Some of these accounts, to be sure, were bound up in an effort to point out how very different the Chinese were from White Americans. But they also show that there was more to the Chinese quarter than groups of 'uncivilized heathens' living in squalor and engaging regularly in prostitution, drug addiction, or gang activity.

The Palatial City

While the Chinese quarter was the most heavily advertised urban 'attraction' in early San Francisco, it was not the only part of the city promoted in the guidebooks or visited by tourists. Among the city's other instantly generated attractions were its hotels, of which San Francisco had sprouted seventy-seven by 1882 – a number that the guidebooks contended exceeded that of any other American city at the time.[51] Many guidebooks, in fact, discussed San Francisco's hotels before any other 'points of interest' – and not just because tourists would need to find accommodation upon arrival. Some of these hotels were constructed with a type of built-in heritage, fashioned as they were after established hotels in Europe and the Eastern United States to make them appear more familiar to the visitors they wanted to attract. Many travellers expressed both surprise and delight at the city's grand hotels, perhaps because they expected the city to be filled with little more than unadorned wooden shacks crowding muddy lanes – a popular impression of the city that lingered from its Gold Rush days.

Hotel construction was frequent in San Francisco's early years. Mostly, these hotels catered for the large number of citizens who needed accommodation on a temporary basis, but they also serviced a growing nineteenth-century visitor population. San Francisco's first hotel of major significance was the Union Hotel. Located on Portsmouth Square, it was completed in September of 1850 and outfitted with a saloon with 'six splendid chandeliers' diffusing light onto 'magnificent Paris mirrors with ornamental gilt frames.'[52] The most luxurious hotels, however, were begun in the 1860s and catered primarily for travellers – only the wealthiest of whom could afford to stay in them for more than a few days. Among the more notable of these were the Lick House (1861), the Russ House (1862), the Occidental Hotel (1860s), the Cosmopolitan Hotel (1860s), the Grand Hotel (1870), the Palace Hotel (1875), and the Baldwin Hotel (1877). The Lick House was noted for its comfort, palatial dining room modelled after that of Versailles, and marble floors; the Occidental Hotel for its gigantic dining room, theatre, and 'modern' conveniences; the Cosmopolitan Hotel for its 'aristocratic feel' and elevators (the first in the city); the Grand Hotel for its mansard roof and central court; and the Baldwin Hotel for its architectural splendour, elegant interior, and theatre. A writer discussing San Francisco hotels in the 1870s noted that Americans thought San Francisco's hotels were 'second to none in the country'. William Doxey's *California Tourists Guide*, published in 1881, argued that the collection of hotels in San Francisco was unsurpassed worldwide.[53]

And yet this collection may have been merely adequate if not for William Ralston and William Sharon's Palace Hotel. While only the guidebooks, strangers' guides, and some travellers' literature discussed the full range of grand hotels in San Francisco, everybody seemed to write about the Palace Hotel – at the time considered 'the most remarkable building of its kind in the world'. Upon completion, the Palace Hotel was by far the city's biggest (seven storeys), largest (755 rooms), and most expensive hotel ($3,250,000), and it dominated its surroundings. It was commonly praised for its structural prowess ('massive' and 'solid') and its ability to withstand fire as much as for its Classical architectural detail. It was an urban landmark, and local businesses frequently promoted themselves by noting their proximity to it. By the early 1890s the Palace Hotel had become famous enough nationwide for Hutchings to use it as a benchmark upon which to indicate the height of Yosemite's El Capitan. 'Within and without,' Doxey's guidebook boasted, 'the kingly structure . . . far surpasses not only in size but in grandeur all the hotels of Europe and America.'[54]

A frequently discussed aspect of the hotel was its circular carriage court with its glass roof, marble-tiled and colonnaded promenade, and tropical garden with exotic plants, statuary and fountains. This court not only added dignity to the hotel's massiveness and solidity, but it enabled hotel guests to step out into the realm of the carefully manicured urban picturesque before they ventured into the sublimity of the urban jungle. But the fantasy world of the Palace Hotel (and all of the luxury hotels) was hardly a permanent escape, and those travellers staying there did not remain sheltered inside for their entire visit – nor was this their desire. There was a whole city outside the door, and they explored it.

A 'World-Class' City

The official travel literature served to whet the tourists' appetite. Guidebooks frequently promoted new public buildings (especially the new City Hall and the US Mint), office buildings, theatres, and public transportation which, by the 1880s, included the cable cars – the 'wonderful wire-rope railroads.'[55] Because of San Francisco's unusual topography and the spectacular ocean and bay views one could obtain from atop its various hills, these guides also publicized viewpoints, such as the observatory atop Telegraph Hill. Less frequently, guidebooks advertised cemeteries, asylums, the Sea Wall, Meiggs' Old Wharf, and Lotta's Fountain on Market Street as visitor attractions in addition to libraries, city and county offices, banks, principal newspapers, places of amusement, and theatres.[56] Despite the myriad urban aspects that were of interest to most travellers, the official literature favoured a portrayal of San Francisco as a civilized retreat on the edge of the continent, dotted with pleasant drives, first-class hotels, fine restaurants, and prominent civic buildings. Thus, guidebooks attempted to transform even the urban into the picturesque.

Other than the hotels, the buildings most heavily promoted as tourist attractions for their architectural grandeur usually were built of 'solid and substantial' materials and featured some derivation of the Classical architectural vocabulary. The guidebooks, and occasionally the travel books, were at pains to point out the details. One of Hittel's guidebooks, for example, promoted the United States Mint branch by describing its 'Doric style' and 'massive fluted columns', and the Merchants' Exchange by noting its 'pillars of polished granite.'[57] In the late 1860s Charles Loring Brace argued that the 'new' public buildings in San Francisco, such as the Bank of California and the Merchant's Exchange, were 'unusually good in effect' and 'better than our new buildings in New York.'[58]

Writers of promotional literature in nineteenth-century San Francisco seemed to base their opinions about whether San Francisco was 'world class' on how closely it resembled the grand cities of Western Europe or the Eastern United States – particularly Paris, London, and New York, but occasionally Boston, Philadelphia, and Vienna. Hittell, for example, commonly referred to San Francisco as the 'Paris of America', while guidebooks and strangers' guides of the 1870s and 1880s supported this claim and consistently pitted San Francisco against its alleged Eastern and European rivals.[59] San Francisco's hotels, for example, were most heavily promoted by the guidebooks if they offered meals on the 'European plan'. A brochure advertising the Palace Hotel prior to its grand opening boasted that its architect had visited hotels of the 'principal cities' of America and Europe before executing the design.[60] However, it was not only the built environment that generated tourists' comparisons to other cities. A.H. Wylie, a London traveller, explained that women in San Francisco dressed almost perfectly, and that 'London and Paris might well take an example from them.'[61]

San Francisco was only one of many areas of the West Coast that journalists, politicians, and other boosters compared to Europe and the East Coast in the late nineteenth century.[62] Whether or not San Francisco did replicate Paris, London or New York in its architecture, planning, or general attitude was not important, however. What seemed to be important was that those staking a claim to this heritage considered these older cities to be repositories of culture and refinement, and by associating San Francisco with them, they could help mark the city's entrance upon the world stage and assure visitors that there was more to San Francisco than its image as a riotous frontier town.

To some extent this promotional tactic worked. Many travellers' accounts included this official view, citing a few examples of monumental architecture, institutions and parks, and noting how far along San Francisco was toward 'civilization'. To one guidebook publisher, the overall combination of architectural attractions meant that San Francisco had emerged as a world-class city:

> The place has many other features besides its hotels, which may well surprise strangers who come to the rapidly-built town expecting to see the roughest evidences of its recent birth, and find, on the contrary, that it is one of the foremost cities in the world in civilization . . .[63]

But tourists' opinions of San Francisco vis-à-vis those more 'established' cities in the East or in Europe was not based on 'civilization' or gentility alone. Like its East Coast and European counterparts, nineteenth-century San Francisco was a distinctively urban place, filled with the often non-genteel aspects of everyday life, work and activity. In their letters and diaries (and, on occasion, in the travel books), tourists consistently remarked on the overall activity of the city – something they did not expect at such a vast distance from Europe or the Eastern Seaboard. According to these travellers, no city west of Chicago offered as much as San Francisco.

The Urban Experience

Tourists rarely experienced San Francisco as a series of discrete sites – the way it was seemingly intended to be experienced if they heeded the guidebooks. Books and diaries written by travellers to San Francisco in the nineteenth century indicate that those who came to San Francisco saw the city not just as a series of sites or a temporary way-station before the next train left for more 'natural' locales inland or along the coast. Thus, while many traveller's books include descriptions of individual sites, the San Francisco tourist experience also included the ordinary spaces of everyday life – the dense residential landscape with innumerable houses featuring 'bow' or 'bay' windows, the shops, the markets, the people in the streets, the restaurants, and the hills – in short, the whole of the cultural landscape.

For many tourists the experience also included that of arrival and the transportation system traversing the city's landscape. Tourists also noted the extraordinary diversity of the city's population and its very 'public' – albeit sometimes rambunctious – nature. They were often surprised and fascinated by the noise and activity generated by a city that in 1849 had consisted of a few loosely scattered shacks on sand dunes. Such urban activity muddied the pristine picture of luxury hotels and grand civic edifices advertised in the guidebooks and depicted in much of the imagery that accompanied them.

Instead of deterring tourists, however, this activity was a source of interest; and it was the very combination of the individual sites – both public and private – and the electricity of human drama in the urban environment that suggested to them that San Francisco had come of age. Leslie explained that the city's frequent afternoon promenades reminded her of fashionable 'openings' elsewhere, when 'the lay figures have suddenly received the life and power of locomotion.'[64] What might have at first seemed an 'uncivilized' colonial outpost far from conventional centres of 'culture' in Europe or on the East Coast (California, after all, had only been admitted to statehood in 1850) seemed to many tourists to feature an urban flavour that was more similar to what they experienced at home.

This is not to contend that early San Francisco was just like Paris, London, or New York, but that it featured – to many travellers, anyway – some of the urban characteristics that they had grown to expect in older and more established cities. San Francisco actually seemed to offer even more, because this urban flavour was packed into a dense and architecturally distinctive residential landscape within a scenic topography surrounded on three sides by water. It also featured a Pacific Rim connection that was not a characteristic of Eastern or European cities. And trans-continental travellers got a taste of this intensified urban landscape immediately upon arrival.

Arriving in this city on the 'other' edge of the continent was apparently an unforgettable experience, and nearly all traveller's accounts and diaries included some discussion of it. By the 1850s most travellers came by rail to Oakland and then boarded a 'large and commodious' ferry that took them across the bay to San Francisco.[65] Others transferred from the railroad to a steamer in Sacramento, and then plied the river system through the Sacramento delta and out into San Francisco Bay.[66]

While the official documentation often described the comfort and efficiency of the Oakland ferries, none of it described the commercial activity of transportation hawking that greeted tourists on the other side (figure 6.3). In his diary – in which he pasted newspaper reproductions of lithographic prints and maps of the sites and the cities he visited – New Yorker Banyer Clarkson made a point of noting the 'continuously ringing'

Figure 6.3. Scene outside the Ferry Building, photographer unknown, 1889. Nearly all tourists experienced a scene somewhat like this one when they arrived in nineteenth-century San Francisco following a transcontinental journey. Cable railways and horses outside the Ferry Building await and distribute passengers to and from the Oakland ferry and points beyond. Cities are listed along the building's frieze, suggesting (from right to left) that Sacramento and San Jose are as easily accessible as St. Louis and Portland. Other cities here include Yuma, Calistoga, Napa, Red Bluff and Santa Rosa. (Photo courtesy of the Bancroft Library, University of California, Berkeley, Hills-Hecht Collection.)

bell of the 'huge floating palace' that transported him across the bay and the 'crowd of hotel runners' at the ferry terminal in San Francisco. Fascinated with the ethnic diversity of the area – particularly the immigrants from Asia – he also noted that the wharf at the foot of Brannan Street was 'always the scene of confusion and also a Babel of languages on the arrival of a steamer from Yokohama or Australia.' Likewise, Leslie, for whom San Francisco represented the 'Mecca' of a long journey, wrote of an 'army' of hotel runners at the wharf.[67]

Everyday activity in urban San Francisco – its sounds as well as its sights – were also an important part of nearly every traveller narrative. While the guidebooks occasionally advertised the city's markets because they gave the out-of-town visitor a chance to see a variety of produce they were not likely to find at home, the tourists often combined this visual feast with an interest in the human activity that surrounded it. Clarkson, for example, was impressed with the 'profusion, size, and perfection' of the fruit sold in the markets. But he was also startled to discover how busy the market activity was – particularly on Sundays. Remembering the prominent sound of church bells from St. Mary's Cathedral tolling on a Sunday, he wrote:

> The melody as the sound comes floating through the air is constantly broken by the noise and rumble of hotel coaches and street cars, for Sunday is unknown in this Western Metropolis. Kearney Street is all alive with buyers, and salesmen do a lively business behind the corner.[68]

Indeed, shoppers found San Francisco a fascinating place, not merely for the markets or 'odd' trinkets available for purchase in the Chinese quarter, but also for the general commercial activity that took place out of doors as well as in. Leslie, for example, mentioned that many of the city's smaller shops were 'open to the street like booths, especially the cigar and liquor establishments, in one of which we saw a man throwing dice for a drink.'[69]

Buildings, too, became part of the busy landscape of commerce experienced by travellers. Travellers frequently described the Merchants' Exchange on California Street, which some thought offered a magnified interior version of San Francisco street life. London visitor Arthur Guillemard filled his travel log with hyperbole to set the scene in the Exchange:

> The crowding, pushing, and heat are almost unbearable, but the scene is sufficiently lively to induce one to become a spectator for a few minutes. The gestures and shouting of the members are quite frantic. A man on my right is proclaiming in stentorian tones his desire to dispose of a certain number of shares in some extraordinarily-named mine. To him through the crush comes a buyer, elbowing and fighting his way, regardless of limb and apparel. He reaches his man after a fierce and prolonged struggle; he seizes him by the

shoulders with both hands, shakes him vigorously, roars in his face, shouts in his ear, and after a short tussle of words, which causes both combatants to turn purple in feature and foam at the mouth with frantic excitement, the conflict ceases, the bargain is concluded, a few notes are pencilled down, and buyer and seller retire amicably to 'cool out' with a brandy smash. Can the London Stock Exchange furnish a scene as exciting as this?[70]

Guillemard's account may not have been a tremendous exaggeration. Leslie noted that the scene 'was one of the wildest excitement,' where 'combat is but a mild term to apply to the jostling, yelling, frenzied, purple-faced struggle, roused into new vigour at each call of a new stock; the bidders crowding to the centre, gesticulating, pushing, ready to tear each other to pieces, or themselves fall down in a fit of apoplexy.'[71]

The moving landscape, too, was of interest to many travellers. For example, the sheer number of streetcars (which travellers often confused for cable cars, although both were in operation in San Francisco in the 1870s), represented an uplifting experience for John Reynolds, who wrote his mother periodically of his personal troubles during a visit to San Francisco in the summer of 1875:

I feel decidedly lonesome though in the midst of the sights and sounds of the busy city, street cars pass the house every 10 minutes. It is 1½ miles to the principal streets of business . . . I never saw as many street cars almost every main street seems to have them and always full too – this is surely a very busy city.[72]

Cable cars, of course, were of major interest to most travellers, along with the city's topography. New Yorker L.D. Luke, among others, was awed by these technological wonders and found it necessary to write about what he knew about their operations:

The word up when applied to many of the streets of this city is very appropriate. The grades of some of the street walks are so rapid that they are ascended all the way by flights of wooden steps. The first object that attracted my attention in the city was some [*sic*] beautiful little palace street cars, passing up and down steep grades without horse or steam or any visible propelling power. Under the paving is an endless cable in motion, and over that is a cleft in the road-bed, two inches wide, and extending over the entire length of the line. A grappling iron extends down through the car and also through the cleft below, where it grasps the cable, and its gentle motion moves the car forward.[73]

Novelist Helen Hunt Jackson, who also wrote widely of her nineteenth-century travels both in America and abroad, stepped out of San Francisco's

Occidental Hotel, boarded a carriage, turned a corner, and found herself moving up 'extraordinarily' steep streets and described the 'small, wooden, light-coloured, and picturesque' houses that lined them. Jackson was surprised at the density and the juxtapositions, noting that a small Chinese laundry building could abut a luxurious house.[74]

Many travellers were simply amazed at the city's energy. One British diarist, after arriving at the Occidental Hotel, wrote of an evening walk where 'we noticed a great number of saloons for drinking & dancing I should think as there was music playing in most of them, some of them were in the cellars & I fancy very fast places; there was such a great noise in the streets.'[75] Albert Richardson, an American traveller who ventured West in the 1860s, was one of several visitors astonished with the speed at which the city had grown in such a short time. Richardson reserved his most glowing praise not for any one attraction in particular, but for the 'teeming life of the great metropolis', the busy harbour with its 'miles of steamers and sailing vessels', and its cosmopolitanism – a more diverse city than any other except New York.[76] Other writers saw San Francisco as *more* diverse than New York – at least relative to their overall populations. This urban energy caused still others to note that San Francisco never slept: to them, it was a vibrant, lively place that was as full of activity by night as by day.[77] Indeed, the liveliness tourists encountered was a cultural creation, it was not merely a visual spectacle constructed by the guidebooks.

Harry Jones was one visitor whose sense of San Francisco's vibrancy was illuminated by its cultural and ethnic diversity. To Jones, Montgomery Street, for example, was a street 'thronged by crowds' that was at once 'the Wall Street and Broadway of the city':

> Every nation and tongue has representatives here. Californians, merchants and miners, Mexicans – I have seen them, with high-peaked saddle and lasso, riding by – negroes, the broadest Irish, Germans, and Chinese make up the multitude. Sudden fortunes bring the miner into the best hotels. The man sitting near you at dinner may be well dressed, but he may have hands horny and brown as a navigator's, and a navigator's appetite. In the same room, perhaps at the same table, are elegant Californian belles. The way in which society, as seen in the streets and inns, is jumbled up here, is very striking.[78]

Boston resident Susie Clark, on an excursion trip in the 1880s, also marvelled at the city's constant activity. For Clark, the city was a 'tumultuous, wide-awake' place – a place to be celebrated for the noise and activity that arguably drove large numbers of residents away from cities in the nineteenth century. If one was to look out of a hotel window in the middle of the night, Clark wrote, one would see 'stores open, houses brilliantly lighted, cable cars with clanging alarm-bells whizzing by, merry strollers whistling under our window, strains of distant music in the air, and the same features of activity that belong to daylight.'[79]

The Vibrant City

The city described by Clark and others suggests that San Francisco not only provided a source of visitor fascination because of its many 'sights', but that the visitor experience was multi-layered and multi-sensory. Neither a generalized understanding of nineteenth-century American history (with an emphasis on the 'huddled masses' in teeming cities and the supposed desire to escape them), nor an understanding of the tourist as simply desiring to escape the everyday, can be loosely applied to tourists in this far-western metropolis. Tourists in nineteenth-century San Francisco were as interested in the monumental, visually-oriented city promoted and manufactured by the guidebooks as they were with the everyday life that made their experience of it into a larger whole – a whole considerably larger than its constituent parts. There was no *single* 'tourist experience'.

Those who ascended the steep slopes of Telegraph Hill for the bird's-eye view of the landscape of commerce with its throngs of people, tightly-packed houses, active industries, and great numbers of water craft entering and exiting the Golden Gate had only risen above the city temporarily; they would soon return to the urban landscape which provided them with so much interest. And it was down in this urban landscape that tourists consumed the city in ways that both followed and ignored the instant heritage manufactured by the guidebooks.

Indeed, tourists to nineteenth-century San Francisco did *experience* the city. They did not wait around for grand City Beautiful planning schemes to reorganize it into broad avenues, stately buildings, and gleaming monuments; nor did they 'see' it as an organized document requiring a single vantage point from which they could remove all that did not fit with the pristine image provided by the official documentation. It would have been too difficult to maintain such an image anyway: while the city displayed flashes of civility to tourists, any sense of it as a completely genteel place was erased the moment they set foot on shore.

The process of heritage manufacturing by the official publications and the consumption of this heritage by tourists was a process of selection and exploration, not of direction and slavish obedience. Guidebooks did not prepare tourists for the possibility – or the likelihood – that their view of the Palace Hotel would include the sights and sounds of busy Market and Montgomery Streets, or that their trip to the Cliff House would require a lengthy ride in a loud streetcar overflowing with a polyglot of passengers. Instead of seeking places of refuge away from the allegedly chaotic and morally decaying nineteenth-century city, tourists revelled in San

Francisco's energy, and considered the energy itself a true mark of the city's 'advancement'. Consequently, tourists helped this alleged urban 'other' – in an extraordinarily early stage of its development – assimilate among the ranks of America's and the world's 'foremost' cities. In turn, tourists established that there was more to tourism in the nineteenth-century American West than extended visits to the region's natural wonders. Although most tourist journeys to the West at this time were not motivated by the lure of cities, San Francisco nevertheless became an integral part of the nineteenth-century tourist landscape.

Notes

1. J.G. Player-Frowd, *Six Months in California*, London, Longmans, Green, and Co., 1872, pp. 1, 16-17. According to a popular guidebook of the 1870s, the Grand Hotel and the Cosmopolitan were considered among San Francisco's 'first-class' hotels. See G.A. Crofutt, *Crofutt's Trans-continental Tourist,* New York, Geo. A. Crofutt, 1874, p. 150. The 'What Cheer House' was described in S. Bowles, *Our New West: Records of Travel Between the Mississippi River and the Pacific Ocean*, Hartford, CN, Hartford Publishing Company, 1869, p. 333, as a second- or third-class hotel, but one with 'excellent meals,' a library, a cabinet of minerals, and a collection of stuffed birds.

2. One account that located the beginning of San Francisco tourism with the Panama-Pacific Exposition is K. Starr, *Americans and the California Dream 1850-1915*, New York, Oxford University Press, 1973. Regarding the 'rationalization' of tourism in the 1890s, where business leaders realized how tourism could assist cities economically and called for the redesign of cities and cost-cutting to appeal to tourists, see N. Harris, 'Urban Tourism and the Commercial City,' in W.R. Taylor (ed.), *Inventing Times Square: Commerce and Culture at the Crossroads of the World*, Baltimore, Johns Hopkins University Press, 1991; or C. Cocks, 'A City Excellent to Behold': Urban Tourism and the Commodification of Public Life in the United States, 1850-1915,' Ph.D. diss., University of California, Davis, 1997, pp. 1-12.

3. John Towner and Geoffrey Wall have argued that the scholarly disregard of tourism in America more generally has had to do with the fact that scholars, given the relative newness of North America (compared to European settlement), have been preoccupied with documenting the 'history of exploration, settlement, resource exploitation, and other aspects of the permanent occupation of North American space.' See Towner and Wall, 'History and Tourism,' *Annals of Tourism Research*, vol. 18, no. 1, 1991, p. 77.

4. There are only a few exceptions. In 'A City Excellent to Behold,' Cocks has made one of the strongest cases for the importance of urban tourism through an analysis of late nineteenth- and early twentieth-century New York City, Washington DC, Chicago, and San Francisco. In 'Urban Tourism and the Commercial City,' Harris argued that middle-class tourism grew in New York City from 1890 to 1920. More problematic is J. Sears, *Sacred Places: American Tourist Attractions in the Nineteenth Century*, New York, Oxford University Press, 1989. Sears included a chapter about prisons, asylums, cemeteries, and parks as nineteenth-century tourist attractions in East Coast cities (pp. 87-121), but he argued that each of these

'urban' attractions were appealing because they provided antidotes to the disorder of city life.

5. The first Western cities to catch tourists' attention were those along or near the main line of the first intercontinental railroad: Omaha, Denver, Salt Lake City, Sacremento, Oakland and San Francisco. Other trans-continental lines opened in the 1870s and 1880s made the cities (or large towns) of Houston, San Antonio, El Paso, Santa Fe, Los Angeles, Minneapolis, St. Paul, and Portland, among others, part of a larger western urban tourist circuit.

6. In *In Search of the Golden West: The Tourist in Western America*, Lincoln, University of Nebraska Press, 1957, p. 23, E. Pomeroy acknowledged that San Francisco was a tourist destination; but, overall, cities played a very small role in his analysis. The influence of this text in promoting an understanding that tourism in the West was based on the attraction of 'natural-wonders' has been immeasurable. One book that follows in this vein is A.F. Hyde, *An American Vision: Far Western Landscape and National Culture, 1820-1920*, New York, New York University Press, 1990.

7. The exception was Salt Lake City, the home city for those business leaders who proposed the initiative. See M.S. Shaffer, 'See America First: Re-Envisioning Nation and Region through Western Tourism,' *Pacific Historical Review*, vol. 65, no. 4, 1996, pp. 559-582.

8. Whether in urban or rural areas, the processes of industrialization were sources of fascination for many travellers. Nineteenth-century traveller's literature, for example, was filled with descriptions and images of the latest 'technological' wonders, including new railroad bridges, tunnels, and mining machinery. In *Sacred Places*, Sears included a chapter about tourists visiting an early coal mine in Mauch Chunk, Pennsylvania. Charles Nordhoff, in *California: For Health, Pleasure, and Residence*, New York, Harper & Brothers, 1874, pp. 20-22, 73, suggested that trans-continental railroad travellers take a detour in Chicago to examine the stockyards, and further detours in California to view gold mining and the 'celebrated Yuba Dam' near Marysville. Regarding factory tours in late-nineteenth- and early-twentieth-century America, see W. Littman, 'The Production of Goodwill: Factory Tours in America,' paper presented at the Vernacular Architecture Forum Conference, May 9, 1998, Annapolis, Maryland.

9. As cited in Pomeroy, *In Search of the Golden West*, pp. 7-13.

10. One guidebook, published by local proprietors in conjunction with the Central Pacific Railroad, explained that the guidebook was to be distributed 'among the Overland passengers at Corlin, en route to San Francisco,' but that 'those *less fortunate* can find them at the principal news depots, or at our office, 207 Kearny Street' (italics mine). In other words, those who could not afford to arrive in San Francisco via railroad would have to purchase a guidebook in the city. See O.F. Tiffany and A.C. Macdonald, *Pocket Exchange Guide of San Francisco*, San Francisco, Tiffany and Macdonald and Central Pacific, 1875.

11. It should be noted that women experienced the Chinese quarter differently from men, that women were only permitted to watch the activity in the Merchants' Exchange from a separate viewing gallery, and that women's writings tended to describe shopping more so than men's. Discussions of shopping in early San Francisco were not, however, exclusive to literature written by women. Men often

described the opportunity to buy the 'bric-a-brac' in the Chinese quarter, and they frequently noted the city's bustling markets.

12. Many scholars have gone to painstaking lengths to define tourism, often breaking it up into different 'types' based largely on the experiences *expected,* rather than experienced. Much of this literature has come from the field of anthropology (and particularly anthropologists studying non-Western, small-scale societies), where tourists are widely rebuked for alleged attempts to find 'timeless' existence, 'authenticity,' and a 'better way of life' in villages remote from the ravages of modern capitalism. See, for example, D. MacCannell, *The Tourist: A New Theory of the Leisure Class*, New York, Schocken Books, 1975; and V. Smith (ed.), *Hosts and Guests: The Anthropology of Tourism*, Philadelphia, University of Pennsylvania Press, 1977, pp. 2-3.

13. See, for example, J. Urry, *The Tourist Gaze: Leisure and Travel in Contemporary Societies*, London, Sage Publications, 1985. One notable exception is the work of Erik Cohen, who since the late 1970s has been attempting to examine tourism according to an experiential rather than a motivational model. See E. Cohen, 'A Phenomenology of Tourist Experiences,' *Sociology*, vol. 13, no. 2, May 1979, pp. 179-201.

14. Regarding the role of photography for this purpose, see P. B. Hales, *Silver Cities: The Photography of American Urbanization, 1839-1915*, Philadelphia, Temple University Press, 1984, pp. 48-57.

15. I.W. Taber, *Hints to Strangers*, San Francisco, I.W. Taber, 1890. For another example of a company promoting itself along with the city, see *Lick House Tourists' Guide: Giving Principal Routes from Chicago and Saint Louis to San Francisco*, San Francisco, Lick House, 1871. Also, in 'Urban Tourism,' p. 66, Harris argued that tourism was a largely commercial activity, where the 'urban tourist was increasingly attracted to and stimulated by business, by merchandisers, both as a consumer and a visitor.'

16. For examples of how nineteenth-century San Francisco photographers idealized the city, see Hales, *Silver Cities*; or D. Harris, *Eadweard Muybridge and the Photographic Panorama of San Francisco, 1850-1880*, Montreal, Canadian Center for Architecture, 1993. For an account of how lithographic bird's-eye views contributed to this idealization process in the nineteenth-century urban American West, see J. Reps, *Cities on Stone: Nineteenth-Century Lithographic Images of the Urban West*, Fort Worth, Amon Carter Museum of Western Art, 1976.

17. The term 'constructed visibility' is borrowed from D. Gregory, 'Colonial Nostalgia and Cultures of Travel: Spaces of Constructed Visibility in Egypt, 1840-2000,' paper presented at Sixth International Conference of the International Association for the Study of Traditional Environments, Cairo, Egypt, December 15-19, 1998.

18. D. Harris, in *Eadweard Muybridge*, p. 41, argued that the numerous photographic panoramas produced in San Francisco's early years were meant to invoke nostalgic responses to a city in the throes of constant change. In 'Urban Tourism,' pp. 70-74, N. Harris made the same argument regarding guidebooks to and photographs of New York City in the late nineteenth century.

19. My use of the term 'picturesque' is borrowed from Sears, *Sacred Places*, pp. 186-188. Following mid-nineteenth-century theories offered by Frederick Law

Olmsted, Sears claimed that the 'picturesque' meant 'quiet and pensive contemplation' intended to promote 'gentility'.

20. It should be noted that many of the published travel books featured carefully edited and rewritten diaries, and that they straddled the line between official and unofficial documentation. Furthermore, since most diaries were kept by wealthy travellers, even the alleged unofficial documentation cannot be considered fully representative of tourists' experiences. Nevertheless, an analysis of many types of source material at least provides a broader picture than a guidebook-focused study. For the methodology of historical tourism research and its limitations, see J. Towner, 'Approaches to Tourism History,' *Annals of Tourism Research*, vol. 15, no. 1, 1988, pp. 47-62.

21. H. Jones, *To San Francisco and Back*, London, Society for Promoting Christian Knowledge, 1878, p. 80.

22. G.A. Sala, *America Revisited: From the Bay of New York to the Gulf of Mexico and From Lake Michigan to the Pacific*, London, Vizetelly & Co., 1886, pp. 441-455.

23. A good analysis of how sight and observation became the focus of touring by the nineteenth (or at least how travelling was promoted) is J. Adler, 'Origins of Sightseeing,' *Annals of Tourism Research*, vol. 16, 1989, pp. 7-29. For an analysis of how the supposed rational order of nineteenth-century American cities was consistently subverted by noises, smells, and certain commercial activities, see D. Upton, 'The City as Material Culture,' in A.E. Yentsch and M.C. Beaudry (eds.), *The Art and Mystery of Historical Archaeology: Essays in Honor of James Deetz*, Boca Raton, FL, CRC Press, 1992, pp. 53-60.

24. See, for example, *The Strangers' Guide to San Francisco*, San Francisco, Jas. B. Bradford, 1875, p. 47.

25. See, for example, 'Excursion to the Environs of San Francisco,' *San Francisco Evening Picayune,* as reprinted in K.M. Johnson (ed.), *San Francisco As It Is: Gleanings from the Picayune*, Georgetown, CA, Talisman Press, 1964, pp. 82-83. For a fictional account of locals who made frequent walking trips away from their downtown apartment buildings to the city's suburbs in the nineteenth century, see F. Norris, *McTeague: A Story of San Francisco*, 1899, reprint, New York, Penguin Books, 1994.

26. R.W. Lotchin, *San Francisco, 1846-1856: From Hamlet to City*, Urbana, University of Illinois Press, 1974, p. 284; and D.H. Huggins (comp.), *Continuation of the Annals of San Francisco: Part 1, From June 1, 1854, to December 31, 1855*, San Francisco, California Historical Society, 1939, p. 42.

27. According to this history, industrial expansion contributed to massive urban immigration (from overseas and the countryside) and subsequent overcrowding, pollution, and moral decay. Industrialization of the cities then drove many people to the suburban fringe or away from the cities altogether. As a result, there was nothing in the cities themselves that pleasure-seeking tourists would desire to explore. To restore peace and moral order to the urban centres, reformers and 'landscape gardeners' such as Calvert Vaux and Frederick Law Olmsted wanted to insert urban parks and greenery into the city. New York's Central Park, for example, was often referred to as helping provide the city with 'lungs.' Many scholars point to efforts at urban reform to justify their claims about a 'rural

sensibility' that seemed to dominate nineteenth-century American life – at times searching for its roots among the picturesque gardens of British gentry in the eighteenth century, the canvases of seventeenth-century French painters Claude Lorrain and Nicholas Poussin, or the poetry of the American transcendentalists. The literature on this 'rural sensibility' is far too voluminous to detail here. For some good examples, see M. Andrews, *The Search for the Picturesque: Landscape Aesthetics and Tourism in Britain, 1760-1800*, Palo Alto, CA, Stanford University Press, 1989; or L. Marx, *The Machine in the Garden: Technology and the Pastoral Ideal in America*, London, Oxford University Press, 1964.

28. Lotchin, *San Francisco 1846-1856*, p. 285.

29. *Ibid*., p. 285.

30. J.S. Hittell, *Bancroft's Pacific Guide Book*, San Francisco, A.L. Bancroft & Co., 1882, p. 66.

31. Nordhoff first wrote about vacation spots along the eastern seaboard, but devoted later books to accounts of the Pacific slope, promoting it as a place with a fabulous year-round climate where invalids could live out their remaining years with renewed health and vigour (a common theme among writers of the nineteenth-century West). Other travel books by Nordhoff include *Peninsular California*, New York, Harper, 1888; *A Guide to California, the Golden State*, Southern Pacific Railroad, 1888; and *Northern California, Oregon, and the Sandwich Islands*, New York, Harper and Brothers, 1877.

32. Nordhoff, *California: For Health, Pleasure, and Residence*, p. 61.

33. *Ibid*., p. 61.

34. *Ibid*., pp. 63, 87.

35. The classification of this literature as 'non-fictional urban sensationalism' comes from G. Foster, *New York by Gas-Light and Other Urban Sketches*, S. Blumin (ed.), Berkeley, University of California Press, reprint ed., 1990, p. 1. This literature arguably dated to 1851 with the publication of H. Mayhew's *London Labour and the London Poor*, 1861, reprint, Harmondsworth, Penguin Books, 1985. It was followed by A. Mearns, *The Bitter Cry of Outcast London*, 1883, reprint, New York, Humanities Press, 1970. In America this writing had its parallel on the East Coast in such works as J. Riis, *How the Other Half Lives: Studies Among the Tenements of New York*, 1890, reprint, New York, Hill and Wang, 1968; and H. Campbell, T.W. Knox, and T. Byrnes, *Darkness and Daylight, or Lights and Shadows of New York Life*, Hartford, CN, Hartford Publishing Company, 1897. In the West, such books as B.E. Lloyd, *Lights and Shades in San Francisco*, San Francisco, A.L. Bancroft Co., 1876; and W. Boyd, *Lights and Shadows of Chinatown*, San Francisco, H.S. Crocker, 1896, featured similar themes.

36. Nordhoff, *California: For Health, Pleasure, and Residence*, p. 91. Nordhoff's analysis of the Chinese turned from vilifying to complimentary but patronizing in his later *Northern California, Oregon, and the Sandwich Islands*, pp. 141-143.

37. Nordhoff's account of 'John Chinaman' had its East Coast counterpart in a chapter about 'John Chinaman' in Campbell, Knox and Byrnes, *Darkness and Daylight*, pp. 549-573. The name 'John' to describe the 'average' Chinese resident in America was not initiated by Nordhoff. I. Saxon, in *Five Years Within the Golden Gate*, Philadelphia, J.B. Lippincott and Co., 1868, p. 39, explained that the

name was bestowed by 'Californians' upon the 'labouring Chinese'. J.H. Beadle, in *The Undeveloped West, or Five Years in the Territories*, Philadelphia, National Publishing Company, 1893, pp. 313-325, also included a chapter entitled 'John'.

38. 'The City by the Golden Gate,' *Scribner's Monthly*, vol. 10, no. 3, July 1875, pp. 266-285.

39. Rev. J.C. Holbrook, 'Chinadom in California,' in J.M. Hutchings, *Hutchings' Illustrated California Magazine*, vol. 4, no. 3, September 1859, p. 129.

40. W. Glazier, *Peculiarities of American Cities*, Philadelphia, Hubbard Brothers, 1886, p. 466.

41. F.W. Green, *Notes on New York, San Francisco, and Old Mexico*, Wakefield, E. Carr, 1886, p. 65.

42. Tiffany and Macdonald, 'An Evening Promenade,' *Pocket Exchange Guide of San Francisco*, p. 188.

43. References to a transplanted 'Canton' in San Francisco came with the earliest Chinese settlement during the Gold Rush. See, for example, 'Young China,' *San Francisco Evening Picayune*, June 27, 1851, as reprinted in Johnson (ed.), *San Francisco As It Is*, p. 170. For similar examples, see *The City and Port of San Francisco*, San Francisco, San Francisco Chamber of Commerce, 1896; and Taber, *Hints to Strangers*, p. 3. For references to San Francisco's Chinese quarter as Peking, see Beadle, *The Undeveloped West*, p. 302; W. Doxey, *Doxey's Guide to San Francisco and Vicinity: The Big Trees, Yo Semite Valley, the Geysers, China, Japan, and Sandwich Islands*, San Francisco, Doxey and Co., 1881, p. 58; or H.H. Jackson, *Bits of Travel at Home*, Boston, Roberts Brothers, 1893, p. 63. As Mongolia, see Boyd, *Lights and Shadows of Chinatown*, p. 1, and 'The City by the Golden Gate,' *Scribner's Monthly*, July, 1875, p. 282; F.M. DeWitt, *An Illustrated and Descriptive Souvenir Guide to San Francisco: A New Handbook for Strangers and Tourists*, San Francisco, Frederic M. DeWitt, 1898, p. 29; or *California As It Is*, San Francisco, San Francisco Call Company, 1882, p. 138. For the Chinese as 'Celestials,' see Holbrook, 'Chinadom in California,' p. 129.

44. R. Cumbrian, *Roving Cumbrian Journal*, October 31, 1871-December 20, 1872, Bancroft Library, University of California at Berkeley.

45. Mrs. F. Leslie, *California: A Pleasure Trip from Gotham to the Golden Gate*, New York, G.W. Carleton & Co., 1877, p. 144.

46. Jones, *To San Francisco and Back*, p. 73.

47. C.L. Brace, *The New West: Or, California in 1867-1868*, New York, G.P. Putnam & Son, 1869, preface. For other sympathetic accounts, see J. Codman, *The Round Trip: By Way of Panama Through California, Oregon, Nevada, Utah, Idaho, and Colorado*, New York, G.P. Putnam's Sons, 1979, pp. 126-134; Mrs. J.G. Smith, *Notes of Travel in Mexico and California,* St. Albans, VT, Messenger and Advertiser Office, 1886, p. 100; or H.H. Vivian, *Notes of a Tour in America*, London, Edward Stanford, 1878, pp. 135-136.

48. See, for example, Leslie, *California: A Pleasure Trip*, pp. 139-140.

49. Vivian, *Notes of a Tour in North America*, p. 137.

50. Leslie, *California: A Pleasure Trip*, p. 217.

51. Many of these hotels were probably boarding hotels, which were a vital part of

early San Francisco life. For descriptions of life in early hotels, see W.L. Macgregor, *Hotels and Hotel Life in San Francisco*, San Francisco, S.F. News Company, 1877.

52. Prior to its opening, the *San Francisco Evening Picayune* noted that the Union Hotel, as a whole, was 'the most superb establishment of this kind in California.' Once it opened, the *Picayune* described it as 'an ornament to our city' that was 'scarcely to be surpassed by any House of the kind in the United States.' One article recounted how a group of travellers from the East Coast, who had recently stayed in New York City's luxurious Irving House, pointed out that San Francisco's Union Hotel 'offers every convenience and luxury to be found in [New York's] elegant resort of wealth and fashion.' 'Union Hotel,' *San Francisco Evening Picayune*, September 12, 1850, and November 26, 1850, as reprinted in Johnson (ed.), *San Francisco As It Is*, p. 87.

53. Doxey, *California Tourists Guide*, pp. 15, 16-17; P. Groth, *Living Downtown: The History of Residential Hotels in the United States*, Berkeley, University of California Press, 1994, pp. 28, 32; A. Johnson, letter to A. Bush, 1879, Bancroft Library, University of California at Berkeley; Lick House, *Tourists' Guide*, p. 18; Macgregor, *Hotels and Hotel Life*, p. 9; San Francisco Call, *California As It Is*, p. 139; Saxon, *Five Years Within the Golden Gate*, pp. 13-14.

54. *California Illustrated: A Guide for Tourists and Settlers*, San Francisco, Carnall-Hopkins Co., 1891; Doxey, *California Tourists Guide*, pp. 15-16; J.M. Hutchings, *Yo Semite and the Big Trees: What to See and How to See It*, San Francisco, J.M. Hutchings, 1894, as noted in Sears, *Sacred Places*, p. 137; San Francisco Call, *California As It Is*, p. 139; Green, *Notes on New York*, p. 57; Taber, *Hints to Strangers*; and B.F. Taylor, *Between the Gates*, Chicago, S.C. Griggs and Company, 1882, pp. 71-73.

55. San Francisco Call, *California As It Is*, p. 140; and Tiffany and Macdonald, *Pocket Exchange Guide*, p. 17.

56. See, for example, *Hotel Visitor and Stranger's Guide*, no. 6, February 7, 1880; *Strangers' Guide to San Francisco*, 1875; or *Disturnell's Strangers' Guide to San Francisco and Vicinity*, San Francisco, W.C. Disturnell, 1883. The only theatres that were promoted with some regularity were those in the Chinese quarter and the one in the Baldwin Hotel.

57. Hittell, *Guide Book to San Francisco*, pp. 28, 30.

58. Brace, *The New West*, p. 38.

59. For nineteenth-century comparisons of San Francisco to Paris as a whole or in specific examples, see Bowles, *Our New West*, p. 314; J.S. Hittell, *The Commerce and Industries of the Pacific Coast*, San Francisco, 1882; Lloyd, *Lights and Shades in San Francisco*, p. 29; or Saxon, *Five Years Within the Golden Gate*, pp. 12-13. For a more recent study making these comparisons, see G. Brechin, 'The City Beautiful,' in P. Polledri (ed.), *Visionary San Francisco*, San Francisco, San Francisco Museum of Modern Art, 1990, pp. 46-48.

60. *The Palace Hotel*, San Francisco, the Palace Hotel, 1885.

61. A.H. Wylie, *Chatty Letters from the East and the West*, London, Sampson Low, Marston, Searle & Rivington, 1879, p. 193.

62. Southern California, for example, was consistently compared to Italy, while Alaska and Colorado became known as 'American Switzerlands'. In turn,

Switzerland became known as the 'European Alaska' or the 'European Colorado' to American travellers in Europe. See Pomeroy, *In Search of the Golden West*, pp. 33-34, 57.

63. San Francisco Call, *California As It Is*, pp. 139-140.

64. Leslie, *California: A Pleasure Trip*, p. 121.

65. For descriptions of the large ferry, see Hittell, *Bancroft's Pacific Guide Book*, pp. 43-44; or Taylor, *Between the Gates*, p. 70.

66. Fewer tourists made the entire trip by boat by this time, and even fewer made the treacherous overland wagon journey.

67. Leslie, *California: A Pleasure Trip*, p. 113.

68. B. Clarkson, 'Overland Journey to California and the Western Territories,' diary, 1874, Bancroft Library, University of California at Berkeley.

69. Leslie, *California: A Pleasure Trip*, p. 118.

70. A.C. Guillemard, *Over Land and Sea: A Log of Travel Round the World in 1873-74*, London, Tinsley Brothers, 1875, pp. 199-200.

71. Leslie, *California: A Pleasure Trip*, pp. 138-139.

72. J.W. Reynolds, letter, September 21, 1875, John W. Reynolds Letters to His Family, 1875-76, Bancroft Library, University of California at Berkeley.

73. L.D. Luke, *A Journey from the Atlantic to the Pacific Coast by Way of Salt Lake City Returning by Way of the Southern Route Describing the Natural and Artificial Scenes of Both Lines*, Utica, NY, Ellis H. Roberts & Co., 1884, p. 35.

74. Jackson, *Bits of Travel at Home*, p. 77.

75. Anonymous, 'Notes of a journey from England to San Francisco and Back,' diary, September 19 to November 17, 1877, Bancroft Library, University of California at Berkeley.

76. A.D. Richardson, *Beyond the Mississippi: From the Great River to the Great Ocean*, Hartford, CN, American Publishing Company, 1867, pp. 448-449. For other accounts noting this growth and cosmopolitanism, see 'The City by the Golden Gate,' p. 272; or S.M. Eardley-West (ed.), *Our Journal in the Pacific*, London, Longmans, Green, and Co., 1873, pp. 29-30.

77. S.C. Clark, *The Round Trip from the Hub to the Golden Gate*, Boston, Lee and Shepard, 1890, p. 73.

78. Jones, *To San Francisco and Back*, p. 69.

79. Clark, *From the Hub to the Golden Gate*, p. 73.

Bibliography

Anon, *California Illustrated: A Guide for Tourists and Settlers*, San Francisco, Carnall-Hopkins Co., 1891.

Anon, *Disturnell's Strangers' Guide to San Francisco and Vicinity*, San Francisco, W.C. Disturnell, 1883.

Anon, *Hotel Visitor and Stranger's Guide*, no.6, February 7, 1880.

Anon, *Lick House Tourists' Guide: Giving Principal Routes from Chicago and Saint Louis to San Francisco*, San Francisco, Lick House, 1871.

Anon, 'Notes of a journey from England to San Francisco and Back,' diary, September 19-November 17, 1877, located in Bancroft Library, University of California, Berkeley.

Anon, *The Palace Hotel*, San Francisco, The Palace Hotel, 1885.

Anon, *The Strangers' Guide to San Francisco*, San Francisco, Jas. B. Bradford, 1875.

Anon, 'The City by the Golden Gate,' *Scribner's Monthly*, July 1875, vol. 10, no. 3, pp. 266-285.

Anon, *California As It Is*, San Francisco, San Francisco Call Company, 1882.

Anon, *The City and Port of San Francisco*, San Francisco, San Francisco Chamber of Commerce, 1896

Adler, Judith, 'Origins of Sightseeing,' *Annals of Tourism Research*, vol. 16, 1989, pp. 7-29.

Andrews, Malcolm, *The Search for the Picturesque: Landscape Aesthetics and Tourism in Britain, 1760-1800*, Palo Alto, CA, Stanford University Press, 1989.

Beadle, John H., *The Undeveloped West, or Five Years in the Territories*, Philadelphia, National Publishing Company, 1893.

Bowles, Samuel, *Our New West: Records of Travel Between the Mississippi River and the Pacific Ocean*, Hartford, CN, Hartford Publishing Company, 1869.

Boyd, William, *Lights and Shadows of Chinatown*, San Francisco, H.S. Crocker, 1896.

Brace, Charles Loring, *The New West: Or, California in 1867-1868*, New York, G.P. Putnam & Son, 1869.

Brechin, Gray, 'The City Beautiful,' in Paolo Polledri (ed.), *Visionary San Francisco*, San Francisco, San Francisco Museum of Modern Art, 1990.

Campbell, Helen, Thomas W. Knox, and Thomas Byrnes, *Darkness and Daylight, or Lights and Shadows of New York Life*, Hartford, CN, The Hartford Publishing Company, 1897.

Clark, Susie C., *The Round Trip from the Hub to the Golden Gate*, Boston, Lee and Shepard, 1890.

Clarkson, Banyer, 'Overland Journey to California and the Western Territories,' diary, 1874, located in Bancroft Library, University of California at Berkeley.

Cocks, Catherine, '"A City Excellent to Behold": Urban Tourism and the Commodification of Public Life in the United States, 1850-1915,' Ph.D. diss, University of California, Davis, 1997.

Codman, John, *The Round Trip: By Way of Panama Through California, Oregon, Nevada, Utah, Idaho, and Colorado*, New York, G.P. Putnam's Sons, 1979.

Cohen, Erik, 'A Phenomenology of Tourist Experiences,' *Sociology*, vol. 13, no. 2, May 1979, pp. 179-201.

Crofutt, George A., *Crofutt's Trans-continental Tourist*, New York, Geo. A. Crofutt, 1874.

Cumbrian Roving, 'Roving Cumbrian Journal,' October 31, 1871-December 20, 1872, located in Bancroft Library, University of California at Berkeley.

DeWitt, Frederic M., *An Illustrated and Descriptive Souvenir Guide to San Francisco: A New Handbook for Strangers and Tourists*, San Francisco, Frederic M. DeWitt, 1898.

Doxey, William, *Doxey's Guide to San Francisco and Vicinity: The Big Trees, Yo Semite Valley, the Geysers, China, Japan, and Sandwich Islands*, San Francisco, Doxey and Co., 1881.

Eardley-West, Sydney Marlow (ed.), *Our Journal in the Pacific*, London, Longmans, Green and Co., 1873.

Foster, George, *New York by Gas-Light and Other Urban Sketches*, Stuart Blumin (ed.), Berkeley, University of California Press, reprint ed. 1990.

Glazier, Willard, *Peculiarities of American Cities*, Philadelphia, Hubbard Brothers, 1886.

Green, Frank W., *Notes on New York, San Francisco, and Old Mexico*, Wakefield, England, E. Carr, 1886.

Gregory, Derek, 'Colonial Nostalgia and Cultures of Travel: Spaces of Constructed Visibility in Egypt, 1840-2000,' paper presented at Sixth International Conference of the International Association for the Study of Traditional Environments, Cairo, Egypt, December 15-19, 1998.

Groth, Paul, *Living Downtown: The History of Residential Hotels in the United States*, Berkeley, University of California Press, 1994.

Guillemard, Arthur C., *Over Land and Sea: A Log of Travel Round the World in 1873-74*, London, Tinsley Brothers, 1875.

Hales, Peter B., *Silver Cities: The Photography of American Urbanization, 1839-1915*, Philadelphia, Temple University Press, 1984.

Harris, David, *Eadweard Muybridge and the Photographic Panorama of San Francisco, 1850-1880*, Montreal, Canadian Center for Architecture, 1993.

Harris, Neil, 'Urban Tourism and the Commercial City,' in W.R. Taylor (ed.), *Inventing Times Square: Commerce and Culture at the Crossroads of the World*, Baltimore, Johns Hopkins University Press, 1991.

Hittell, John S., *Bancroft's Pacific Guide Book,* San Francisco, A.L. Bancroft & Co., 1882.

Hittell, John S., *The Commerce and Industries of the Pacific Coast*, San Francisco, 1882.

Holbrook, Rev. J.C., 'Chinadom in California,' in J.M. Hutchings, *Hutchings' Illustrated California Magazine*, vol.4, no.3, September 1859.

Huggins, Dorothy H. (comp.), *Continuation of the Annals of San Francisco: Part 1, From June 1, 1854, to December 31, 1855*, San Francisco, California Historical Society, 1939.

Hutchings, James Mason, *Yo Semite and the Big Trees: What to See and How to See It,* San Francisco, J.M. Hutchings, 1894.

Hyde, Anne F., *An American Vision: Far Western Landscape and National Culture, 1820-1920*, New York, New York University Press, 1990.

Jackson, Helen Hunt, *Bits of Travel at Home,* Boston, Roberts Brothers, 1893.

Johnson, Angie, letter to Abigail Bush, 1879, located in Bancroft Library, University of California, Berkeley.

Johnson, Kenneth. (ed.), *San Francisco As It Is: Gleanings from the Picayune*, Georgetown, CA, The Talisman Press, 1964.

Jones, Harry, *To San Francisco and Back*, London, Society for Promoting Christian Knowledge, 1878.

Leslie, Mrs. Frank, *California: A Pleasure Trip from Gotham to the Golden Gate*, New York, G.W. Carleton & Co., 1877.

Littman, William, 'The Production of Goodwill: Factory Tours in America,' paper presented at Vernacular Architecture Forum Conference, May 9, 1998, Annapolis, Maryland.

Lloyd, Benjamin E., *Lights and Shades in San Francisco*, San Francisco, A.L. Bancroft Co., 1876.

Lotchin, Roger W., *San Francisco, 1846-1856: From Hamlet to City*, Urbana, University of Illinois Press, 1974.

Luke, L.D., *A Journey from the Atlantic to the Pacific Coast by Way of Salt Lake City Returning by Way of the Southern Route Describing the Natural and Artificial Scenes of Both Lines*, Utica, NY, Ellis H. Roberts & Co., 1884.

MacCannell, Dean, *The Tourist: A New Theory of the Leisure Class*, New York, Schocken Books, 1975.

Macgregor, William L., *Hotels and Hotel Life in San Francisco*, San Francisco, S.F. News Company, 1877.

Marx, Leo, *The Machine in the Garden: Technology and the Pastoral Ideal in America*, London, Oxford University Press, 1964.

Mayhew, Henry, *London Labour and the London Poor*, 1861; reprint, Harmondsworth, Penguin Books, 1985.

Mearns, Andrew, *The Bitter Cry of Outcast London*, 1883; reprint, New York, Humanities Press, 1970.

Nordhoff, Charles, *A Guide to California, the Golden State*, Southern Pacific Railroad, 1888.

Nordhoff, Charles, *California: For Health, Pleasure, and Residence*, New York, Harper & Brothers, 1874.

Nordhoff, Charles, *Northern California, Oregon, and the Sandwich Islands*, New York, Harper and Brothers, 1877.

Nordhoff, Charles, *Peninsular California*, New York, Harper, 1888.

Norris, Frank, *McTeague: A Story of San Francisco*, 1899; reprint, New York, Penguin Books, 1994.

Player-Frowd, John G., *Six Months in California*, London, Longmans, Green and Co., 1872.

Pomeroy, Earl, *In Search of the Golden West: The Tourist in Western America*, Lincoln, University of Nebraska Press, 1957.

Reps, John, *Cities on Stone: Nineteenth-Century Lithographic Images of the Urban West*, Fort Worth, Amon Carter Museum of Western Art, 1976.

Reynolds, John W., letter, September 21, 1875, John W. Reynolds Letters to His Family, 1875-76, located in Bancroft Library, University of California, Berkeley.

Richardson, Albert D., *Beyond the Mississippi: From the Great River to the Great Ocean*, Hartford, CN, American Publishing Company, 1867.

Riis, Jacob, *How the Other Half Lives: Studies Among the Tenements of New York*, 1890; reprint, New York, Hill and Wang, 1968.

Rothman, Hal, *Devil's Bargains: Tourism in the Twentieth-Century American West*, Lawrence, University Press of Kansas, 1998.

Sala, George A., *America Revisited: From the Bay of New York to the Gulf of Mexico and From Lake Michigan to the Pacific*, London, Vizetelly & Co., 1886.

Saxon, Isabelle, *Five Years Within the Golden Gate*, Philadelphia, J.B. Lippincott and Co., 1868.

Sears, John, *Sacred Places: American Tourist Attractions in the Nineteenth Century*, New York, Oxford University Press, 1989.

Shaffer, Marguerite S., 'See America First: Re-Envisioning Nation and Region Through Western Tourism,' *Pacific Historical Review*, vol. 65, no. 4, 1996, pp. 559-582.

Smith, Mrs. J. Gregory, *Notes of Travel in Mexico and California*, St. Albans, VT, Messenger and Advertiser Office, 1886.

Smith, Valene (ed.), *Hosts and Guests: The Anthropology of Tourism*, Philadelphia, University of Pennsylvania Press, 1977.

Starr, Kevin, *Americans and the California Dream 1850-1915*, New York, Oxford University Press, 1973.

Taber, Isaiah West, *Hints to Strangers*, San Francisco, I.W. Taber, 1890.

Taylor, Benjamin F., *Between the Gates*, Chicago, S.C. Griggs and Company, 1882.

Tiffany, O.F. and A.C. Macdonald, *Pocket Exchange Guide of San Francisco*, San Francisco, Tiffany and Macdonald and Central Pacific, 1875.

Towner, John and Geoffrey Wall, 'History and Tourism,' *Annals of Tourism Research*, vol. 18, no. 1, 1991.

Towner, John, 'Approaches to Tourism History,' *Annals of Tourism Research*, vol. 15, no. 1, 1988, pp. 47-62.

Upton, Dell, 'The City as Material Culture,' in A.E. Yentsch and M.C. Beaudry (eds.), *The Art and Mystery of Historical Archaeology: Essays in Honor of James Deetz*, Boca Raton, FL, CRC Press, 1992.

Urry, John, *The Tourist Gaze: Leisure and Travel in Contemporary Societies*, London, Sage Publications, 1985.

Vivian, Henry Hussian, *Notes of a Tour in America*, London, Edward Stanford, 1878.

Wylie, A.H., *Chatty Letters from the East and the West*, London, Sampson Low, Marston, Searle & Rivington, 1879.

Chapter 7

Re-Presenting and Representing the Vernacular: The Open-Air Museum

PAUL OLIVER

'History,' the car manufacturer Henry Ford is reputed to have said, 'History is bunk.' Bunk, or bunkum, means verbal rubbish, tinged with deception. Actually, what Ford said was 'History is bunk, as it is taught in schools.'

Ford's concern was so genuine that he founded the Henry Ford Museum and Greenfield Village near Dearborn, Michigan, to introduce Americans to the material facts of their history.[1] However, Ford could be cavalier in his methods. Wishing to demonstrate that many American families came from rural England, in the mid-1920s he tried to purchase a row of Cotswold cottages to transport and rebuild in his open-air museum of Greenfield. But, alarmed at the impending demolition of Arlington Row, Bibury, the local community alerted the Gloucestershire Archaeological Trust, who succeeded in preventing Ford from going through with his plan. In 1929 Arlington Row was bought by the Royal Society of Arts, and twenty years later it was given to the National Trust for safe keeping.

Even though he failed to transport Arlington Row, Ford later bought a Cotswold house and blacksmith's forge from another village, and had all 500 tons of stone and timber shipped to Michigan. Whether Americans learned much more of their history as a result is open to question, even if, by default, Bibury was the richer for the preservation effort set in motion by Ford's plan.[2]

The Skansen Movement

Henry Ford's Greenfield collection demonstrates how a single influential, affluent and motivated person, with a certain perception of history, can arrange for the location, demolition, transfer and re-erection of a collection of buildings (in Greenfield, about a hundred) ostensibly for the benefit of present and future generations. A number of open-air museums started this way. Artur Hazelius in Sweden was responsible for assembling the buildings which, as 'Skansen' on Djugarden Island, Stockholm, opened in 1891. Hazelius believed that 'getting to know and falling in love with the past is the essential basis for all kinds of new production – a tree is the stronger the deeper are its roots.'[3] *Skansen* in Swedish means 'fort', and though the word referred to the former function of Hazelius's site, it came to symbolize the enclosed, contained character of the 'open-air' museum. Others soon followed, with the Lillehammer dentist Anders Sandvig building a collection which was adopted by the municipality and opened in Maihaugen in 1904. 'In my opinion Maihaugen should be a collection of homes, where it is possible to come close to the people who lived in them, to learn to know their way of life, their tastes, and their work,' Sandvig wrote, though he had more comprehensive ideas in mind. 'My aim is not merely to preserve a haphazard collection of old houses. No, I would like to include the whole parish, as a complete entity, in my picture-book.'[4]

Known henceforth as *skansens* in Scandinavia and Eastern Europe, some forty-four open-air museums had been established by World War I, many at the instigation of passionate collectors, but others under museological or institutional, provincial or national auspices. The Seurasaari Open-Air Museum, for example, situated on Seurasaari Island near Helsinki, Finland, was founded in 1909 by Professor Axel Olai Heikel, whose declared aim was 'to collect typical buildings from the different regions of Finland in order to display folk architecture and how Finnish people lived.'[5] His purpose was at least in part nationalistic, symbolically using the museum to distance Finland from its Russian and Swedish neighbours. The site, which was part of a public park, was rented from the City of Helsinki by a newly formed company in 1911; however, two years later it was acquired by the state and operated as a national museum under the auspices of the National Archeological Commission (later, the National Board of Antiquities).

Other such collections were similarly adopted, for example the Nederlands Openluchtmuseum, Arnhem, which was founded by a small private group in 1912 and opened in 1918, operating independently for thirty years until the state assumed overall responsibility in 1941. Its declared purpose was to 'present a picture of the daily life of ordinary people in this country as it was in the past and has developed in the course of time.' In producing a 100-page guide, the group 'concentrated on the social and economic background rather than overburdening the visitor with technical terms or local names, or giving detailed accounts of the past history of the buildings, which in any case have mostly found their way to the museum more or less by chance.'[6]

Such museums grew in number, and many in size, between the wars, with Muzeul Satulut in Bucharest, Romania, which opened in 1939, eventually displaying well in excess of 300 buildings. After World War II open-air museums continued to proliferate, with Norway establishing an association to unite more than a hundred in that country. Typical was the Trøndelag Folkemuseum, Sverresborg, Trondheim, which was privately founded in 1913. Staffed voluntarily and subsisting on private funds and later by grants, it has retained its independent status. By 1980 the Polish historian of the *skansen* movement, Jerzy Czajkowski, could write that 'there are approximately 500 big and small, ready or being built, skansen museums in Europe at present.'[7]

Skansen, Village Museum, Freilichtmuseum, Openluchtmuseum, Musée de Plein Air, Open-Air Museum, whatever the term used, they all conserve buildings within outdoor museum compounds. Or so it appeared, until notices of a Freilichtmuseum near Stadthagen in north Germany led to a park inhabited by full-size concrete dinosaurs! There is no agreement on terms, and often little on what is preserved, how buildings are relocated and restored, what form the museum may take, and even why they are assembled. This may be regarded as inevitable, bearing in mind the vast range of terrains, resources, materials, periods, traditions, functions, and building types which may be considered to merit preservation.

During the past century the incursions made into the rural landscapes of countries on all continents have been of a scale which is almost beyond measure. The expansion of cities, the growth of suburbs, the migration of peoples, the construction of superhighways, the decline of small agricultural economies, and the growth of agribusiness are among the manifold factors which have contributed to the irreparably changed face of the world's rural landscapes. Such a summary may be platitudinous, but it is undeniable, nonetheless. Falling victim to such physical changes have been countless fields, farmlands and forests, among which have been untold rural buildings, sometimes of considerable age, that have been lost forever. Sometimes this loss has gone virtually unnoticed, as farm properties have fallen to the auctioneer's hammer along with the farmlands. But in certain other instances their destruction has been part of deliberate policies, motivated by industrial greed or political ideologies, of which the planned programme of village elimination and 'rural systemization' of the Ceaucescu regime in Romania was the most notorious.[8] Their loss has been paralleled by the destruction of innumerable urban buildings of all kinds and often of considerable quality, that have obstructed industrial or commercial expansion and the exploitation of land values.

Though the safeguarding of certain examples of architecture has been assured through the efforts of individuals and conservation groups since early in the century, the period following World War II has witnessed a redoubling of efforts to do so. The 'listing' and grading of buildings, devising of heritage trails, establishment of eco-museums, declaration of World Heritage Sites, and other measures have been taken to ensure the protection of significant buildings. Some are of archaeological importance; others are of unquestioned architectural merit – singular buildings, whose scale, architectural quality, prominence in their urban or landscape context, historic record, and, in most cases, religious significance are justification enough for their conservation. Yet many buildings are conserved for their historic associations rather than their intrinsic value as architecture. For example, it is its function as a national symbol and the home of the first President of the Union, George Washington, that is the primary reason for the preservation of the seventeenth-century Virginian mansion, Mount Vernon. Its conservation might also be argued on aesthetic grounds, but this could not be said of the modest homestead at Johnson City near Fredericksburg, Texas. Yet the conservation of the Sam Early Johnson Log House, the reconstructed birthplace and restored Boyhood Home of Lyndon Johnson, whose memory is preserved with the building while emphasizing his rural background, could be justified as examples of regional, vernacular architecture.

Distinguished from free-standing châteaux, churches and country houses, monasteries and mosques, by their rural and largely domestic character, the majority of open-air museums display examples of what their publications variously term 'regional', 'folk', 'traditional', 'peasant' or 'vernacular' architecture. Such assemblies were suggested as early as 1790, and a number of precursors date from the nineteenth century. Though some of these consisted of only a few structures, as was the case with the buildings relocated on Norway's Bygdoy Peninsula on the orders of King Oscar II in the 1880s, a few were larger. Among these early efforts was the Ethnographic Village of twenty-four houses, a church, and several farm buildings from the Carpathian Basin re-erected in 1896 as part of the Hungarian Millenary Exhibition. Dismantled the following year, it nevertheless inspired the formation of a number of open-air museums in Hungary a lifetime later.

Museums in the Open Air

Though it has been in international use since the 1950s, it remains uncertain at what date and to which location the term 'open-air' museum was first applied. The vagueness of the term means that it can embrace a variety of situations, though it is generally employed to identify a museum of buildings located in a territory often approximately equivalent to that of a provincial zoo. The unifying concept of the 'open-air' museum refers to the dispersal of the buildings within the territory and not to the inherent nature of the exhibits themselves. This permits the display not only of domestic buildings but of cart sheds, hayricks, farm outbuildings, and unique structures such as the chain-maker's shop at Bromsgrove (England) or the water-powered laundry at Arnhem.

Certain such complexes could be on open access, but for the purposes of this discussion the open-air museum is defined as one in which diverse buildings have been relocated in a physically, if artificially, defined landscape setting to which access is gained by payment of an entrance fee. Encompassed within the concept are village museums, such as Old Sturbridge Village, Massachusetts; folk artefact collections, like the Welsh Folk Museum, St. Fagan's; museums of building, such as Avoncroft near Bromsgrove, England; ethnographic museums, for instance, the Park Etnograficzny W Sanoku, Poland; and heritage centres like the Isle of Arran Heritage Museum, among several other more or less synonymous designations.

Within its territorial confines the open-air museum may conform to one of a small number of specific types. In much of Europe the term 'park museum' denotes an arrangement in which buildings are dispersed at intervals that are sufficiently distant so as to preserve their distinct identities, whether in function, period, method of construction or region of origin. The Niedersächsisches Freilichtmuseum at Cloppenburg, north Germany, is a representative of this type. By contrast, the term 'village museum' is applicable to those where a number of buildings have been clustered in village form, simulating, for instance, the relationship of domestic buildings to church, windmill, smithy, and village green, as at the Provincial Open-Air Museum, Bokrijk, in Flanders, Belgium. The disposition of elements may depend upon the intentions of the museum authorities. For example, the last-named was founded within the Bokrijk Provincial Domain in 1958, with its objectives 'based on Culture, Nature and Recreation'. Commenting that 'earlier open-air museums erected their buildings with little thought to their actual inter-relationship,' the 'museum village' of Bokrijk was constructed 'so that visitors feel that they are stepping into the past and entering a snapshot of life in former times.'[9] It is unusual in that it includes a simulated Brabant 'Old Town' comprising buildings transferred or partially reconstructed from Antwerp, now located around the 'Antwerp Square', with additional streets of buildings from Leuven and Diest. Future plans include clusters of urban buildings from Limburg and Flanders.

A regional approach to the organization of the open-air museum is particularly evident in 'national' collections that seek to illustrate the local or provincial traditions within the frontiers of a country. Such is the Swiss Open-Air Museum at Ballenberg near Brienz, founded in 1978, where some eighty farmsteads from all cantons are 'arranged in thirteen architectural units, on the basis of their regional origin.' About 80 hectares have been devoted to buildings in clearings on a mainly forested total site of some 200 hectares. Here the visitor can 'discover a realistic representation of the peasant style of living. Ballenberg does not just consist of ancient dwellings that have been rebuilt, but represents a vivid reconstruction of rural Switzerland of the past centuries. Truly a dream . . .'[10]

Founded earlier, in 1967, but still far from completion, the Hungarian Open Air Museum in Szentendre, near Budapest, was more ambitious. Within a space of 46 hectares, its master plan envisaged 'the relocation of about 300 buildings to the Museum for re-erection in nine groups, each representing a region in Hungary.' The regional groupings, of which four are completed, 'show[s] the pattern of settlement characteristic of each region,' and with furniture and demonstrations of customs, give visitors 'a comprehensive picture of life in the villages and market towns of Hungary in a defined segment of time.' Similarly, while the director of the museum at Szentendre is committed to the authenticity of the depiction of the chosen periods, he also emphasizes that 'authenticity also means authenticity of material and structure. The buildings chosen are usually not exhibited in the form they were found. They have to be restored to the form, material and structure they had at the time to be represented.'[11]

Smaller museums may also opt for the 'village' structure, either by arranging the component buildings that they have acquired in a village format and obtaining others that would complete the notional ensemble, or by assembling from scratch the representative buildings necessary to create the preconceived village considered to be typical of its locale. Old Sturbridge Village, 'a bit of past history come to life,' is pre-eminently such a museum. It began when Albert and Cheney Wells of Southbridge, Massachusetts, purchased a tract of some 250 acres on which stood just two buildings. Relocating and re-erecting over thirty additional buildings, including several around a green, they opened the village in 1936. Private and non-profit, it was 'chartered by the Commonwealth of Massachusetts to carry on its work of instructing and entertaining visitors, giving them a brief glimpse into rural New England that existed in 1800 or in 1820.'[12] More eclectic still is the Harold Warp Pioneer Village at Minden, Nebraska, the buildings of which were drawn mainly from Nebraska, but the contents of which and other exhibits were chosen and purchased in the United States, Canada and Mexico to illustrate 'Man's Progress'. In spite of its name, a Nebraska village form is not simulated, the buildings being arranged in a cartwheel plan permitting the visitor to 'see everything, by walking less than one mile.'[13]

Open-air museums whose stated intentions are fundamentally architectural are fewer than might be expected. In England the Avoncroft Museum of Buildings was initiated by a group of people, led by F.W.B. Charles, the architect specializing in timber framing, who endeavoured to prevent the demolition of a fifteenth-century timber-framed building in Bromsgrove. Though they were unsuccessful in this, they saved the framing timbers and re-erected the building on a 10-acre site in 1967. Other projects were commenced with the aim 'to encourage interest in buildings of architectural and historical value, where possible to prevent the demolition of such buildings when they are threatened, and to give advice on their restoration.'[14] Buildings that could not be restored in situ have been rebuilt at this museum, their value being the criterion for inclusion rather than their relationship to others within an ensemble. Perhaps the most notable rescue is the early fourteenth-century roof originally of the Guesten Hall at Worcester Cathedral. Of similar date and intention is the Weald and Downland Open Air Museum at Singleton, near Chichester, England. Founded by J.R. Armstrong in 1967, and opening four years later, its buildings today range from a fine Wealden house to a man-operated tread-wheel. Its founders aimed 'to establish a centre that could rescue representative examples of vernacular buildings from the South-east of England, and thereby to generate an increased public awareness of the built environment.' An introductory exhibition displays 'regional building materials and methods,' and a collection has been gathered of 'artefacts representing the country crafts and industries, the building trades, and agriculture.'[15]

Many museums serve a single function, preserving the architecture of a particular cultural or religious group such as the Shaker Museum at Hancock, Massachusetts, or an industry or craft like that of the leatherworkers displayed at Das Museum im Lederhaus, Purgstall, Germany. Curators have differing perceptions of the purpose of such museums, and of their clientele, and these influence the manner in which they present the architecture in their care. But ask a director whether a museum is there to preserve buildings, to educate children, to inform the public, to communicate a past way of living, to aid architects, anthropologists or ethnologists in their work, and the answer is invariably, 'all of these'. Beyond this, museum directors generally define their objectives in terms that reflect their respective views on quality, relevance, historic significance, status, and architectural merit. And to support these views, they may have teams of ethnographers and architectural historians, engineers and technologists; they may also be subject to an administering institution or university. In 1966 an Association of European Open Air Museums was established, which has agreed on standards of conservation, display, facilities, and other common aspects of their work, though many museums do not subscribe to the organization.[16]

Like Italy and Spain, France came late to the idea of the open-air museum, but at Cuzals on the Célé river is to be found 'le plus beau Musée de Plein Air du Quercy, unique en France.'[17] It is of interest in a number of ways; for instance, it has a château on the site. Châteaux are common enough in France, but this was different from most, rebuilding on a former site having been commenced in the 1920s and discontinued by the late 1930s. Its anachronistic hydrological service system, with a pumping station and gravity feed to the château, was restored in the museum. Other buildings at Cuzals include a farm, allegedly built in 1910, and a reconstruction of a farmhouse from the pre-Napoleonic period. Examples of roof types and a row of stone walls demonstrate different methods of construction. A collection of oil presses, another of tractors, a toy museum, a basket-making exhibit, physick garden, children's carousel, an apiary, a bakery, and a 'museum of fire' have much that entertains while having little sense of direction, or evidence of co-ordination. In 1998 two corbelled stone *cabanes* were being built by architectural students from La Villette, Paris – in an area where many stone builders still live and work. Within its 50 hectares, Cuzals demonstrates both the successes of open-air museums and the problems confronting them.

Presentation, Re-Presentation and Representation

As the foregoing examples have shown, the creation of an open-air museum is almost without exception, the result of a motivation on the part of an individual or of a group of people sharing a common concern or interest, to preserve a number of buildings for posterity. This means that the essential drive is one of a perceived need, but the distribution of the museums in different countries, their prevalence in some and virtual omission in others, indicate their unco-ordinated establishment and development. Some have been adopted as national or as regional collections, but most are subject to ideas of classification; only in Norway is it possible to see local museums that demonstrate the vernacular traditions of most of the country, permitting comparative study without undue preconditioned selection. In the majority of open-air museums the buildings on display have been appropriated by a founder, an organization, a provincial or other authority so as to project an overriding concept, whether this is of a presumed local tradition, a style of living in a favoured period, the values of a religious sect, the home of a memorialized individual, the labours of former workers, or some other historical construction. Statements of founders, extracts from charters, or quotations from published guides reveal a preoccupation with the visual image, often expressed in romantic terms: to create a 'picture-book', to 'present a picture', to enter a 'snapshot of former times'. The image is seldom of buildings alone but of a 'comprehensive picture of life in the villages'. Though it is unusual to acknowledge that the resultant picture is 'truly a dream', the emphasis on the presentation of 'life' in the past is ubiquitous.[18] But these constructed environments and picturesque images raise many issues that bring the validity of the open-air museum into question.

The issues are ones of presentation, re-presentation, and representation. Rarely are buildings that have been presented to a museum, rescued from demolition, or 'saved for the nation', or which have 'found their way to the museum more or less by chance', in a state where they can be simply transported and re-erected. Removal of a building is not in itself a simple process, but one which may involve legal, contractual and planning approvals. Eventually, if this task is to be responsibly completed, it requires detailed surveys, the examination of the building's condition, and the drawing, recording and numbering of its component parts for dis-assembly and re-assembly. In turn, the latter necessitates much physical but specialized work and sensitivity and method in the sequences of dismantling, handling, transporting, storing, and protecting from damage or deterioration.

Neither will the building be in a condition that permits straightforward re-erection. Extensive repairs may be required; materials that are considered inappropriate may need to be replaced (corrugated galvanized metal sheeting by thatch, for example); extraneous elements (add-ons) may need to be removed; and missing parts may need to be replaced or constructed. This is done in the name of 'authenticity', a concept which, however, normally relates 'to the time to be represented' – like the Japanese museum of buildings in which a representation of William Shakespeare's birthplace is declared to be 'more authentic' than the Stratford-upon-Avon house because its new construction is comparable with the state of the original in the sixteenth century.[19] Such work cannot be undertaken lightly or spontaneously: it requires forethought and planning; selection and design decisions; professional skills in history, archaeology, architecture and building; and not inconsiderable investments of money, time and effort. Some may be given voluntarily, but whatever the circumstances, the processes of re-erection and reconstruction are demanding and time consuming.

They also require explanation, though the terminology employed is not consistent and can be misleading. 'Erection' of a building and 're-erection' are often used synonymously, while 'reconstruction' may mean 'rebuilding' in one museum, but largely obscure the fact that a structure is a new simulation in another. The use of the term 'reconstruction' to mean 'new construction made to represent a former building' (such as the Napoleonic farmhouse at Cuzals) is now widely adopted but less widely made known to the visitor. Similar ambiguities relate to 'repair' and 'renewal', to 'renovate' and 'restore', to 'reproduce' and 'reproduction', and, notoriously, to 'preservation', 'conservation' and 'restoration'. Though some or all of these terms are generally used in guides and museum handbooks, a glossary that explains them, as distinct from one which gives historic or indigenous terms, is extremely rare. The looseness and ambiguity of the terminology assists museums in conveying their 'pictures' of 'life in past times' by laying a thin cloud of obscurity across the simulated villages and their furnished but lifeless buildings.

The buildings in an open-air museum cannot be casually obtained nor merely presented as they stand. Many are selected – or rejected – for what they are believed to represent within the chosen context, and good (if possible, excellent) examples are sought. As such, they are represented on specially selected sites and displayed as representative of the underlying theme of the museum. In forming part of a larger and more complex whole, their interrelationship with other buildings may be invented rather than replicated in the process of re-presentation. Representation in the chosen context is dependent on an understanding of the building-producing and -occupying culture in history, and its interpretation by the museum's curatorial management. How these are achieved and for what purpose will depend on the message that is intended: no museum re-presentation or representation is value free.

Outside and Inside

That the buildings are gathered within a compound is made explicit at the entrance to the open-air museum which, in the majority of instances, has a bureau and ticket office, administrative offices, and gated access. These are necessary and often extensive, but frequently, as, for instance, at the Nederlands Openluchtmuseum, Arnhem, or the National Folk Museum, St. Fagan's, Wales, the offices of administration and conservators are placed close to the entrance. Apparently, this is to assure the visitor of the extent and intensity of the work involved in preserving the buildings, even if the conservator's premises are generally closed to the interested tourist. Their design often presents problems. Thus, having failed to buy Independence Hall in Philadelphia, Henry Ford had it replicated for the entrance to Greenfield. The entrance at Frilandsmuseet, Lyngby, Denmark, is of similar scale and palatial impressiveness, notwithstanding the museum's declared purpose 'to portray the living conditions of the countryman.'[20] The forms and details of these entrance buildings emphasize the Classical tradition rather than the vernacular, while the defensive walls of the museum compounds distance the visitors from the buildings and artefacts within, with a security that merits the term *skansen* (fort). That the contents represent the 'other', the peasant community of the past, could not be more emphatically expressed. The Niedersächsisches Freilichtmuseum at Cloppenberg, north Germany, is one of the few open-air museums which uses a vernacular building as its entrance. Albeit a large one, the Austellungshalle, 'Münchausen-Scheune', dating from 1561, is skillfully employed to allow the building to indicate the character of the museum without large notices or announcements.[21] More inviting perhaps, but in an essentially modern group of buildings, the entrance to the Hungarian Open Air Museum, Szentendre, is a veritable market of pavilions, shops, souvenir stalls, and cafes.

Beyond the entrance the visitor enters the invented environment and artificially constructed 'open-air' landscape of the museum. Few are geometrically planned, and many group the houses around an open space or 'green'. Often, the arrangement of buildings appears to be determined at least in part by the circulation pattern prescribed for visitors and the convenient placing of restaurant or other facilities. In some museums, buildings may be clustered according to their affinity with others in arrangements which can appear convincing but which are not necessarily accurate. Such is the case at Sirigoino open-air museum in Yugoslavia, where all the buildings of a simulated farm were placed in parallel on a sloping site and facing the same direction even though the internally facing farm settlements located on contour patterns were clearly evident in the immediate locality.

Some collections are assembled *ad hoc* as buildings have been salvaged, bequeathed or made available, rather than by planned development. Often this has meant their relative isolation, as in the 'park' type of museum. Such serendipitous accumulation has its benefits, but purposeful expansion depends on the potential availability of buildings considered important enough to save and restore. Both the clustering of unrelated buildings and the distant dispersal of others can be seen at the Frilandsmuseet, Denmark. In many instances a compromise is effected by creating the semblance of the village form of a region, which is linked to another with a path or roadway, symbolizing considerable physical distance. This is the case at Szentendre, where a group of houses from the Upper Tisza region was located nearest the entrance, selected in the Communist era to demonstrate by the contrast of their styles of living the oppression of the serfs by the 'minor gentry', and of them in turn by the nobility. Now made less blatant, the museum bears a revisionist message.

Many fine buildings have been conserved in literally hundreds of museums, but when selection has been possible, museologists may bring professional criteria to exemplify, as in this case, a socio-political theme. But they may also bring criteria of quality, for example, in craftsmanship, decoration, sheer size, or the uniqueness of the building concerned. While this can be meritorious in itself, it can result in an overall distortion of the extent to which they are representative of buildings in their local context, emphasizing the special rather than the commonplace. In the early years of many museums, collecting was coloured by romanticism, as was the case at the Openluchtmuseum, Bokrijk, Belgium, founded in 1953 by the artist C. Wellens, who rebuilt the Wellenshoeve from Kempen 'in accordance with his own ideas' which included the addition of a dog-powered butter mill.[22] This was a salutary warning to later curators who have been assiduous in their restoration, though they cluster buildings by their province or region rather than by their functional relationships.

Not least of the problems confronting museologists is one of period. There is a general archaeological tendency to restore a building as it was first constructed, without the additions, adaptations and appurtenances that accrued over time. Only historic alterations, such as the insertion of a solar floor in an English hall house, are sometimes spared. This is done in the name of authenticity and does not take into account the modification of buildings to meet changing needs and functions. Comparison of many photographs taken both before and after dismantling, transferral, conservation and re-erection confirms the widespread practice, though there are numerous exceptions. A building may remain in largely unaltered form, even if the roofing thatch has been replaced by slates over several centuries. But interiors are more subject to change, not only in the removal or building of partitions and other minor alterations, but also in the furniture, artefacts and furnishings utilized. Certain of these, such as wooden bowls or andirons and crooks, have changed comparatively little in form, but ceramics, textiles, painted decorations, stoves and many items of household equipment are more period specific.

Curators may choose to keep the interiors vacant, as is occasionally the case, but though an empty room may hold the interest of an architect, it can have little appeal for the general public. Usually curators have to decide what period their recreated interior assumes. This is largely conditioned by what is available from the museum store – few buildings arrive with artefacts intact. Sometimes the reserves are considerable – for example, Anders Sandwig left a legacy of 30,000 objects to Maihaugen. Inevitably, most furnished interiors appear as stage sets for specific historical drama productions, undisturbed by any actors. Conscious of this, some museum directors have arranged tableaux of wooden or waxwork figures, such as groups of purchasers and shopkeepers, or schoolchildren and their teacher in a classroom. These and some fifteen other period settings with characters in arrested action are displayed at the Musée des Arts et Traditions Populaires, Wattrelos, near Lille, northern France. Others have followed the lead of Colonial Williamsburg and dressed some of their staff in period costume, a practice which brings with it further dimensions of the problem of representation, creating not only an artificial ambience but also fictitious characters to people it. Some measure of accuracy is introduced when the actors are involved in a craft or occupation, such as firing pottery or baking cornbread, but such events are customarily to be seen at scheduled times only, perpetuating the impression of museum as theatre.

In the majority of museums the buildings have been restored and, where necessary, rebuilt, so that they have an ensured life of a few generations. Their crisp edges, tight bonds and joints, colour-washed walls, unimpaired brick and stonework, painted doors, and sharp mouldings are often at variance with the condition of similar buildings standing beyond the museum enclosure. The exteriors bear no evidence of the dirt and damage of cattle, draught oxen and horses, of lurching loads and lumbering carts which leave their impressions on the buildings of working farms. Across the threshold, the interiors are usually immaculate, the tiled floors, the treen and pewterware, ceramics and hardware, plain and painted furniture, clean and polished. Devoid of mud and grease, they frequently present an idealized and antiseptic version of the peasant environment.

This distancing from the hardship and, not infrequently, the discomfort of rural life is made still more detached by the inhibitions against entering rooms or handling of objects. Suspended ropes, even glass walls and doors, separate the visitor from some of the living spaces, which remain mute, inflexible and undisturbed. Though the museum is not one of conventional glass cases, this curatorial fixation persists. Protection of the artefacts against damage or theft is, of course, the most readily offered reason – the objects in the Frilandsmuseet, Lyngby, for instance, being wired into position and alarmed. Moreover, there are serious problems of safety for both objects and visitors, as indicated by the fire at Arnhem in 1970 and, even more drastically, the fire at the museum of Sanok, Poland, in 1994, when thirteen large buildings and 140 exhibits were destroyed.[23] At the opposite extreme, it is unusual for a museum to be the outcome of a disaster, but such is the case with the Buried Village of Te Wairoa, Rotorua, New Zealand, a museum of a Maori settlement which was engulfed by a volcanic eruption in 1886 and subsequently recovered, Pompei-like, by excavation.[24] Here is one museum whose still emptiness is the essence of the site.

Departures from the hygienic standards shared by the majority of museums are rare. At Arnhem a single exhibit, a farmhouse from Krawinkel, has been re-erected as it was found: 'Beyond the smart facade, the old huddle of buildings, lean-to's and sheds remained unaltered.'[25] But the smell of the fodder that once boiled in the fireplace is undetectable, and even the dung-heap seems sanitized. Of listing walls, loose joints, peeling plaster, and rotting timbers there is little to be seen; verisimilitude in representation does not extend that far. A more remarkable exception is the museum at Howick, New Zealand, an open-air museum which is a recreation of a 'fencible' (defendable) military settlement of the 1840s.[26] More than thirty buildings have been recovered, largely due to the efforts of the Howick Historical Society and, as at Maihaugen, the initiative of a local dentist. He sought to keep every building in the condition in which it was obtained, with layers of soiled and peeling wallpaper and paint, broken and rusty locks, nails as wall hooks, and splintered floorboards. The society was in conflict with regional authorities who, with the safety of visitors in mind, wished to avoid infection, contamination, or minor injuries. Apparently, this was not considered to be a problem at Cuzals, where the farmhouse, the cattlebyre, and the cowman's room within it were left as they were found, in a welter of dishevelled clothing and bedding, with mud-caked boots and socks in the manure on the earth and cobble floors, so that the living presence of the former occupants was almost tangible. This is rarely the case: objects are displayed with the meanings and aesthetics of the mid-twentieth century; with qualities of orderliness, cleanliness, purity and simplicity clearly apparent. Restoration of the buildings to their presumed genotypes, the precision and refinement of the skills applied to them, the undisturbed interiors with their ascetic frugality and perfection in the placement of artefacts, speak of the values projected on them rather than the values of their former inhabitants.

Guided and Misguided

To take the visitor through their collection, virtually every open-air museum publishes information. This may simply be in the popular twice-folded A4 leaflet form by which most museums attract browsers in hotel lobbies and information centres. Couched in phrases that evoke nostalgia and sentiment, awareness of history, and, above all, regional and national pride, they are published in their tens of thousands. Most museums publish a separate and more fully descriptive guide, though they vary in the amount of information they carry. All buildings are numbered as if they were in a display case, and in so presenting them a circulation route is implied. In most guides they are illustrated with photographs, now in colour and with current graphic styles, though less frequently with the perspectives, isometrics, or other orthographic projections which formerly conveyed architectural information. The imagery employed indicates the museum's expectations of the market, which may now be larger than in the 1970s. Arnhem, which succeeds in including photographs, sketches and plans in its guide, claimed in excess of half a million visitors in 1990.[27]

Some museum guides use reproductions of prints, plans, sections and projections of selected buildings, and sometimes pictures that record both the buildings *in situ* and after removal. The guide to the Australian House Museum, Geelong, Victoria, uses most of the above, though its drawings are sketch isometrics. Early photographs and reproductions of documents and media reports give it a distinctive character.[28] Texts may range from the elementary to the informed, from a catalogue to a history, or a carefully edited combination of these. As few visitors read the guide in detail while entering buildings, a summary text is helpful. Inevitably, the approach varies with the type of museum and the expected clientele. Cover photographs tend to emphasize the bucolic, in lands where the sun always shines. The layout may be lively while the typography remains conventional, signifying both open-mindedness and seriousness of purpose. In some cases booklets or game sheets are available for children, with line drawings to colour, items to name, or blank pages for their own drawings. More frequently, these are issued to assist teachers with their classes, for school groups constitute the largest bodies of visitors to some museums, though adult bus tours are also numerous. The attention span of such groups is frequently brief, and many museums produce a basic map guide issued with the ticket to simplify the circuit.

To offset the impression that the simple guides can give, it is important for many museums to underpin their work with publication of their research. In part this is to establish their credibility with conventional museums, many of whom, after a century, still regard open-air museums of building as dubious. Academic papers on archaeological surveys, specific building types, technical problems, and socio-historical accounts are numerous. Some publish monographs like those of the Stichting Historisch Boerderij-Onuderzoek, the research wing of the museum at Arnhem, while others may have a regular publication, such as *Haz Ès Ember* (*House and Man*), the yearbook of the Hungarian Open Air Museum, Szentendre.[29] Though some independent museums publish such studies, Old Sturbridge Village Museum producing its 'Booklet Series' since the 1950s, they are typically the productions of museums that have institutional, provincial or national interests behind them.[30] This reflects both the economic and the academic bases of these museums. But it is also indicative of the motivation on the part of the latter to reinforce the regional and historical provenance of the buildings, promoting identification with former folk communities. Sometimes this is overtly expressed, as in the case of the Orkresne Vlastivedne Muzeum (the District Motherland Museum), Stara Lubovna, in the Slovak Republic.[31]

In the corner of a display of agricultural equipment, and close to the bakery at Cuzals, someone had posted a notice:

> QUELLES SONT LES DIFFERENCES ENTRE UNE ÉCOLE, UNE PRISON, UN ZOO ET UN MUSÉE DE PLEIN AIR?

Freely translated, it reads 'What are the differences between a school, a prison, a zoo and an open-air museum?'[32] Whether the question was written by a perplexed visitor, by a teacher for her class, by a doubting museum employee, or a member of the management could not be ascertained; Cuzals was hardly burdened by officials. Behind the rhetorical question, it seemed that there was a recognition of the didactic, idealistic, puritanical, contained, herded, classified, near extinct/surviving, wanted/unwanted, re-presented and represented features which are common to the forms and philosophies of all open-air museums. In effect, the question may well be asking by implication: is history as it is taught in open-air museums, also bunk?

Open Air and Overseas

The foregoing observations have been based on visits to open-air museums in over twenty countries and more than a dozen US states, though for this chapter the examples cited have been more limited. After a century of existence the nature and purpose of circumscribed and gated museums of buildings need to be re-examined. Such places can, and often do, serve useful purposes, but there is too little agreement on their functions, the basis of the selection of buildings, the degree to which they should attempt to evoke a sense of place and period, the planning of the environments in which they are sited, and the movement and freedom of visitors. The nature and quality of the publications produced, the research which they permit or encourage, and the extent to which they serve local or national political interests are frequently constrained. This last condition is clearly evident at the museum of Chinese Regional Architecture at Guilin, whose imposing entrance, lined with the flags of many nations, underscores its propaganda role. The entrance building on its plinth, approached by a full-width flight of steps, leads to a geometrically formal garden around which are newly built and austere versions in replica of the indigenous architecture of some of the peoples of south China, including the Dai and Hani. There has been no attempt at the relocation of existing traditional buildings, nor any evidence of how they are situated in their authentic contexts.

Similar issues become highlighted when the small but growing number of open-air museums in the industrializing and developing world are considered. For example, the differences in quality and objective become readily apparent when the Museum of Traditional Nigerian Architecture at Jos is considered. This extensive and remarkable museum was the inspiration of Professor Zbigniew Dmochowski of Gdansk Technical University, Poland. His detailed research in the 1950s and 1960s led to his attempting to represent the diverse architecture of Nigeria's peoples on the 65-acre site. With a few exceptions, he chose less to preserve than to replicate, and he greatly benefited by the employment of indigenous craftsmen to build the houses.[33] Though less thoroughly documented, Bomas of Kenya, a museum of traditional building near Nairobi, was also built by the members of tribal groups, the dwelling types being current, though less monumental than many that were reconstructed at Jos. Both museums were created by motivated expatriates, but the Village Complex in the National Handicrafts and Handlooms Museum, Delhi, was built for the Asiad '72 Exhibition. Representing some fifteen regions, its huts being 'built in facsimile with authentic construction by the villagers themselves. In every hut items of day-to-day life are displayed in order to recreate the cultural contexts in which such objects were actually used.'[34] The modest administrative building was designed by the Indian architect Charles Correa. More contentious, however, is the regular plan to which the displayed huts and houses conform.

Though construction by indigenous builders is less likely to misrepresent traditional structures, the Nicobarese builders at Delhi's Village Complex were obliged to reduce their traditional meeting house to an eighth of its customary volume. Such distortion of scale or quality is also evident at Taman Mini, Jakarta, where each Indonesian province has been required to provide a representative building. Thus, Bali built a typical domestic compound, while Central Sumatra chose the 'aristocratic' Minangkabau house with its rich carving and painting and its collection of regal wedding costume. Sulawesi was represented by the Toraja, who produce one of the most spectacular building types of any vernacular tradition. They had no need to embellish or exaggerate, though they did miniaturize their grave figures by building a rock shelter in perspective. But northern Sumatera, represented by the Batak Toba, built an immense and over-scaled version of their domestic house, doubtless to rival other regions. Both the Indonesian Taman Mini and the Indian Village Complex are well patronized by their own nationals as well as by tourists. The management of the Delhi Museum hope that their complex 'will symbolize the urgency for the preservation of rural technology and traditional aesthetic values in rapidly industrializing India.'[35]

Already the small but growing number of open-air museums in industrializing countries are encountering the problems of re-presentation and representation; these problems are not necessarily the same as those in Western museums, but they are clearly comparable. As their numbers increase, issues of intention, of veracity, guidance and publication will become more problematic. Associations of museums may be instituted, and advice based on the experience of a century in the West may be sought. Whether or not this is the case, the time is surely overdue for a serious reconsideration of the functions and forms of open-air museums. Many are deceptive, their invented, rather than authentic, environments being based on sentimentality, nostalgia, and the falsification of 'life' in selected periods by sanitizing, insulating and idealizing the buildings and their contents. By rejecting the changes that have been wrought by generations, by deleting all references to the modifications and influences of subsequent periods, they fossilize them in mythical time.

There is little doubt that many open-air museums are popular, and many are used for such educational purposes as allowing schoolchildren to 'step into the past'. It may be argued that by so doing they bring history alive and make future citizens aware of their heritage. It may also be contended that by preserving vernacular domestic and functional buildings, they are helping to generate an understanding of regional architecture which will ensure respect for it in the future. Such was not the case in Romania, however, where one of the most respected of museums and oldest of scientific academies concerned with conservation was powerless to prevent the destruction of whole villages in the name of progress. In fact, the open-air museum may be counterproductive in this respect. By saving 'representative' examples of the vernacular for contemplation in idealized surroundings, they can relieve those intent on the destruction of such buildings in the world beyond the museum's walls of any responsibility to protect them. Again, by creating 'a picture' that associates vernacular buildings essentially with the past, they also isolate them from the modern world, an isolation which is emphasized by their enclosure within the museum confines. As many of the buildings are presented or selected for their uniqueness or special qualities of scale, craftsmanship or decoration, by implication they low-rate the simpler and less singular buildings which nevertheless contribute unspectacularly to their environmental contexts.

These considerations are important with regard to the European and North American open-air museums, whose continued existence in their current forms and serving their present purposes is seriously open to question. They are even more important in the developing world where self- and community-built buildings based on vernacular traditions will continue to be essential if the housing demands of expanding populations are to be met in the twenty-first century. Rather than romantic depictions of old buildings with invented histories in spurious and contrived settings, vernacular architecture in the developing world needs respect and support, with encouragement given for its continued use of renewable resources, passive climatic modifications, spatial organization based on social structures and scale according to need. Admittedly, this too is part of an agenda, but it is one which, together with instruction in building skills where they are declining and more forethought in the choice of building examples and their presentation, could give new purpose and relevance to the open-air building collections of the future.

Notes

1. N. Zook, *Museum Villages USA*, Barre, MA, Barre Publishers, 1971, p. 55.

2. Information on a conservation notice: Arlington Row, Bibury, Gloucester, National Trust, n.d.

3. J. Czajkowski, 'An Outline of Skansen Museology in Europe,' in Czajkowski (ed.), *Open-Air Museums in Poland*, Poznan, Biblioteka Muzeum Narodowego, Roloictwa, 1981, p. 13.

4. F. Valen-Sendstad, *The Sanvig Collections: Guide to the Open-Air Museum*, Lillehammer, Gjovik, nd, pp. 7-8.

5. R. Ailonen and R. Kinnunen, *Seurasaari Open-Air Museum Visitor's Guide*, Helsinki, National Board of Antiquities, 1986, p. 3.

6. *Netherlands Open-Air Museum*, Arnhem, Netherlands Openluchtmuseum, 1993, p. 7.

7. Czajkowski, 'An Outline of Skansen Museology in Europe,' p. 18.

8. D.C. Giurescu, *The Razing of Romania's Past* (IJS/ICOMOS), London, Architecture and Technology Press, 1989, pp. 20-23.

9. M. Laenen, *Provincial Open-Air Museum, Bokrijk, Guide*, Brussels, Ludion-Lannoo, nd, p. 6.

10. D. Melli, M. Gschwend and C.Schutt, *Guide to the Swiss Open-Air Museum, Ballenberg*, Brienz, Switzerland, 1987, p. 3.

11. M. Cseri and E. Füzes, *Hungarian Open-Air Museum*, Szentendre, 1997, p. 9.

12. S. Chamberain, *A Tour of Old Sturbridge Village*, New York, Hastings House, 1969, p. 4.

13. *The Harold Warp Pioneer Village*, leaflet and map, Minden, NB, n.d.

14. *The Avoncroft Museum of Buildings*, Bromsgrove, Avoncroft Museum, 1973.

15. R. Harris (ed.), *Weald and Downland Open-Air Museum Guidebook*, Singleton, Weald and Downland Museum, 1981, p. 57.

16. For a summary of the approach of museologists to vernacular architecture, see A. De Jong, 'Museological Approach,' in P. Oliver (ed.), *Encyclopedia of Vernacular Architecture of the World*, Cambridge, Cambridge University Press, 1997, pp. 49-52.

17. *Cuzals: Musée de Plein Air Du Quercy*, Sauliac-sur-Célé, Lot, France, n.d. This leaflet is the substitute for a guide to the museum.

18. Phrases from guides noted here are drawn from quotations previously cited.

19. J. Hendry, Personal communication.

20. K. Uldall, *Frilandsmuseet: the Open-Air Museum*, English Guide, Lyngby, Denmark, Nordlundes Bogtrykkeri, 1972, p. 2.

21. H. Kaiser and H. Ottenjann, *Museumsdorf Cloppenburg, Niedersächsisches Freilichtmuseum*, Cloppenburg, Museumsdorf, 1997, pp. 8-10. This 200-page guide is one of the most comprehensive in Europe.

22. Laenen, *Provincial Open-Air Museum, Bokrijk, Guide*, p. 37.

23. J. Czajkowski, *Park Enograficzny W. Sanoku W. Ogniun*, Sanok, Sierpien, 1994.

24. *The Buried Village of Te Wairoa*, guide leaflet, Rotorua, Smith, n.d.

25. *The Netherlands Open-Air Museum*, p. 113.

26. *Howick Historical Village*, Auckland, Howick and District Historical Society, 1995. This leaflet substitutes for a detailed guide

27. *The Netherlands Open-Air Museum*, p. 9.

28. F. Campbell, *Guide to the Australian House Museum*, Victoria, Deakin University, Geelong, nd.

29. C. Miklos and F. Endre, *Haz Ès Ember*, Yearbook no. 11, Szentendre, Hungarian Open Air Museum, 1997.

30. E.P. Hamilton, *The Village Mill in New England*, Sturbridge, Old Sturbridge, Inc., 1964. Some twenty-four titles were published in the Booklet Series by 1964, some on the collections of material culture.

31. J. Stika and J. Langer, *Ceskoslovenska Muzea V Pirode*, Vydalon Vydavatelstvo Osveta Martin, 1989. Lists all curatorial, ethnological and architectonic authorities attached to open-air museums in the former Czech Republic.

32. Noted at Cuzals Museum, July 1998.

33. Z.R. Dmochowski, *An Introduction to Nigerian Traditional Architecture*, three vols., London and Lagos, Ethnographica and National Commission for Museums and Monuments, l990.

34. J. Jain, *National Handicrafts and Handlooms Museum*, Delhi, Crafts Museum, c.1989, p. 4.

35. *Ibid.*

Bibliography

Anon, *Cuzals: Musée de Plein Air Du Quercy*, Sauliac-sur-Cele, Lot, France, n.d.

Anon, *Howick Historical Village*, Auckland, Howick and District Historical Society, 1995.

Anon, *Netherlands Open-Air Museum*, Arnhem, Netherlands Openluchtmuseum, 1993.

Anon, *The Avoncroft Museum of Buildings*, Bromsgrove, Avoncroft Museum, 1973.

Anon, *The Buried Village of Te Wairoa*, guide leaflet, Rotorua, Smith, n.d.

Anon, *The Harold Warp Pioneer Village*, leaflet and map, Minden, NB, n.d.

Ailonen, R. and R. Kinnunen, *Seurasaari Open-Air Museum Visitor's Guide*, Helsinki, National Board of Antiquities, 1986.

Campbell, F., *Guide to the Australian House Museum*, Victoria, Deakin University, Geelong, nd.

Chamberain, S., *A Tour of Old Sturbridge Village*, New York, Hastings House, 1969.

Cseri, M. and Endre Füzes, *Hungarian Open-Air Museum*, Szentendre, 1997.

Czajkowski, Jerzy, 'An Outline of Skansen Museology in Europe,' in Czajkowski (ed.), *Open-Air Museums in Poland*, Poznan, Biblioteka Muzeum Narodowego, Roloictwa, 1981.

Czajkowski, Jerzy, *Park Enograficzny W. Sanoku W. Ogniun*, Sanok, Sierpien, 1994.

De Jong, A., 'Museological Approach,' in P. Oliver (ed.), *Encyclopedia of Vernacular Architecture of the World*, Cambridge, Cambridge University Press, 1997.

Dmochowski, Zbigniew R., *An Introduction to Nigerian Traditional Architecture*, three vols., London and Lagos, Ethnographica and National Commission for Museums and Monuments, l990.

Giurescu, D.C., *The Razing of Romania's Past* (IJS/ICOMOS), London, Architecture and Technology Press, 1989.

Hamilton, E.P., *The Village Mill in New England*, Sturbridge, Old Sturbridge, Inc., 1964.

Harris, R. (ed.), *Weald and Downland Open-Air Museum Guidebook*, Singleton, Weald and Downland Museum, 1981.

Jain, Jyotindra, *National Handicrafts and Handlooms Museum*, Delhi, Crafts Museum, c.1989.

Kaiser, H. and H. Ottenjann, *Museumsdorf Cloppenburg, Niedersächsisches Freilichtmuseum*, Cloppenburg, Museumsdorfd, 1997.

Laenen, M., *Provincial Open-Air Museum, Bokrijk, Guide*, Brussels, Ludion-Lannoo, nd.

Melli, D., M. Gschwend and C.Schutt, *Guide to the Swiss Open-Air Museum, Ballenberg*, Brienz, Switzerland, 1987.

Miklos, C. and F. Endre, *Haz Ès Ember*, Yearbook no.11, Szentendre, Hungarian Open Air Museum, 1997.

Stika, J. and J. Langer, *Ceskoslovenska Muzea V Pirode*, Vydalon Vydavatelstvo Osveta Martin, 1989.
Uldall, K., *Frilandsmuseet: the Open-Air Museum*, English Guide, Lyngby, Denmark, Nordlundes Bogtrykkeri, 1972.
Valen-Sendstad, F., *The Sandvig Collections: Guide to the Open-Air Museum*, Lillehammer, Gjovik, nd.
Zook, N., *Museum Villages USA*, Barre, MA, Barre Publishers, 1971.

Chapter 8

Making the Nation: The Politics of Heritage in Egypt

TIMOTHY MITCHELL

One of the odd things about the arrival of the era of the modern nation-state was that for a state to prove it was modern, it helped if it could also prove it was ancient. A nation that wanted to show it was up to date and deserved a place among the company of modern states needed, among other things, to produce a past. This past was not just a piece of symbolic equipment, like a flag or an anthem, with which to organize political allegiance and demonstrate a distinct identity. As every recent theorist of nationalism has pointed out, deciding on a common past was critical to the process of making a particular mixture of people into a coherent nation.[1]

The idea of the nation presents a way of living the locality of social relations by imagining them to extend back over a continuous period of time. The political community can then understand its present historically. This projection into the past may help make the present seem natural, disguising some of the arbitrariness, injustice and coercion upon which it depends. Historical thinking achieves this not just by projecting a past, but by organizing that past as the life of a self-directing object – the 'nation' or 'society'. Contemporary political arrangements then acquire a degree of inevitability by appearing as the genetic destiny of this historical being.

Recent writings on nationalism have also pointed out that to produce a past a state has to produce a place. If the nation depends on extending present social relations back through time, this can only be done by defining their geographical extent. The self-contained coherence of a society developing through history depends upon fixing the geographical domain that gives the particular society its limit. Benedict Anderson has argued that the idea of the nation came about when modern forms of writing enabled the social worlds of individual citizens to expand. For example, innovations such as the modern novel and newspaper enabled people to imagine unknown others as members of the same community.[2]

Yet in many parts of the world the idea of the nation required people not only to expand their sense of community in novel ways, but in equally novel ways to constrict it. People's sense of religious community or tribal cognation, their networks of trade and migration, communities of learning and literature, and patterns of imperial power and allegiance were in many places much more diverse than the narrow boundaries of modern nation states. Ernest Renan famously remarked that the idea of the nation required that people learn to forget certain aspects of their past.[3] Many people also had to learn to forget, or at least to reconsider, their sense of place. They were supposed to reduce the significance of those interconnections, exchanges, genealogies, hegemonies, moral systems, and migrations that had defined a social landscape whose horizons reached beyond what became the new boundaries of the nation – or even to forget their existence altogether.[4]

Let me illustrate these questions of how a nation's past is made in the case of modern Egypt. One might suppose that the lower Nile valley, compared to many other parts of the world, offered a well-defined geography within which to imagine a self-contained society. It should have been relatively easy to picture Egypt as a self-sufficient nation, to minimize the wider relations people may have had with other regions, and to give its particular mixture of communities a singular and self-contained past. Moreover, the survival of monuments from more than 5,000 years before – indeed, the powerful image of 'ancient Egypt' as the cradle of civilization – would seem to offer modern Egyptian nationalism a neat and uncontroversial way to lay together superincumbent images of people, place and past.

Yet constructing the past is never so straightforward. In the first place, ancient monuments do not automatically belong to one's own past. As someone from England, I can admire the imaginative scale and ancient precision of Stonehenge, but I cannot feel those stones as part of my own past. In order to belong to one's history, monuments must connect with some aspect of one's social identity. Something similar seems to be true of the way the monuments of ancient Egypt figure in the politics of Egyptian nationalism. Periodically, efforts have been made to present the Pharaonic past as a source of modern Egyptian national identity. The idea that modern Egypt is a society whose ancestry goes back in a continuous line to a Pharaonic beginning is also the view of the nation's history one finds in Egyptian school textbooks. However, such uses of the past have generally been of limited political use in the country's modern politics.

The most sustained effort to invoke the glories of ancient Egypt as the source of modern Egyptian identity came in the second quarter of the twentieth century following the discovery of Tutankhamun's tomb in the Valley of the Kings, near Luxor, in 1922. When the British archaeologist Howard Carter unearthed the riches of the first royal tomb to be found intact in modern times, the event attracted worldwide attention. The discovery also occurred in the same year in which Egypt won partial independence from the British military occupation established in 1882, and provided the new nationalist government with a powerful expression of the nation's identity. The government refused to allow the British archaeologists to take possession of 50 per cent of the discovered treasure, the practice followed with earlier finds.[5] Its determination to keep control of the treasure provided a useful demonstration of the country's newly acquired autonomy.

Beyond this incident, however, the Pharaonic past has played only a minor and diminishing role in Egyptian nationalism. For another decade or so after the discovery of King Tut's treasures, a group of conservative writers with cultural ties to Europe continued to insist on the significance of the nation's Pharaonic origins. But they did so as part of an argument against Europeans who insisted on the Oriental and therefore backward character of Egypt, and against local intellectuals who insisted on the exclusively Islamic character of their society. Their concern was to show that Egypt was a modern, Western nation, a view to be proven by the fact that the West's own past lay within Egypt. Yet the significance of the past for these writers was not so much that it gave the nation a distinct and authentic identity, but that it showed that the nation belonged to the larger community of modern civilization. The role of the past was to serve as a sign of the modern.[6]

In the same period a right-wing populist party, Misr al-fatâh (Young Egypt), also began to emphasize the importance of the Pharaonic past, finding there an expression of its belief in leader worship, militarism, and an Egyptian imperialism stretching from the Mediterranean to the equator. However, this, too, was short lived. By the 1930s most political argument in Egypt had reverted to themes that connected more readily with people's everyday experience and self-conception, principally the themes of Islam, Arabism, and anti-imperialism. These political identifications did not necessarily refer to the confines of the Nile valley, and gave local politics a much wider resonance than a purely Egyptian nationalism.[7]

The difficulties and ambiguities in the production of a nation's past can be more fully understood if one shifts from the history of nationalism, as it is conventionally written, to a political process I would call 'making the nation'. I find it useful here to think in terms of Homi Bhabha's distinction between the nation as pedagogy and the nation as performance.[8] The history of nationalism reconstructs the more or less coherent story of how the nation emerges as a pedagogical object. It pieces together the official nation that is invoked in the ideology of political parties, the propaganda of government programmes, the rhetoric of school textbooks, the memoirs of public figures, and the news reporting and opinion-making of the mass media. These sources constitute the formal archive examined by any standard history of the emergence of twentieth-century Egyptian nationalism.[9] What such an account generally overlooks is the more mundane and uncertain process of producing the nation. I have in mind the variety of efforts, projects, encounters and struggles in which the nation and its modern identity are staged and performed. The difference between performance and pedagogy is not a question of looking at the practical rather than the ideological, or the local rather than the national. Both involve the making of meaning, and both take place in particular sites among particular parties. What is different about making or performing the nation is that it always involves the question of otherness.

In the nation as pedagogy, the emergence of the national community is understood as the history of a self that comes to awareness, or of a people who begin to imagine their collectivity. History takes the form of the growing self-awareness or imagination of a collective subject. This imagination is thought to develop through a gradual revealing of the subject to itself, shaped by those powers of communication, reason and consciousness that define our understanding of an emergent self. There is no encounter with otherness, except as part of the general discovery of a world beyond the self. In the performative making of the nation, on the other hand, otherness plays a constitutive role. The nation is made not out of a process of self-awareness, but out of encounters in which this self is to be made out of others – or rather, is to be made by making-other. The nation is made out of projects in which the identity of the community as a modern nation can be realized only by distinguishing what belongs to the nation from what does not, and by performing this distinction in particular encounters. Unlike conventional accounts of the emergence of the nation as pedagogy, understanding of such encounters cannot be governed by the consciousness of a collective subject that produces the meaning of the nation; this collective subject – the nation – is not the author of the performance, only its occasional, unstable effect. Moreover, one can bring into view the forms of difficulty, uncertainty, violence and subversion that the making of the nation can involve.

In this chapter I take as examples of this process two episodes from recent Egyptian politics. One is a campaign launched in the 1930s and 1940s to define and preserve an indigenous cultural heritage, pursued through a struggle to create a distinctively Egyptian vernacular architecture. The other is a dispute over the protection and presentation of the heritage of ancient Egypt, in particular the Theban Necropolis where Howard Carter earlier unearthed the treasures of King Tut. In 1945 these two different efforts to produce and defend a national heritage came together in plans to demolish and rebuild a village in southern Egypt. In the 1990s, more than half a century later, the village remained the site of an unresolved struggle over the question of national heritage.

Making the Nation in the 1940s

In 1945 the Egyptian government commissioned the Cairo architect Hassan Fathy to design and build a model village to rehouse the inhabitants of the village of Gurna, which lies on the west bank of the river Nile opposite the town of Luxor, 400 miles south of Cairo. The village consists of a string of hamlets stacked along the desert escarpment beyond the sugarcane fields at the valley's edge, amid the ancient rock-cut tombs and temples known to archaeologists and tourists as the Theban Necropolis. A year or two earlier the Department of Antiquities had been embarrassed by the removal of an entire wall of one of the tombs under its guard. Needing a scapegoat, it blamed the local inhabitants and demanded their removal.

Fathy designed the model village of New Gurna using mud-brick construction and elaborate vaults and domes. His intention was to develop an affordable and aesthetically pleasing popular architecture. The building of the village marked the birth of a new vernacular style, one that was to become internationally famous for announcing the rejection of Western modernism and seeking to re-appropriate the styles and materials of a local heritage. The event also marked, as Kees van der Spek has noted, the moment of this new vernacular's untimely death.[10] 'The Village', as locals still refer to Fathy's project, was constructed over the following three winters in the corner of a large, irrigated tract of sugarcane fields. In 1948, with only a fraction of the 50-acre village constructed, Fathy was forced to abandon the project, partly because of bickering between government departments, but mostly because one night that winter, men from the old hamlets of Gurna, whose families opposed the planned eviction and resettlement, had cut the dyke and flooded the low-lying village.

Fathy's account of these events, published twenty years later, expressed his disappointment at the failure of his plans 'to revive the peasant's faith in his own culture,' and his bitterness towards the 'suspicious and strict' inhabitants of Gurna who had refused to co-operate and 'were not able to put into words even their material requirements in housing.'[11] It is easy to criticize Fathy today, whether for his paternalism and arrogance towards villagers who stood in the way of his architectural vision, or for the cosmopolitanism that led him to propagate this vision in books published in English and French and admired around the world, but which cut him off from those who preferred to read in Arabic.[12] My concern here is not with Hassan Fathy, however, but with those events in the 1940s in Gurna, where the attempt to define and preserve a national heritage was simultaneously born and destroyed. It is this relationship between producing a national heritage and its subversion that I want to explore. I will examine how the manufacture of the modern vernacular, the attempt to preserve a national culture, as well as the protection of a more ancient, archaeological past, seemed to depend upon a relationship of force and a structure of antagonism. The aim is to learn something larger from the fact that the birth of a national heritage movement in New Gurna was also the moment of its violent demise.

The history of the model village intersects with a continuing effort to present and preserve another national heritage: the monuments of ancient Egypt. The road past New Gurna today is filled with tourist buses, which stop beyond the village at the Colossi of Memnon before proceeding to the Valley of the Kings and other ancient sites. None of the buses stops at the model village, which is barely visible behind the shops and tourist signs that line the street (figure 8.1). New Gurna is a thriving community, but Fathy's houses have been overbuilt with additions, infills, and extra floors (to the extent that domed roofs allow), or in some cases pulled down. Fathy's village school – the one public building immediately put to use fifty years ago as planned – was demolished in the late 1980s and replaced with a larger school built with a concrete frame and baked red brick. The handmade mud bricks of Fathy's original school provided the rubble to make the new building's driveway.

Figure 8.1. The Model Village of New Gurna, built in the 1950s, as photographed in 1999. (Photo by author)

One thing, however, survives intact after more than fifty years: the unfulfilled desire to evict the inhabitants of Old Gurna. After several intervening failures, in the years 1992 to 1994 new plans were drawn up, as part of a master plan for Luxor funded by the United States Agency for International Development, to 'depopulate' the seven or eight hamlets on the Gurna escarpment, from Sawalim in the north to Gurnat Mar'i in the south, plus the neighbouring hamlet of Medinat Habu.[13] Over the following four years new villages again were built, this time located in the desert 5 to 10 kilometres north of Old Gurna, and again the households of Gurna largely refused to move and see their village demolished. On January 17, 1998, after several earlier skirmishes, a government bulldozer accompanied by two truck-loads of armed police moved into Gurna to carry out demolitions. A group of about three hundred villagers gathered, later swelling to several thousand, and drove the police back with stones, pushing the retreating bulldozer into a canal. The police opened fire on the villagers with automatic rifles, killing four and leaving more than twenty injured.[14] This incident set back the relocation plans, but by the end of the same year the head of Luxor City Council, Major General Selmi Selim, confirmed that the plans to depopulate what he referred to as 'nine shanty areas known as Old Gurna' would go ahead as part of a vision to turn the area into 'an open air museum and cultural preserve.'[15] 'You can't afford to have this heritage wasted because of informal houses being built in an uncivilised manner,' he explained to the press.[16]

The major general's use of the term 'uncivilized' to justify the evictions echoed the earlier language of Hassan Fathy. Fathy's account of the events of the 1940s told the history of New Gurna as a story of the progress of culture and intelligence impeded by the ignorance and lawlessness of the natives. The families of Gurna lived mostly as tomb robbers, Fathy said (an accusation to which I will return), and it was to preserve this lawless way of life that they sabotaged the project. (In the plans for the model village there were to be several public buildings, including a theatre and an exhibition hall, intended to create the kind of public spirit that Fathy felt was missing in ordinary villages and was needed if Egypt's vernacular heritage was to be revived; but there was also to be another public building never previously found in villages – a police station.) This violence and lawlessness provided the pretext for building the new village. It was only by addressing the problems of ignorance and the absence of civilization that an architect interested in a programme to create a modern vernacular could find an opportunity to work. However, hidden within the project was a larger violence, not visible in the plans, but making them possible.

As a way to uncover some of this larger violence, I begin with what may seem a minor event in Fathy's account, what he referred to as the 'malaria epidemic' of 1947. He noted in passing that the epidemic 'killed about a third of Gurna's inhabitants,' but he concentrated more on the restrictions imposed on travel from Cairo and other delays the epidemic caused to his project.[17] It seems startling today that Fathy would not discuss any larger objections to uprooting and relocating a community in the midst of such suffering. But in fact there was more to this oversight. Writing twenty years later, Fathy had collapsed together two separate epidemics. And these events were not just an obstacle to his plans but the condition that made them possible.

The 1947 epidemic was actually an outbreak of cholera, not malaria, and affected mostly Lower Egypt. However, three years earlier, in 1942-44, an epidemic of malaria had occurred in the Luxor region, an outbreak of gambiae malaria, the disease's most lethal form, brought from the south by recent irrigation work designed to expand the sugarcane plantations, and by increased wartime traffic with Sudan. It was this earlier epidemic, along with the famine that resulted from wartime food shortages and men too sick to harvest the wheat crop or earn wages cutting cane, that killed more than a third of the people of Gurna.[18] It is estimated that from 100,000 to 200,000 people may have died in Luxor and neighbouring regions, with the heaviest casualties in places like Gurna and other sugarcane plantations, where perennial irrigation enabled the *gambiae* mosquito to reproduce. In May 1944 the manager of a plantation near Gurna estimated that 80-90 per cent of the local population had contracted the disease, and the doctor in the local town of Armant reported 80-90 deaths a day.[19] Even in the 1990s old men in Gurna remembered those times, when there were not enough healthy men even to bury the dead, and corpses were taken to their graves on the back of donkeys.

The *gambiae* malaria epidemic provoked a political crisis in Cairo in 1944-45. Opposition politicians blamed the enormous death toll on the severe poverty of the Luxor region and the rest of the extreme south (Qena and Aswan Governorates), where a handful of owners controlled most of the land in sugar plantations of tens of thousands of acres each, and the majority of the population was landless and worked for starvation wages. A deputy in Parliament argued that living conditions in the Soviet Union were far better. However, the governing Wafd party, which represented the interests of large landowners, was anxious to defuse this radical threat to the principle of landownership. It argued that the cause of the epidemic was not poverty and inequality but the unsanitary living conditions in the villages. Instead of land reform and the redistribution of wealth, it proposed a plan to demolish the country's traditional villages and replace them with well-ventilated, sanitary and attractive model villages.[20] It was in the midst of this political crisis that Fathy negotiated and won government funding for the construction of his model village at Gurna.

The government paid for the purchase of 50 acres of sugarcane land from Boulos Hanna Pasha, who owned thousands of acres in the Gurna region and was one of the largest landowners in Upper Egypt. The 50 acres were to provide space for the village with its generously proportioned houses and its numerous public buildings, a fresh-water pond for swimming, and a public park for recreation – but not a single acre on which to grow food. Rather than open the question of rights to the land, Fathy helped establish a textile workshop, employing twenty child weavers, to provide some income for the village. Later a visiting government official noticed that the children in the workshop 'looked thin and hungry,' and suggested that they be given a bowl of lentil soup every day. 'It was a sensible and practical suggestion,' Fathy admitted. But no money could be found to provide the food.[21] Brushing aside such problems, Fathy saw his village as a pilot project launching a 'National Program for Rural Reconstruction' that would lead 'to the complete regeneration of the Egyptian countryside through rebuilding its villages.'[22]

This approach to social problems, which saw in the recovery of national heritage the recovery of social energy and purpose, reflected the hubris of an architectural politics that was coming to believe in the powers of planning as a means to the construction of new national subjects and an alternative to more popular and effective proposals that threatened the social order of large landownership. Fathy felt that the very participation of villagers in the planning process – a revolutionary idea – would provide the means for them to recover their lost individuality. They would develop into subjects of the nation by rediscovering their power to make decisions. 'Ideally,' Fathy wrote, 'if the village were to take three years to build, the designing should go on for two years and eleven months.'[23] The limitations of this view – the inability to consider that villagers might prefer to stay in the houses they had already designed and built themselves – reflected the fact that Fathy himself came from the Egyptian landowning class (his father was the absentee owner of several large estates).[24] Whatever his own disagreements with this class, all his architectural commissions were connected with such a landowning background. Besides the village of New Gurna, he also built a model farm for the Royal Agricultural Society and rest houses for the Anglo/American-controlled Chilean Nitrate Company – two institutions promoting commercialized, large-scale agribusiness. And most of his earlier architectural designs were country houses for the proprietors of large estates.[25]

If Fathy saw the villages of Gurna as violent, lawless, and unable 'to put into words even their material requirements in housing,' when one puts his project into a larger social context, it is the project that begins to seem violent, and its author who is perhaps 'not able to put into words' its material context. The sugar plantations of the Gurna region had originally been village land. But from the 1860s, the ruling household in Cairo had begun to seize the land, paying little or no compensation, as new irrigation schemes made it possible to prevent the annual Nile flood and plant the year-long cane crop. Then, after the country's Ottoman Turkish ruling household was declared bankrupt by its British and French bankers in 1875 and the British army invaded and occupied Egypt, the foreign bankers managed the estates and then auctioned them off, not returning them to the original village cultivators, but selling them intact to barons like Boulos Pasha. Thus, when Fathy and the government neglected to provide New Gurna with land to grow its own food, or even bowls of lentil soup for child workers, this was not an innocent oversight: it was the continuation of a process of appropriation carried out over the preceding sixty or eighty years through the depredations of a Turkish-Egyptian elite and their British backers. And it was to counter the new threats to this political order, around 1944-45, that men were dreaming up plans for model villages and Fathy was proclaiming the architect's 'unique' ability 'to revive the peasant's faith in his own culture.'[26]

In projects of this sort one can see the difficulties of making the nation. To perform the nation, groups must be included by first declaring them excluded by their lack of civilization, villages destroyed in order to preserve them, pasts declared lost so that they may be recovered. Fathy desired to 'revive' an indigenous culture as a means of developing an Egyptian national heritage. To perform this revival, he needed the people of Gurna – yet he needed them as a community outside the nation, whose elimination would help bring the nation and its past into being. The Gurnawis were to be portrayed as ignorant, uncivilized, and incapable of preserving their own architectural heritage. Only by constructing them in this way would the architect have an opportunity to intervene, presenting himself as the rediscoverer of a local heritage that the locals themselves no longer recognized or understood. As the spokesman bringing this heritage into national politics, the architect would enable the past to speak and play its role in giving the modern nation its character. Thus, the people of Gurna could only enter into national politics by submitting to an act of violence. And to preserve their heritage, the architect had to first destroy it. Old Gurna was to be pulled down and rebuilt – and not just because it was built over antiquities, for if the project succeeded, every other village of Egypt could also be demolished and rebuilt. The preservation of the past required its destruction, so that the past could be rebuilt. Likewise, performing the nation required that every one of its rural inhabitants be declared outside the nation, uncivilized and unhygienic, so that in rendering them civilized and clean, the nation could be made.

One can add an ironic conclusion to the story of New Gurna. The violence of the plantation system, and the large irrigation schemes that made it possible, were the source of the opportunity to build the model village, as an urgent response to the system's crisis in the mid-1940s, one that sought to draw upon the forms and materials of the vernacular. But the irrigation schemes also meant that the fields were no longer flooded, and there was no longer an annual deposit of Nile silt that allowed renewal of the rich alluvial mud out of which Fathy's mud-brick architecture was built. By the 1980s the government was forced to ban the use of alluvial muds for brick-making in an attempt to limit the further loss of fertile soil. The celebration of a vernacular based on centuries of local mud was launched at precisely the moment when (and for the same reasons that) the mud became for the first time in history no longer in renewable supply.

At the same time, the violence and exclusion employed in manufacturing heritage and making the nation also provide moments when things do not go according to the official script. Nation-making is a performance that remains open to improvisation and restaging. Such restagings are not the subversive acts of outsiders, but the imaginative response of those in whose lives the nation is performed. After all, it was thanks to the violent history of the irrigation schemes that the villagers had the power to thwart Fathy's plans: the model village was protected by a dyke, and the dyke could be cut. In more ways than one, the vision of preserving and reviving a vernacular peasant culture was haunted by the violence that made it possible.

Making the Nation in the 1990s

Fifty years later, in the mid-1990s, the government was still trying to evict the population of Old Gurna, and still describing them as lawless and uncivilized. To the old arguments about tomb-robbing, official statements added new claims: that waste water from the houses of Gurna was damaging the tombs; that their 'living conditions are poor, unhygienic, and spoil the view';[27] that the houses blocked access to Pharaonic tombs and impeded the expansion of the tourist industry; and that the presence of this large population in what was now a UNESCO World Heritage-listed site prevented its archaeological preservation and its development as an open-air museum for tourism. From the Ministry of Culture and American development experts in Cairo, to the Luxor City Council and local contractors and tourism investors, there was now a coalition of forces working to transform Gurna into a site that was clean, well lighted and signposted, with wide roads and ample parking, and people-free: in a word, not just an ancient heritage site but a modern one.

Making the nation in the 1990s occurred in a different context from the 1940s, with new actors and audiences. The staging of national heritage now involved new groups of outsiders, including UNESCO, the World Bank, the United States Agency for International Development, international tour operators and tourists, and new forces claiming to speak in the name of the nation, including local officials and entrepreneurs. The alliances and interactions among these forces shaped the local dynamics of the tourism industry and the local production of the nation. The script still spoke of uncivilized natives, whose ignorance and lawlessness threatened a national and global heritage. But once again, the performance of the nation had a more uncertain outcome.

The new plans to evict the population dated back at least to a UNESCO document of 1980, and were formalized in a planning and relocation study carried out in 1992-94. The new relocation site, first identified and surveyed in the 1950s, lay several kilometres to the north of Gurna, beyond the outlying hamlet of Tarif, and had come to be known as New Tarif (figure 8.2). Repeating themes first articulated by Hassan Fathy, the Terms of Reference for the relocation study, funded by USAID, emphasized the need for 'community participation' in the planning process and for detailed architectural, social and cultural surveys of the old village. This now included the making of an ethnographic film about the community that was to be destroyed. Such surveys were undertaken, and the resulting house designs once again represented Cairo architects' interpretation of local aesthetics. They made few concessions to the actual layout of local houses (bathrooms, for example, were designed to be placed inside the houses, and adjacent to the kitchens, rather than in a rear courtyard – where local residents had always placed them, in part because one often has to go for days without running water). And 'community participation' largely took the form of constructing different model houses, made of solitex, plywood, and butterfly cloth, which villagers could visit to select their preferred design.[28] Some of these houses were later built, but a Cairo consulting firm also discovered that the Luxor City Council had handed over 45 per cent of the relocation area to a private contractor, who threw up minuscule and nondescript concrete-and-red-brick boxes. Several hundred villagers eventually agreed to move to these new settlements, in most cases those living in extreme overcrowding (since 1978 the government had banned further building in Old Gurna). However, since villagers were able to exchange one old house for several new ones, only a few dozen old houses

Figure 8.2. Unoccupied housing. New Tarif, 1999. (Photo by author)

were available for demolition. And when the government tried to force other villagers to move, the result more than once was violent resistance, culminating in the riot and shootings of January 1998.

The US Agency for International Development, the World Bank, and the Egyptian government spent hundreds of millions of dollars during the 1990s alone planning and attempting once again the eviction of the people of Gurna.[29] Despite this vast employment of architects, planners, ethnographers, bureaucrats and bulldozers, there was little investigation of the actual need for the evictions or their possible impact. The Terms of Reference for the planning and relocation study, drawn up with USAID funds, called for detailed analyses of the aesthetics, topography, population and culture of Old Gurna, as a means of preserving something of the community that was to be destroyed. But there was to be no investigation of the actual problems these people were said to be creating, which might have raised questions about the need for the evictions and for the employment of so much expertise.

The alleged problems can be examined one by one. First, it was said, the people of Gurna were tomb robbers. This accusation has been repeated so often that even many critics of the eviction assumed it to be true. The image of tomb robbers is a standard element in local media representations of Gurna – from Shadi `Abd al-Salam's famous film of 1969, *Al-Mumiya* (*The Mummy*), to a popular television serial aired during the middle of the struggles in 1996-97, *Hilm aj-Janubi* (*The Southerner's Dream*), whose plot turned on the conflict between an evil tomb robber in the Luxor area and an educated hero who sought to defend and rediscover Egypt's heritage.[30] Occasionally, the authorities reinforced these images by staging a raid on a Gurna house, as in 1996 when Muhammed al-Adhim, 63 years old, came home and found that the authorities had discovered a tomb cut into the rock behind the wall of his great grandmother's bedroom. The tomb was just an empty tunnel, but this did not stop the authorities from arresting the old man, who worked as an assistant in a local dentist's office, and making a public example of him. 'I am completely stunned. I never knew there was a tunnel . . . ,' he said. 'I think the tourist authority just made a balloon to attract foreigners. Tomorrow they will say these slippers I am wearing came from Ramses II.' Tomb robbers, he pointed out, are supposed to make lots of money. 'But can you tell me where is my Mercedes, where is my six-storey house?'[31]

It is no doubt the case that over some two hundred years, until the middle of the twentieth century when international conventions against the trade in antiquities were put in place, the people of Gurna formed a small part of the global networks that moved the treasures of ancient Egypt to the great museums and private collections of Europe and North America. Yet it is curious that those Gurnawis are today considered 'tomb robbers', while it is still difficult to describe the British Museum in London or the

Metropolitan Museum in New York as collections of stolen goods. It is also true that a small illicit trade in Egyptian antiquities still continues, driven by the demand from private collectors in the West. Occasionally, these trading rings are broken; however, news reports then show that the source of stolen goods is almost invariably places such as the Valley of the Kings, where no Gurnawis live, or storerooms and other sites under the control of the government.[32] Anecdotal reports from Gurna suggest the same thing.[33] These problems would be best addressed by measures such as better pay and training for local employees of the antiquities authority, and above all by more vigorous international action against Western dealers. In the early 1990s, for example, scholars on the Cultural Property Advisory Commission in the United States tried to persuade the Egyptian government to ask the United States to ban the importation of artefacts illegally exported from Egypt, but no ban was introduced. But it is easier to demand the eviction of villagers from a hillside in southern Egypt than to investigate how the trade in antiquities is actually organized and run and to collaborate on measures against international dealers. Development agencies, architects, planners and academics can then repeat without evidence the claim that Gurnawis are tomb robbers. I have no idea who in particular is involved in the antiquities economy, and I do not want to make accusations without evidence. My point is that the planners and developers have no idea either, which means that it is not clear what impact the evictions would have on the alleged problem of illegal trade.

It is possible that it could make the problem worse. The official image of the people of Gurna – as destitute, uncivilized, and poverty-stricken peddlers and dealers – conceals the complex nature of the local economy. There have been cursory attempts to study this economy, by the World Bank and other agencies. Such studies have, however, been narrowly based on local government records and formal questionnaires, so typically they have uncovered only a small part of people's economic lives – and very little of the complicated relationship between the villagers and the heritage industry.

I will give one example of this complexity. In 1978 the heritage managers imposed a ban on any further building in the antiquities zone. They also required that all buildings be of mud-brick construction. This meant that villagers who extended their houses, or added a concrete frame inside to support a third or fourth storey, or tried to open a tourist shop in the ground floor, were required to make a series of payments, often amounting to tens of thousands of pounds, to a variety of local officials to persuade them to make exceptions to the law. Such 'variances', as they are called in the more regularized practice of the United States, continue to represent an unofficial, yet pervasive element in the local economy. To depopulate the region would cause this system of payments to collapse. Those who campaign against the evictions in support of the villagers often mention how the villagers depend upon the local heritage industry for their incomes. But the financial relationship also operates in reverse. If the depopulation of Gurna were to deprive local officialdom of a major source of unreported income, might another source of income, the antiquities trade, have to expand in compensation? I do not know, nor does anyone else. But it is certainly plausible that the depopulation would have the opposite effect to the one intended.

If one turns to the second, and more sophisticated, set of arguments for removing the villagers from the heritage site, one finds that once again things may be closer to the reverse of what is claimed. Whether or not they are robbing its tombs, it is argued, the villagers of Gurna are damaging the Theban Necropolis by their very presence. 'Living conditions are poor, unhygienic, and spoil the view,' the authorities claim, and, more seriously, the waste water from the Gurna houses is damaging the surrounding tombs. Houses built over tombs, moreover, prevent the development of tourism.

Again, it is not clear what the evidence is for these claims. The hamlets of Gurna are not allowed to have running water or to dig wells. They must fetch all the water they need in wheeled oil drums pulled by donkeys. The only running water on the Theban hillside is in the accommodations of the European archaeological missions. While moisture damage is a problem, there has been no comprehensive geological survey of the Gurna site, with its alternating layers of Theban limestone and Esna shale, to assess the impact of habitation (versus, for example, the impact of the general raising of the water table and humidity levels since the building of the Aswan High Dam, or the increasing number of flash floods attributable to global climate change), or to identify which locations can support human occupation without damage to the tombs.[34] Once again, despite the hundreds of millions of pounds spent on outside consultants, these basic studies have not been done.

Moreover, while it is true that a handful of ancient tombs of Old Gurna have houses built over their entrances, there are many hundreds of other tombs that are not concealed by houses, yet have not been opened up to tourism. Some of these are used by the authorities for other purposes: for example, as storerooms. The tombs that are concealed by the Old Gurna houses the authorities want to demolish are arguably better off than all the rest. First, while tombs of no archaeological significance may be simply cave-like extensions of the houses built against them, the few of archaeological merit are closed off from houses by interior gates and controlled by the antiquities department. This relationship between household and tomb may represent a more historically interesting aspect of local heritage than many of the empty tombs cleared out and opened up as tourist sites. Indeed, one or two archaeologists working in the area have started to dig not in uncleared tombs but in the piles of debris cleared out by earlier excavations. Earlier excavators were interested only in Pharaonic treasure, or at most in the art and artefacts of the Pharaonic period. Yet many of the tombs served as human habitations over subsequent centuries, and the debris of earlier excavations contains rich evidence of this long period of Coptic and early Islamic local life. The communities living among the tombs today may date back a mere four of five hundred years, but the relationship they represent between a dead past and a living community is arguably an integral part of the history of the Theban Necropolis (figure 8.3).[35]

Figure 8.3. A hamlet of Old Gurna, 1999. (Photo by author)

It is also important to consider the likely consequences of the larger aim of the depopulation programme, whose goal is an enormous increase in the numbers of tourists visiting the area. While no studies have been done on any actual damage the villagers of Gurna do to the archaeological sites, there is detailed information about the damage that tourists do, and especially the damage done by tourists' waste water. If a typical tomb in the Theban Necropolis is occupied by twelve visitors, in one hour their sweat increases the relative humidity by an average of 5 per cent. At the peak of the tourist season, as many as 4,500 tourists visit the Necropolis every hour. More than one-third of them, between 1,500 and 2,000, visit the three most popular tombs, causing the humidity in them to increase by up to 100 per cent, a level at which one-fifth of the wall painting can be lost. While villagers can be denied running water to reduce the problem of waste water, there appears to be no equivalent way to stop tourists from sweating. The master plan for Luxor, of which the depopulation of Gurna is a major part, envisions quadrupling the number of tourists within twenty years, from the rate of one million each year in the 1990s to four million. And just about every one of those three million extra visitors will want to squeeze themselves, dripping with perspiration, into and out of the tombs of Gurna. Far from eliminating the problem of waste water, the plans for Gurna are going to add to it significantly.

Considering these potential impacts, therefore, it may be more appropriate to view planners and officials the way Hassan Fathy wanted people to see the villagers, as unable to put certain things into words. What current official silence seems to hide, and thus gives away, is the destructive and unsustainable plans for the local development of tourism. In the 1940s archaeologists and government officials claimed the depopulation of Gurna was required to preserve the nation's archaeological heritage. In the 1990s depopulation has been linked not just to arguments about archaeological preservation, but to demands to create a proper tourist experience. National heritage is now to be shaped by the forces and demands of a worldwide tourist industry. Yet once again these forces have been open to local forms of subversion.

The tourism developments of the 1990s go back to 1982, when the World Bank hired the US consulting firm Arthur D. Little to draw up a programme for increasing tourism revenue in Luxor. The consultants revived the proposal for the depopulation of Gurna (and also revived another of Fathy's proposals from half a century earlier: the establishment of a co-operative to improve the quality of locally made souvenirs). According to the consultants, the increase in tourism revenue once the local population was removed would come from a number of areas: better visitor management (new roads, bus parks, and other facilities) to increase the flow of tourists; a new airport terminal across the river in Luxor; and increased water and electric supplies and other infrastructure to enable the development of the luxury sector of four- and five-star hotels and Nile cruise ships. Since there was a limit to the number of tourists who could be squeezed each hour in and out of King Tut's tomb, the consultants took the approach that income growth would come partly from a shift toward wealthier tourists.

The funds for the development programme itself have also had little local impact. World Bank documents show that more than half the budget for the 1980s development projects – $32.5 million out of a total of $59 million – was to be spent abroad, to pay for foreign contractors, consultants and equipment.[36] The Egyptian government borrowed this

$32.5 million from the World Bank, following the usual practice with Western development assistance to countries like Egypt according to which the assistance is actually money paid by Egypt to the West.[37] The balance of the budget, the local costs, were paid directly by the government. The contractors building the new roads initially agreed to employ some of the villagers. But as Ayman, one of the villagers who did this work, later told me, when they discovered that for more than twelve hours a day of heavy labour they were to be paid only 5 pounds (about $1.50), they refused to work, and the contractors brought in cheaper labour from other, more impoverished villages.[38] The major local construction contract, building a new river embankment at Luxor, which might also have employed local workers, was awarded (following perhaps the World Bank's new free-market principles) to a military workforce from China.

These investments made possible a phenomenal growth in tourism over the following decade. From 1982 to 1992 the number of visitors to Egypt and their estimated expenditures more than doubled (although attacks by Islamic militants caused numbers to dip again in the 1990s).[39] In Luxor most of the growth, as planned, was in luxury hotels and cruise ships. Across the river in Gurna, those who had established small hotels or other tourist enterprises before the development ban was imposed also did well. And they typically put their profits into importing small air-conditioned tour buses from Germany, or buying land and putting up apartment buildings in Luxor.

But for many villagers there was almost no way of breaking into the tourist business, except for those who found unskilled hotel jobs in Luxor at below-subsistence wages. A few dozen young men did better, by finding a foreign tourist to marry – usually a much older woman, who would visit each winter for a few weeks and, with luck, would be wealthy enough to set the husband up in business. One woman, an enterprising Californian divorcée named Happy, began to build a small hotel on the edge of the desert south of the Theban Necropolis. The building was stopped by the authorities, of course, and after six years and many payments the hotel was still not quite finished. Most of the husbands have settled for something less, such as an imported car to run as a tourist taxi. Cruising past those working in the sugarcane fields in their air-conditioned Peugeots, these men seem to underline the separation of the tourist world from the village.

The World Bank's programme for the development of Luxor tourism was designed to increase this separation. The Bank's consultants, Arthur D. Little, conducted a survey of tourists' experience in Luxor and reported that the biggest problem concerned the visitors' contact with the local population. Tourists complained of being bothered continually by people trying to take them somewhere or sell them something. The consultants recommended that no further peddler's licenses be issued. More significantly, the visitor-management scheme they devised was planned to minimize unregulated contact with the tourists and increase their physical separation from the local community. A separate river ferry and bus facilities were developed to better isolate the movement of tourists from local traffic. An enclosed visitor centre with its own restaurant and shops was to be built, to enclose the tourists waiting for transportation. And in Ayman's village the plans called for an elevated walkway to be erected through the middle of the village, so that tourists could cross from the bus park to the Pharaonic temple without touching the village itself.

Enclave tourism, as this kind of arrangement is called, is now the typical pattern of tourist development in regions outside Europe and North America. It appears to be required by the increasing disparity between the wealth of the tourist and the poverty of those whose countries they visit. Today the Egyptian Ministry of Tourism appeals to foreign capitalists considering putting money into hotels or other tourist enterprises in Egypt with the claim that investors 'are enjoying outstanding profits in the tourism field,' thanks to the easy repatriation of those profits and to 'labor costs that are more than competitive on a world-wide scale.'[40] In the late 1980s the Ministry calculated that each tourist spent on average $100 a day in Egypt, which was more than most hotel employees earned in a month. A decade later the disparity was far greater.[41] The difference in wealth is so pronounced that the tourists' enjoyment can only be secured by physical separation from the host community.

However, there is a larger reason for the creation of enclave tourism. As the industry becomes concentrated in the hands of luxury hotels, mostly under the management of US- or European-based international chains, as well as half a dozen enormously wealthy Egyptian entrepreneurs, the hotel managers have sought to increase their profit by channelling more and more tourist expenditure within their own establishments.[42] The grand Egyptian hotels that used to provide little more than spacious accommodations and an elegant dining room have been replaced by hotel complexes that offer three or four different restaurants and cuisines, several bars, shopping arcades, a swimming pool and fitness club, cruises and excursions, business facilities, and evening lectures and entertainments. The Nile cruise ships and the walled 'tourist villages' popular where space is plentiful, such as along the Red Sea coast, are even more self-contained.

Except for a small elite, the local population is excluded from these enclaves, kept out by the prices charged and the guards posted at the gate. The result is a system of almost total segregation. Most Luxor tourists live, eat and sleep in their enclave hotels, travel in separate air-conditioned taxis and buses, and go to separate entertainments. The few occasions when organized tourists encounter the local street, whether half an hour set aside for shopping in the Luxor bazaar or a five-minute walk from the cruise ship to an archaeological site through a strip of village, become frenzied scenes in which local peddlers, merchants and entrepreneurs try to secure some small share of the tourist business.

Driven by the planning of international hotel chains and local entrepreneurs, this process of segregation has been further encouraged by government and World Bank policy. In the 1980s the World Bank directed Egyptian public funds into building the infrastructure for tourist development, with projects like the one in Luxor. In the 1990s the World Bank began pushing for the profits from this public investment to be switched into private and, especially, foreign hands. Supported by a former Egyptian banker turned Minister of Tourism, Fuad Sultan, in 1992 the World Bank paid the consultants Coopers and Lybrand Deloitte to draw up plans to sell off the country's luxury hotels – which, although managed by international hotel chains, were still owned by the state. The hotels were highly profitable, providing returns of up to 50 per cent or more of revenue.[43] As the consultants acknowledged, the investors enjoyed prospects for windfall profits from the future resale of undervalued properties.

Whatever the windfall, the increased control of Luxor tourism by outside capital will have two likely consequences. First, it will send not just the profits from tourism abroad, but tourist expenditure in general. Increasing international integration of the tourist industry decreases the proportion of tourist expenditure that remains in the host country or region.[44] The integration of the hotel industry was accompanied in the 1990s by that of the foreign tour operators. In Britain, for example, by 1998 just four companies controlled over 75 per cent of the overseas tour industry. The largest company, Thomson, with its own retail chain, tour operator, and airline, was itself controlled by the North American conglomerate Thomson Travel Group.[45] Second, as those who purchased Egypt's tourist assets increase the pressure on local managers to build their share of a limited market, the process of segregating the tourists within their luxury enclaves will intensify. For Ayman and the other young men of Gurna and neighbouring villages seeking employment, both developments will likely decrease the proportion of tourism income available to the local community.

Yet even as the process of segregation has developed, the lives of the local community have been increasingly affected by the tourist presence. Because of the kind of industry tourism is, its development involves more than a simple process of segregation. A conventional industry, whether based in manufacturing or agriculture, involves organizing people to produce. Mass production relies upon all the well-known methods of recruiting and disciplining a workforce, organizing their use of time, their movement, and their arrangement in physical space, and developing systems of instruction, supervision and management. Mass tourism, by contrast, involves organizing people to consume. It relies upon similar methods of managing flows and timetables, arranging physical space, and instructing and supervising to maximize the process of consumption. Tourism is usually defined from the tourist's point of view as a form of leisure, and therefore in contrast to work. But it is better seen as another form of industry, organized around the maximization not of production but consumption.

Tourism is an industry of consumption, and the consumption not of individual goods but of a more complex commodity, experiences. No object of consumption is ever just a thing. The purchase of food, clothing or cars is always the purchase of a certain taste, lifestyle or experience. One pays not just for the thing but for what it signifies. With tourism, this consumption of what things signify is taken to the extreme. The tourist industry sells not individual objects of signification but entire worlds of experience and meaning.

In Luxor the tourism industry markets the consumption of ancient Egypt. The experience is created out of the archaeological sites, but also by organizing the contemporary society to appear as a reflection and extension of the past. The 1982 World Bank report on visitor management explained that 'the creation of an *overall environment* is needed on the West Bank in order for Luxor to reach its full market potential.'[46] This meant turning Gurna into an 'open-air museum', its population moved out, and its houses to be destroyed – except for a few left standing as examples of local architecture and used to house artisans and craftsmen producing tourist artefacts.

In 1981 half a million tourists visited Luxor, and each stayed for an average of only 2.1 nights. A decade later the number of visitors in a good year was more than double that, but the length of stay had declined, to an average of less than one night.[47] The local tourist industry has less than 24 hours within which to maximize the tourist's consumption. This requires a meticulous planning of meals, drinks, sleeping and entertainment, plus the requisite trips to Karnak and Luxor temples, the sound-and-light show, the *felucca* ride, and Luxor bazaar, followed by King Tut's tomb and other sundry tombs and temples of the Theban Necropolis across the river.

This mass-production of experience produces a curious common interest between tourism's over-organized heritage consumers and some of the local community. In the 1982 World Bank survey, alongside the complaint about the behaviour of peddlers and local merchants, the most frequent tourist request was for more meaningful contact with the local population. Many tourists to Luxor are anxious to escape the routine and meet 'real Egyptians'. Many of the local population, interested in diverting tourist expenditure back toward their own needs, are keen to help. Ayman's aunt, Zaynab, for example, has a house directly in front of a parking area for tour buses. Her children hang around the buses, out of sight of the tour guides, and catch the eye of tourists lagging behind the main group as it heads off toward the temple. They then invite them into the house to watch their mother baking bread at the earthen oven. The kids expect a tip of a pound or so, and some of the tourists even pay their mother.

The mass consumption of heritage includes countless small encounters of this sort, in which the logics of exclusion, impoverishment and eviction are briefly suspended. Such events operate like a local eco-tourism, almost invisible to the large-scale tourist industry, performing, like Zaynab's kids, behind its back, yet for many individual tourists often representing the highlight of their day, far more memorable than all that sweaty Theban heritage. These encounters very occasionally develop into longer exchanges, including the foreign women who as tourists find a local husband in the village. None of this is necessarily an eco-tourism to celebrate, for it is usually constructed on considerable inequalities and misunderstandings. But it does serve as a reminder that the manufacture and consumption of heritage produces encounters beyond the control of heritage managers, where the act of consumption briefly undermines the place of things in the heritage system.

Making Hidden Violence Visible

Let me conclude by bringing this analysis of tourism and the heritage industry back to the question of producing the nation, and to the people of Gurna. In November 1996 the heads of more than seventy households threatened with eviction and demolition signed a petition to the authorities. 'We the people of Gurna,'[48] the petition begins,

> . . . have become threatened in our homes, we have become agonized with fear, while our houses are demolished above our heads and we are driven from our homeland. Sirs, you know the feelings suffered by the refugee driven from his home, the exile from his land, who becomes a stranger in his own country. We have begun to wonder whether we are Egyptians.

The petition describes the fear and violence of relocation, connecting it to other, more brutal expulsions of a sort that Egyptians in recent history have not had to face. The villagers then invoke for themselves the question of the nation: 'We have begun to wonder whether we are Egyptians.' This simple question opens up the contradictions of nation-making. Their eviction has been justified as a project of producing the nation. To preserve the national heritage, and to turn a lawless and uneducated population into honest citizens of the state, they must be expelled from their homes. To produce the nation requires a local act of violence, and in revealing this violence, its victims bring to light the forces and instabilities that nation-making brings into play. The petition continues:

> The pretext for all this is that we damage and do harm to tourism and that we threaten the safety of the monuments. We do not understand who has fabricated these rumors. We come from the monuments and through the

monuments we exist. Our livelihood is from tourism. We have no source of sustenance beyond God except for our work with tourism . . . We are married to the tourists . . .

Against the popular official portrayal of them as backward, unclean, ignorant, and an obstacle to the development of a modern heritage site, they declare 'we are married to the tourists.' Both a metaphor for their close involvement with the tourist industry and a reference to the fact that foreign women have in fact married local men, this claim gently but insistently subverts the official rhetoric.

Given that the authorities have been periodically attempting to evict the people of Gurna for more than five decades, and have on their side today all the resources of bulldozers, armed police forces, tourism investors, and United States and World Bank consultants, it is important to take seriously this power to subvert the violent plans of the heritage industry and the local performances of nation-making. This subversion, I have tried to show, is not the pure resistance of an indigenous community opposed to the plans of the authorities. It is a subversion that operates within, and opens up to view, the contradictions within the projects of heritage and nation-making. The manufacturing of a national heritage attempts to divide the world into consumers of tradition and the dead, depopulated heritage they are to consume. But on numerous levels and in multiple ways, neither the consumers nor those facing eviction agree to this programme. And in their minor acts of disruption, they bring its hidden violence into view.

Notes

An earlier version of this paper was presented at the Sixth Biennial Conference of the International Association for the Study of Traditional Environments, 'Manufacturing Heritage/Consuming Tradition,' Cairo, December 15, 1998. I want to thank Nezar AlSayyad, Caroline Simpson, Kees van der Spek, Muhammed Hamdouni Alami, Rania Fahmy, Kristine McNeil, Alice Diaz-Bonhomme, Jennifer Bell, Boutros Wadie, Siona Jenkins, Michael Jones, and David Sims. None is responsible for the views presented here.

1. The standard views on nationalism and the invention of the past now include those of E. Hobsbawm and T. Ranger (eds.), *The Invention of Tradition*, Cambridge, Cambridge University Press, 1983; E. Gellner, *Nations and Nationalism*, Oxford, Oxford University Press, 1983; and B. Anderson, *Imagined Communities: Reflections on the Origins and Spread of Nationalism*, 2nd ed., London, Verso, 1991.

2. Anderson, *Imagined Communities*.

3. E. Renan, *Qu'est-ce qu'une nation? What is a nation?*, introduction by C. Taylor, English version by W.R. Taylor, Toronto, Tapir Press, 1996.

4. For a further discussion of some of the problems about conceptions of space raised by Anderson's argument, see my essay, 'The Stage of Modernity,' in T. Mitchell (ed.), *Questions of Modernity*, Minneapolis, University of Minnesota Press, 2000.

5. Such treasure hunting had provided the main incentive for Western archaeology, and its major means of support. Its ending led to a sharp reduction in Western archaeological excavations in Egypt. They did not expand again until the late 1950s, when funds from UNESCO and other non-profit sources became available in response to the imminent destruction of ancient sites caused by the building of the High Dam at Aswan.

6. I have borrowed the notion of the past as 'the sign of the modern' from N.B. Dirks, 'History as a Sign of the Modern,' *Public Culture*, vol. 2, no. 1, Spring 1990, pp. 25-32.

7. On Egyptian nationalism in this period, see the two books by I. Gershoni and J.P. Jankowski: *Egypt, Islam and the Arabs: The Search for Egyptian Nationhood*, Oxford, Oxford University Press, 1986; and *Redefining the Egyptian Nation, 1930-1945*, Cambridge and New York, Cambridge University Press, 1995. Also see criticisms of their work in my review of the latter book, *American Political Science Review*, June, 1996; and in the excellent review essay by C.D. Smith, 'Imagined Identities, Imagined Nationalisms: Print Culture and Egyptian Nationalism in the Light of Recent Scholarship,' *International Journal of Middle East Studies*, vol. 29, 1997, pp. 607-622. For further discussion of the intellectual debates of this period, see, among others, C.D. Smith, *Islam and the Search for Social Order in Modern Egypt: A Biography of Muhammad Husayn Haykal*, Albany, SUNY Press, 1983; and J. Beinin and Z. Lockman, *Workers on the Nile: Nationalism, Communism, Islam, and the Egyptian Working Class, 1882-1954*, Princeton, Princeton University Press, 1987.

8. H.K. Bhabha, 'DissemiNation: Time, Narrative, and the Margins of the Modern Nation,' in *The Location of Culture*, London and New York, Routledge, 1994, p. 145.

9. For a representative example, see Gershoni and Jankowski, *Redefining the Egyptian Nation.*

10. K. van der Spek, 'Dead Mountain vs. Living Community: The Theban Necropolis as Cultural Landscape,' paper presented at the UNESCO Third International Forum, 'University and Heritage,' Deakin University, Melbourne and Geelong, Australia, October 4-9, 1998.

11. H. Fathy, *Gurna: A Tale of Two Villages*, Cairo, Ministry of Culture, 1969, reprinted as *Architecture for the Poor: An Experiment in Rural Egypt*, Chicago, University of Chicago Press, 1973, pp. 40, 43, 51.

12. *Architecture for the Poor* was translated into Arabic only in the 1980s. See N. AlSayyad, 'From Vernacularism to Globalism: The Temporal Reality of Traditional Settlements,' *Traditional Dwellings and Settlements Review*, vol. 7, no. 1, 1995, pp. 13-24. Cosmopolitanism, however, may have been central to Fathy's achievement. Perhaps it gave him a certain distance from the narrower materialism and less generous paternalism of the landowning class from which he came, and opened him to the influence of other inventions of the modern vernacular, such as French colonial architecture in Morocco. For example, the Habous neighborhood built in

the late 1930s in Casablanca, and copied some years later in Rabat, was widely discussed in this period as a reaction against the Modernist movement and Le Corbusier (Muhammed Hamdouni Alami, personal communication).

13. Luxor City Council, 'El Gurna Region Resident Relocation Study and New El Tarif Village Planning through Community Participation: Terms of Reference,' Luxor, Egypt, October 1992.

14. *Al-Ahram Weekly*, February 12-18, 1998.

15. *Al-Ahram Weekly*, May 7-13, 1998; and *Middle East Times*, November 22, 1998.

16. M. Tadros, 'A House on the Hill,' *Al-Ahram Weekly*, April 1-8, 1998.

17. Fathy, *Architecture for the Poor*, p. 60.

18. For the history of the malaria and cholera epidemics, see N. Gallagher, *Egypt's Other Wars: Epidemics and the Politics of Public Health*, Syracuse, NY, Syracuse University Press, 1990. Toward the end of his book, Fathy correctly distinguishes the two epidemics. See *Architecture for the Poor*, p. 166.

19. Gallagher, *Egypt's Other Wars*, pp. 32-35.

20. *Ibid*., pp. 60-66.

21. Fathy, *Architecture for the Poor*, pp. 63-64.

22. *Ibid*., pp. 113, 127, 134.

23. *Ibid*., p. 39

24. *Ibid*., p. 1

25. *Ibid*., p. 5.

26. *Ibid*., p. 51.

27. *Al-Ahram Weekly*, May 7-13, 1998.

28. Luxor City Council, 'El Gurna Region Resident Relocation Study,' p. 21.

29. S. Jenkins, 'Lifting Roots and Moving Home,' *Al-Wakaleh* (Cairo), March 1997, pp. 36-37.

30. See L. Abu-Lughod, 'Television and the Virtues of Education: Upper Egyptian Encounters with State Culture,' in N. Hopkins and K. Westergaard (eds.), *Directions of Change in Rural Egypt*, Cairo, American University in Cairo Press, 1998, pp. 147-165.

31. S. Bhatia, 'Villagers Cursed by Tombs of the Pharaohs,' *The Observer*, 1996.

32. See, for example, Bhatia, 'Villagers Cursed by Tombs of the Pharaohs.'

33. A Cairo-based archaeologist tells me that antiquities he has seen offered for sale in Gurna have been engraved with identifiable excavation register numbers, indicating that they come from government storehouses, not undiscovered tombs hidden among the houses of Gurna.

34. See van der Spek, 'Dead Mountain vs. Living Community.' An archaeologist working for the U.S.-funded Egyptian Antiquities Project has established that there has been a notable increase in the number of flash floods in the Theban Necropolis over the last 200 years, and that the disturbed climate pattern of the last twenty years has produced severe problems, notably a flash flood in 1994 which filled one

tomb alone (KV13) with 95,000 gallons of water. Presentation at the ARCE annual meeting, Chicago, April 1999.

35. See van der Spek, 'Dead Mountain vs. Living Community' for a similar argument.

36. The World Bank, 'Staff Appraisal Report: Arab Republic of Egypt Tourism Project,' typescript, Washington, D.C., April 26, 1979, pp. 19-22.

37. See T. Mitchell, 'The Object of Development: America's Egypt,' in J. Crush (ed.), *Power of Development*, New York, Routledge, 1995.

38. Ayman and other local names used in this chapter are pseudonyms.

39. Government of Egypt, Ministry of Tourism, Tourism Development Authority, Information Management Department, '1992 Tourism Data Bulletin,' January 1993. The number of tourist arrivals dropped from 3.21 million in 1992 to 2.51 million in 1993, recovering to 3.31 million in 1995 and 3.90 in 1996, before dropping again in 1997 and 1998. Economist Intelligence Unit, *Egypt: Country Profile 1997/98*, p. 68.

40. Government of Egypt, Ministry of Tourism, 'Taba Touristic Development Company,' typescript, Cairo 1991, pp. 54-55.

41. Business International, 'Egypt: Profile of a Market in Transition,' Geneva, Business International S.A., 1989, p. 75.

42. See T.G. Freitag, 'Enclave Tourism Development: For Whom the Benefits Roll?' *Annals of Tourism Research*, 1994, pp. 538-553.

43. Coopers and Lybrand Deloitte, 'Egyptian Hotels Privatisation Study: Interim Report,' typescript, June 19, 1991. The management companies typically take 15-20 per cent of profit.

44. J. Urry, *The Tourist Gaze: Leisure and Travel in Contemporary Societies*, London, Sage, 1990, p. 64. Attempts to measure the exact proportion of tourist expenditure that remains in the host country are inconclusive, in part because circumstances differ so much from one economy to the next. See E.P. English, *The Great Escape? An Examination of North-South Tourism*, Ottawa, The North-South Institute, 1986, pp. 17-45. The measurements are also inconclusive because the very nature of the industry, organized around the consumption of experience (see below), makes conventional economic measurement impossible.

45. *The Observer*, July 16, 1995, Business Section, p. 1; and August 9, 1998, p. 4.

46. Arthur D. Little, *Study on Visitor Management and Associated Investments on the West Bank of the Nile at Luxor*, April 1982, p. VII-9, emphasis added.

47. Arthur D. Little, *Study on Visitor Management*, p. VIII-2; and Government of Egypt, '1992 Tourism Data Bulletin.' Since 1981 the average visitor length of stay has declined steadily for Egypt at a whole, except for a jump in 1986-88 caused largely by long-stay summer tourists from the Gulf, very few of whom visit Luxor.

48. Petition from the people of Gurna, signed 'Ahali al-Qurna, 'anhum Abd al-Sallâm Ahmad Sûlî, al-Qurna, Naj' Al-Hurûbât, Al Uqsur.' Typescript, in the author's possession, November 1996.

Bibliography

Abu-Lughod, Lila, 'Television and the Virtues of Education: Upper Egyptian Encounters with State Culture,' in Nicholas Hopkins and Kirsten Westergaard (eds.), *Directions of Change in Rural Egypt*, Cairo, American University in Cairo Press, 1998.

AlSayyad, Nezar, 'From Vernacularism to Globalism: The Temporal Reality of Traditional Settlements,' *Traditional Dwellings and Settlements Review*, vol. 7, no. 1, 1995. pp. 13-24.

Anderson, Benedict, *Imagined Communities: Reflections on the Origins and Spread of Nationalism*, 2nd ed, London, Verso, 1991.

Arthur D. Little, *Study on Visitor Management and Associated Investments on the West Bank of the Nile at Luxor*, Cambridge, MA, 1982.

Beinin, Joel and Zachary Lockman, *Workers on the Nile: Nationalism, Communism, Islam, and the Egyptian Working Class, 1882-1954*, Princeton, Princeton University Press, 1987.

Bhabha, Homi K., 'DissemiNation: Time, Narrative, and the Margins of the Modern Nation,' in *The Location of Culture*, London and New York, Routledge, 1994, pp. 139-170.

Dirks, Nicholas B., 'History as a Sign of the Modern,' *Public Culture*, vol. 2, no. 1, Spring 1990, pp. 25-32.

English, E. Philip, *The Great Escape? An Examination of North-South Tourism*, Ottawa, The North-South Institute, 1986.

Fathy, Hassan, *Gurna: A Tale of Two Villages*, Cairo, Ministry of Culture, 1969; reprinted as *Architecture for the Poor: An Experiment in Rural Egypt*, Chicago, University of Chicago Press, 1973.

Freitag, Tilman G., 'Enclave Tourism Development: For Whom the Benefits Roll?' *Annals of Tourism Research*, 1994, pp. 538-553.

Gallagher, Nancy, *Egypt's Other Wars: Epidemics and the Politics of Public Health*, Syracuse, NY, Syracuse University Press, 1990.

Gellner, Ernest, *Nations and Nationalism*, Oxford, Oxford University Press, 1983.

Gershoni, Israel and James P. Jankowski, *Egypt, Islam and the Arabs: The Search for Egyptian Nationhood*, Oxford, Oxford University Press, 1986.

Gershoni, Israel and James P. Jankowski, *Redefining the Egyptian Nation, 1930-1945*, Cambridge and New York, Cambridge University Press, 1995.

Hobsbawm, Eric and Terence Ranger (eds.), *The Invention of Tradition*, Cambridge, Cambridge University Press, 1983.

Luxor City Council, 'El Gurna Region Resident Relocation Study and New El Tarif Village Planning through Community Participation: Terms of Reference,' Luxor, Egypt, October 1992.

Mitchell, Timothy, 'The Object of Development: America's Egypt,' in Jonathan Crush, (ed.), *Power of Development*, New York, Routledge, 1995.

Mitchell, Timothy, 'The Stage of Modernity,' in Mitchell (ed.), *Questions of Modernity*, Minneapolis, University of Minnesota Press, 2000.

Renan, Ernest, *Qu'est-ce qu'une nation? What is a nation?*, introduction by Charles Taylor, English version by Wanda Romer Taylor, Toronto, Tapir Press, 1996.

Smith, Charles D., 'Imagined Identities, Imagined Nationalisms: Print Culture and Egyptian Nationalism in the Light of Recent Scholarship,' *International Journal of Middle East Studies*, vol. 29, 1997, pp. 607-622.

Smith, Charles D., *Islam and the Search for Social Order in Modern Egypt: A Biography of Muhammad Husayn Haykal*, Albany, SUNY Press, 1983.

Tadros, Mariz, 'A House on the Hill,' *Al-Ahram Weekly*, April 1-8, 1998.
Urry, John, *The Tourist Gaze: Leisure and Travel in Contemporary Societies*, London, Sage, 1990.
van der Spek, Kees, 'Dead Mountain vs. Living Community: The Theban Necropolis as Cultural Landscape,' paper presented at the UNESCO Third International Forum, 'University and Heritage,' Deakin University, Melbourne and Geelong, Australia, October 4-9, 1998.
World Bank, 'Staff Appraisal Report: Arab Republic of Egypt Tourism Project,' typescript, Washington, D.C., April 26, 1979.

Chapter 9

The 'New-Old Jaffa': Tourism, Gentrification, and the Battle for Tel Aviv's Arab Neighbourhood

MARK LEVINE

The city has been widely recognized as a central site for the unfolding of the project of modernity and its sister discourses, colonialism and the nation-state. Scholars have similarly affirmed its axial position in the present era of globalization and (arguably) receding state power.[1] What remains to be determined is the extent to which the contemporary phenomena of globalization has altered the dynamics of urbanization and the struggle for what Henri Lefebvre has called the 'right to the city'. Of particular importance in this regard is the powerful intersection of market-based postmodern architectural and planning discourses and a (re)articulation of communal identities away from identification with the modern nation-state.[2]

This chapter presents a case study of contemporary urbanization in the city-turned-neighbourhood of Jaffa – the economic and cultural capital of pre-1948 Arab Palestine, and now a mixed Arab-Jewish quarter in the city of 'Tel Aviv-Yafo'. It reveals a fundamental continuity during this period of transition from the nation-state to the 'global' era in the century-long Zionist/Israeli (that is, nationalist, and thus exclusivist) imagination of Jaffa and Tel Aviv, the official planning and urbanization discourses produced through that imagination, and the architectures of Arab-Palestinian identity constructed in resistance to them. Jaffa, it will be seen, remains a powerful and poignant example of how the interplay of the discourses of nationalism, modernity, architecture, tourism and gentrification can influence the transformation of an urban space.

From the establishment of Tel Aviv as a garden suburb in 1909 through its evolution into the 'White City' – the world centre of International Style architecture during the 1930s and 1940s – architecture and planning played crucial roles in these processes by visually and discursively separating 'modern, Jewish' Tel Aviv from 'ancient, Arab' Jaffa, and by marking the former city as the pre-eminent symbol of the Zionist rebirth of Palestine. Today Tel Aviv's International Style architecture symbolizes the city's, and the country's, longstanding modernity. Yet the overriding focus on Tel Aviv and its architectural heritage has obscured Jaffa's equally impressive architectural heritage – both its early influence on the design of homes in Tel Aviv, and the frequent deployment of the International Style by the city's bourgeoisie before 1948 to declare their, and Jaffa's, modernity.[3]

I have elsewhere examined the role of architecture and planning in the pre-1948 conflicts surrounding the development of Jaffa and Tel Aviv.[4] In this chapter I examine the battle for Arab Jaffa during the late 1980s and 1990s, a time when Jaffa once again became an object of 'development', both as a site for tourism, and as a new, chic neighbourhood for the burgeoning Jewish elite of 'global Tel Aviv'. More specifically, I will examine how, in the face of creeping dislocation, accompanied (and supported) by daily media and television portrayals of Jaffa as both poor and crime-ridden, *and* chic, exotic and romantic (and thus the ideal tourist site), Arab residents have attempted to re-imagine their 'city' and open up new spaces for agency and empowerment. Through such actions they may ultimately be able to articulate a more autochthonous synthesis of the city's history and its architectural traditions – one that will allow them to remain on the land and develop Jaffa for the benefit of the local, as well as the international, community.

Jaffa and Tel Aviv after the 1948 War

In 1947 United Nations General Assembly Resolution 171 partitioned Palestine into Jewish and Arab states. Despite being surrounded by Tel Aviv and other Jewish towns, the city of Jaffa was included in the territory of the Arab state because of its majority Arab population and its status as the cultural and economic capital of Arab Palestine – its 'Bride of the Sea' (figure 9.1). Fighting in Jaffa began in December 1947, and continued until the surrender of the city to Zionist/Israeli forces on May 13, 1948, following the flight of all but 3,500 of the city's prewar Arab population of 70,000.

At the end of the war all of the twenty-six Arab villages in the Jaffa subdistrict were emptied or destroyed, and Jaffa itself had 'totally collapsed'.[5] For the new Israeli Prime Minister, Ben Gurion, Jaffa was to be resettled entirely by Jews: 'Jaffa will be a Jewish city . . . War is war.'[6] Subsequently, on April 24, 1950, Jaffa was officially united with Tel Aviv. According to one soldier-turned-architect who participated in the capture of the neighbouring village of Salameh: 'from the beginning the

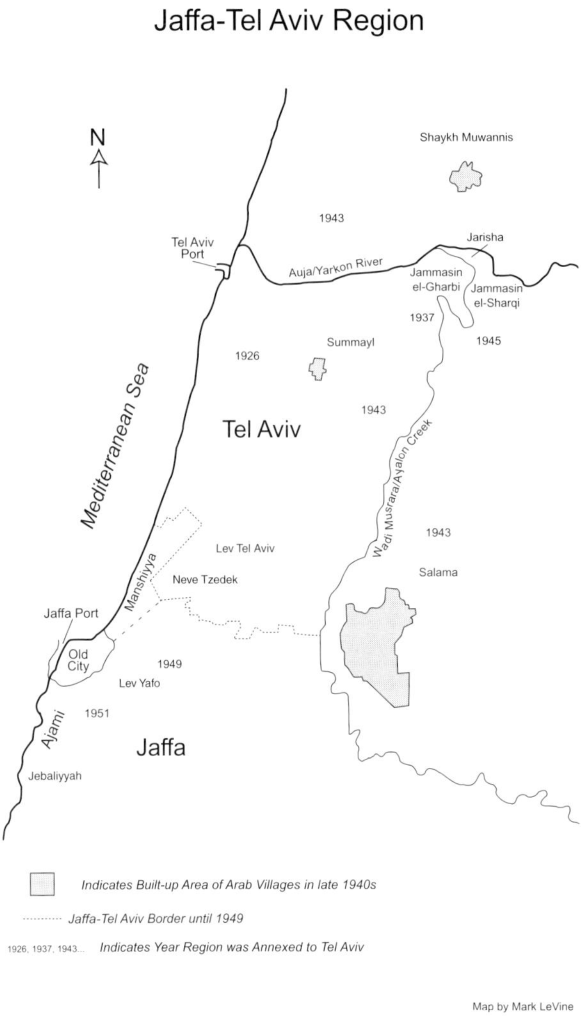

Figure 9.1. Relative locations of Jaffa and Tel Aviv, showing the years in which various surrounding regions were annexed. (Map by author.)

Municipality decided to *erase* [*limhok*] historic Salameh and build in its place something completely new.'[7]

I have elsewhere discussed how the discourse of 'erasure and reinscription', as James Holston has termed the guiding force behind modernist planning, was a major theme in the planning and architecture of Tel Aviv.[8] In fact, such an erasure of the existing Arab presence was a precondition for the symbolic and physical development of Tel Aviv. And ultimately, the transformation of the area was given biblical justification, so that today a passage from Amos greets visitors to the Tel Aviv Museum, located in the home of the city's first mayor, Meir Dizengoff: 'I will restore the fortunes of my people Israel, and they shall rebuild the ruined cities and inhabit them.'[9] But erasure of the former Arab presence was not just physical; it was also discursive. Thus, the Municipality of Tel Aviv changed almost all the Arabic street names in Jaffa into numbers. The idea was that such a system would be maintained until they could be given Hebrew names, the etiology of which was discussed at length in the short-lived Hebrew Jaffa paper, *Yediot Yafo* [*News from Jaffa*].[10]

While the municipality was initially reluctant to annex Jaffa because of the cost of postwar rehabilitation, ultimately the cities were united because the national government saw this as vital to achieving 'the disintegration of Jaffa and the demarcation of the boundaries of a united city of Tel Aviv and Jaffa.'[11] Their rebirth as 'Tel Aviv-Yafo' was announced on April 24, 1950, with 'Tel Aviv' symbolizing Jewish settlement renewing itself in Israel, and 'Yafo' being attached to preserve the historical name.[12]

Background to the Present Socio-Economic Situation

The post-1948 remnants of the Arab community of Jaffa were the poorer Arabs from the surrounding villages and a few Jaffans who remained. Jewish immigrants, mainly from the Balkans, were settled in empty Palestinian properties in the early 1950s. Later, when many of these immigrants moved to newer neighbourhoods in the Tel Aviv region, Palestinians resumed renting and buying properties in Jaffa. Then, after a precipitous drop in the Tel Aviv metropolitan region's population during the period 1972 to 1983 period, the city entered a second phase of transformation. The 'post-industrial era' has now witnessed the relocation of most of the major financial and industrial corporations of Israel to the city, and with them, numerous young-professional ('yuppy/dinkie') couples.[13] This movement was augmented by a new wave of immigration of primarily Soviet Jews beginning in 1989. Meanwhile, within Jaffa proper, the Arab population has almost trebled since 1972 while the Jewish population of the city's two predominantly Arab neighbourhoods, Ajami and Lev Yafo, has fallen dramatically – down to less than 3 per cent in the case of Ajami. Overall, the change in the population of Jaffa and Tel Aviv during this period is shown in the accompanying chart (table 9.1).[14]

Discrimination has played a continuous role in the social life of Jaffa's Arab residents. Among other instances, it is evident in wage differentials, access to jobs, and educational attainment differences between Arabs and Jews.[15] For Arab residents, such conditions have been exacerbated recently by the large increase in the quarter's Arab population, as well as by a decade-long influx of Russian immigrants, who compete with them for jobs and housing. Thus, despite claims by the Municipality of Tel Aviv-Yafo that conditions in Jaffa have actually improved during the past decade, in fact, they have 'meaningfully deteriorated' in recent years, to the point where Arab Jaffa has become the most depressed and disadvantaged community in the entire country.[16] Some indicators of this situation are shown in the accompanying table (table 9.2).

Table 9.1. Change in population of Jaffa and Tel Aviv.

Year	Arab population of Jaffa and percentage of total population	Total population of Tel Aviv-Yafo
1961	5,782 / 1.5 per cent	386,070
1972	6,351 / 2 per cent	363,750
1983	9,455 / 3 per cent	327,265
1992	15,005 / 4.2 per cent	356,911
1997	15,800 / 4.5 per cent	~355,200; 1.908,600 total for Tel Aviv metro area

Table 9.2. The Condition of Employment of Jaffa's Arabs in Relation to the Remainder of the Population.[17]

	Jewish Males in Tel Aviv	Arab Males in Jaffa	Arab Males in Mixed Cities	Arab Males in Arab Towns
Number of years of schooling	12.90	10.04	10.96	9.03
Professional status	48.33	32.97	40.39	36.47
Monthly income in shekels	3,722	2,293	2,306	2,406
Percentage of Academics	24.2	4.2	3.2	5.8
Percentage of Wage-earners	75.7	86.4	79.4	80.6
Percentage employed in the public sector	16.8	4.6	22.8	17.3

The Symbolic Functions of Tel Aviv and Jaffa

The symbolic and discursive functions of Tel Aviv and Jaffa within the Zionist enterprise have always been as important as their economic and political functions, and they currently exercise a determinative influence on the political-economic situation in Jaffa. On the one hand, 'modern', 'clean', and 'well-planned' Tel Aviv has from the start been contrasted with 'backward', 'dirty', and 'unplanned' Jaffa. At the same time, the 'first modern Hebrew city in the world' has since 1948 also been contrasted with Jerusalem, the religious capital of pre-Zionist Jewish Palestine. This dichotomy has continued to be a major theme in Israeli and Western imaginations, in no small part due to postmodernist trends that have encouraged cities to distinguish or differentiate themselves through their architecture, particularly through the selling of image.[18] Thus, the *New York Times* recently explained that 'to many Israelis, the battle of the left-wing and secular Tel Aviv against the nationalist and religious Jerusalem is a struggle for the soul and destiny of Israel.'[19]

Other American and European publications have likewise contrasted 'secular', 'normal', 'cosmopolitan', 'unabashedly sybaritic', and (most importantly) 'modern' Tel Aviv with 'holy' and abnormal Jerusalem.[20] 'A visitor wanting to see what the 50-year-old Jewish state is really all about would do well to plunge into the casual, self-consciously secular and thoroughly modern metropolis on the sea back where the dunes used to be.' This implies that Jerusalem and the seemingly interminable conflict it symbolizes are, in fact, a mirage on the 'Sahara Desert' upon which Tel Aviv was imagined and then built.[21] In a similar vein, the chief architect of Tel Aviv recently titled a book on International Style architecture in the city *Houses from the Sands* [*Batim Min Ha-Hol*].

Such a 'discourse of the sands' can be intimately tied to that 'aesthetic of erasure and reinscription' upon which most modernist planning ideologies, particularly Zionist/Israeli planning, are based.[22] And not surprisingly, the discursive erasure epitomized by the symbolism of sands and the changing of street names has lasted until today. As the *Economist* explained in comparing Tel Aviv and Jerusalem, 'Unlike Jerusalem, Tel Aviv contains hardly any Arabs. It has swallowed the old Arab port of Jaffa, but in the main it was built by Jews, for Jews, on top of sand dunes, not on top of anybody else's home.'[23] The purported absence of Arabs from the land on which Tel Aviv was built is an important reason why Tel Aviv is not considered a 'national' space in the way that the *New York Times* conceives Jerusalem. This is an ironic development considering that Tel Aviv was created as the living embodiment of a Zionist – that is, Jewish national – utopia.[24]

Such renditions of Tel Aviv's creation mythology by the Western media have had a profound impact on the way Jaffa has been imagined by both Israelis and foreign writers during the past ninety years. This is because from the birth of Tel Aviv, the landscape of Jaffa has remained central to the Tel Avivan definition of self – and thus its definition of the 'other' as well. If Arabs were discursively (and ultimately physically) erased from Tel Aviv, the process was even more determined in Jaffa.

Two contemporary depictions of Jaffa, one negative and the other quaint and 'aggressively restored', have framed its envisioning.[25] On the one hand, Jaffa has been, and continues to be, visualized as poor and crime infested. For example, it has served as the setting for many crime or war movies and television shows since the 1960s, because 'it resembles Beirut after the bombardments – dilapidated streets, fallen houses, dirty and neglected streets, smashed cars.'[26] But this image has also been reinforced by media and government depictions and, to a lesser extent, by its reality as a major centre for drug-dealing in the Tel Aviv metropolitan region.

The other image of Jaffa, specifically designed for tourist consumption, is based on the city's character as 'ancient', 'romantic', 'exotic' and 'quaint'. 'Old Jaffa . . . is the jewel of Tel Aviv', is how an official brochure described it.[27] Such depictions are linked to a re-imagining of Jaffa as a historically Jewish space, one that was 'liberated from Arab hands', as the museums and tourist brochures inform visitors.[28] These visions of Jaffa are connected to Jaffa's place as an historic, archaeological, and thus touristic, site within Tel Aviv. According to the Israeli Ministry of Tourism: 'A port city for over 4,000 years and one of the world's most ancient towns, Jaffa is a major tourist attraction, with an exciting combination of old and new, art galleries and great shopping . . . Great care has been given to developing Old Jaffa as a cultural and historical center . . .'[29]

Without a past or a history of its own, the 'City of the Sands' (as Tel Aviv has long been known) has required Jaffa to complete its identity. According to the Tel Aviv Municipality: 'Once Tel Aviv became Tel Aviv-Yafo the young city all at once acquired itself a past – the 3000 years of ancient Yafo . . . [and] was ready for the great leap forward which transformed it into a metropolis. Yafo . . . one of the oldest cities in the world, acquired a future and renewed youth, with widespread progress streaming its way from its youthful neighbor.'[30]

Not surprisingly, Arab Jaffans have protested how their city has since 1948 become little more than 'a margin on the name of Tel Aviv'.[31] One reason is that pre-1948 Jaffa was considered the 'jewel' of Arab Palestine, and was continually depicted in the Palestinian press as the country's most beautiful and important city. As *Falastin* described it in 1946: 'No one doubts that Jaffa is the greatest Arab city in Palestine, and it is inevitable that visitors to Palestine will stop by to see the model of Palestine's cities.'[32] In other words, Jaffa was a symbol, and perhaps the epitome, of Arab Palestine's urban landscape.

Notwithstanding such views, the erasure of Jaffa has now been accepted by many diaspora Jaffans, particularly those returning to visit the city in recent years, who have come to regard present-day Jaffa as a 'figment of the imagination'.[33] And in some ways Tel Aviv has displaced Jaffa in the Palestinian imagination. For instance, when the facilitator of a peace mission in Palestinian-controlled Nablus asked people what their vision of peace was, a Palestinian artist replied 'visiting Tel Aviv and watching the sun set.'[34]

On the other hand, the attachment of the remaining Arab population to Jaffa has grown significantly during the past two decades. In part this has corresponded with the larger trend toward increasing Palestinianization of Israeli Arabs in the wake of the reunification of all of Mandatory Palestine after the Six Day War and the outbreak of the Intifada in 1987.[35] However, this nationalistic re-imagining of Israeli Arab identity has also added greater relevance to the question of territoriality.[36] In fact, there were several violent protests in Arab Jaffa during the 1990s, and Arab community leaders have called for Jaffa's municipal independence (a demand that has won some support among Jewish residents of Jaffa, who also see themselves as excluded from the larger municipality's plans for their neighbourhoods).

Moreover, in response to continued attempts by the Municipality of Tel Aviv to evict long-time Arab residents, Jaffa's Arab community's leadership has threatened a 'housing *Intifada* in the streets . . . declaring with a loud voice that we are planted here and that they will not be able to uproot us from our homes the way they uprooted the orange and olive trees.'[37] This focus on rootedness is deeply imbedded in the Jaffan – and the Palestinian – psyche, as evidenced by the painting by Jaffan artist Suheir Riffi depicting a mother nursing her child rooted into the earth and connected through it to her dilapidated home (figure 9.2).[38]

Globalization, Architecture and Planning in Tel Aviv-Yafo

The specificities of contemporary Jewish and Arab imaginings of Jaffa have influenced the way the Jaffa-Tel Aviv region has experienced globalization and attempts by Israel's leadership (and the leaders of the Municipality of Tel Aviv, in particular) to transform Tel Aviv into a 'world' city. This drive to 'globalize' Tel Aviv may be understood as part of an effort by city leaders to shape and deploy a unique identity, separate from the rest of the country – especially from Jerusalem. The apparent success of this effort has left planners, architects and commentators to wonder 'what to do with a world city that is so different from the rest of the country in which it is located.'[39]

Israeli social scientists have also conceived of and analysed Tel Aviv as a global city, focusing on its entrance into international markets, the increasing disparities between rich and poor, the 'marketization' of social services such as the education system, and the influx of increasingly illegal migrant guest workers (upwards of 100,000 of whom are now said to live in the Tel Aviv metropolitan region).[40]

Likewise, most architects working in Tel Aviv have refused to criticize the municipality's planning policies for 'global Tel Aviv', which call for building high-rises throughout the city to maximize the market value of its land. The eminent Dutch architect Peter Kook, who has worked in Tel Aviv, has described the present Tel Aviv 'style' in a manner that contextualizes it within the political psychology of a significant proportion of the country's Jewish population. For Kook, contemporary architecture in Tel Aviv consists of

> . . . paranoia on the one hand, and the world-wide trend of the worship of money on the other. The paranoia is reflected in the fact Israeli architects are

closed to any outside styles, they only see what the Housing Ministry does, and not what's going on in the wider world. The power of money rules here in a dominant way on both aesthetics and on urban planning . . . Also, there is a psychological factor. Israeli architects take the fortress as their model . . . the security room in their apartments. They are afraid to do more elegant architecture here, with more feeling, because maybe something will [destroy] the building.[41]

Figure 9.2. Painting by Jaffan artist Suheir Riffi, exhibited in the 'To Live Within a Picture Exhibition,' Jaffa, 1997. (Courtesy by author with courtesy of artist.)

The political, economic and discursive roles of architecture in Jaffa and Tel Aviv during this century bear out Michel Foucault's belief that 'architecture and its concomitant theory never constitute an isolated field to be analyzed in minute detail; they are only of interest when one looks to see how they mesh with economics, politics, or institutions.'[42] Certainly, both Jaffa and Tel Aviv, in particularly Tel Aviv, did use town planning as a tool in the 'war over land' during the Mandate period.[43] Yet, not surprisingly, much of Israeli planning literature has avoided any discussion of the Arab minority that would disturb the carefully apolitical suppositions upon which it is based. Instead, such writing has focused on planning as 'change-oriented activity,' in order to 'shift attention away from the document – the plan – to the political process whereby intentions are translated into action.'[44] Thus, for example, in a recent edited volume on planning in Tel Aviv, a chapter on 'Conflict Management in Urban Planning in Tel Aviv-Yafo' consisted of a case study of underground parking in stores in central Tel Aviv.[45] In another chapter, Tel Aviv's chief municipal engineer, Baruch Yoscovitz, explained that there has been very little true planning in the Jaffa-Tel Aviv region since the work of the eminent Scottish planner Patrick Geddes in the mid-1920s: 'Instead of comprehensive planning, these days we have 'pragmatic planning'.'[46]

What Yoscovitz failed to mention in his lengthy analysis, however, was Geddes's specification that 'with all respect to the ethnic distinctiveness and the civic individuality of Tel Aviv, as Township, its geographic, social and even fundamental economic situation is determined by its position as Northern Jaffa . . . The old town, the modern Township, must increasingly work and grow together . . . for Greater Jaffa.'[47] Moreover, however 'pragmatic' the dynamic of planning in Tel Aviv, the chief engineer himself has been an important actor in ongoing battles between the municipality and the Arab residents of Jaffa over the development of Ajami and the Jaffa port.

It is clear, then, that it is precisely the documents, or texts, that are pivotal to understanding the larger discourse of planning, particularly when planning takes place in 'frontier' regions such as Jaffa's Arab neighbourhoods.[48] Within frontier regions under the sovereignty of post-independence settler colonization movements such as Israel, spatial policies are often used as a powerful tool to exert territorial control over minorities. On an urban scale, majority-controlled authorities exercise more subtle forms of spatial control through land use and housing policies, and in so doing, they create segregation between social groups.[49] This is particularly true when, as in Israel, the government has taken almost all planning and development powers out of the hands of local Arab communities.

The discussion has thus far suggested that the goal of any analysis of planning in the Jaffa-Tel Aviv region should be to clarify the complex web of relations between governmental, semi-governmental, and pseudo-governmental organizations and institutions that control the planning system in Israel. The number of institutions involved, and the complexity of their relations, indicates that, despite claims to the contrary, planning is highly politicized and ideological.[50] But what is new in this equation in Israel today is the increasingly prominent role that private interests are beginning to play, in Jaffa in particular, and how this shift has altered the internal boundaries within the land and planning system, while maintaining the traditional Israeli focus on permanent Jewish ownership of as much land as possible.[51]

Fuelled by a larger discursive, even epistemological, shift in Israeli society, the strategic shift toward privatization in city planning has led to a situation in which planners chart a course of development focused on middle- and upper-class Israelis, implemented through private developers, which implicitly pits Jews against their Palestinian co-citizens. Thus, Arab land has been expropriated; the construction of new, privately-developed Jewish housing has been approved; and the new Jewish 'owners', who have invested time and money in their new homes, naturally take the lead in fighting against the claims of the previous (now 'illegal') Arab inhabitants. This is how the government, working through private developers, has brought the economic interests of liberal Israelis in line with perceived 'national' interests *vis-à-vis* increasing Jewish ownership, control, and presence on the land.[52]

How and why have such policies become so embedded in the city's (and the country's) political economy that they have been rendered nearly invisible, or at least unremarkable – especially when a stroll through Jaffa, or a glance at a map, will show that from both an architectural and a planning perspective, Jaffa's development, and Ajami's in particular, have closely mirrored that of Tel Aviv?[53] A review of the history and discourses of post-1948 planning in Tel Aviv and Jaffa may provide some insight into these matters.

In the postwar/unification planning of the early 1950s Jaffa and the surrounding villages were considered 'slums'. As such, they were scheduled for rehabilitation, the goal of which was to redevelop the 'ancient city of Jaffa' and the surrounding neighbourhoods under the slogan 'today slums, tomorrow seashore parks'[54] (figure 9.3). However, by the early 1980s a new generation of 'renewal' efforts had begun in the older neighbourhoods of Neve Tzedek and Lev Tel Aviv, prompted by a structural reorganization of the city's economy that had begun in the previous decade. This effort sought to 'reviv[e] the region as a space for living in the center of the city by drawing a mainly young population to it.' Both Neve Tzedek and Lev Tel Aviv featured architecture that made them attractive for gentrification. Lev Tel Aviv, having already undergone extensive reconstruction in the 1930s, featured the International Style buildings that had placed Tel Aviv

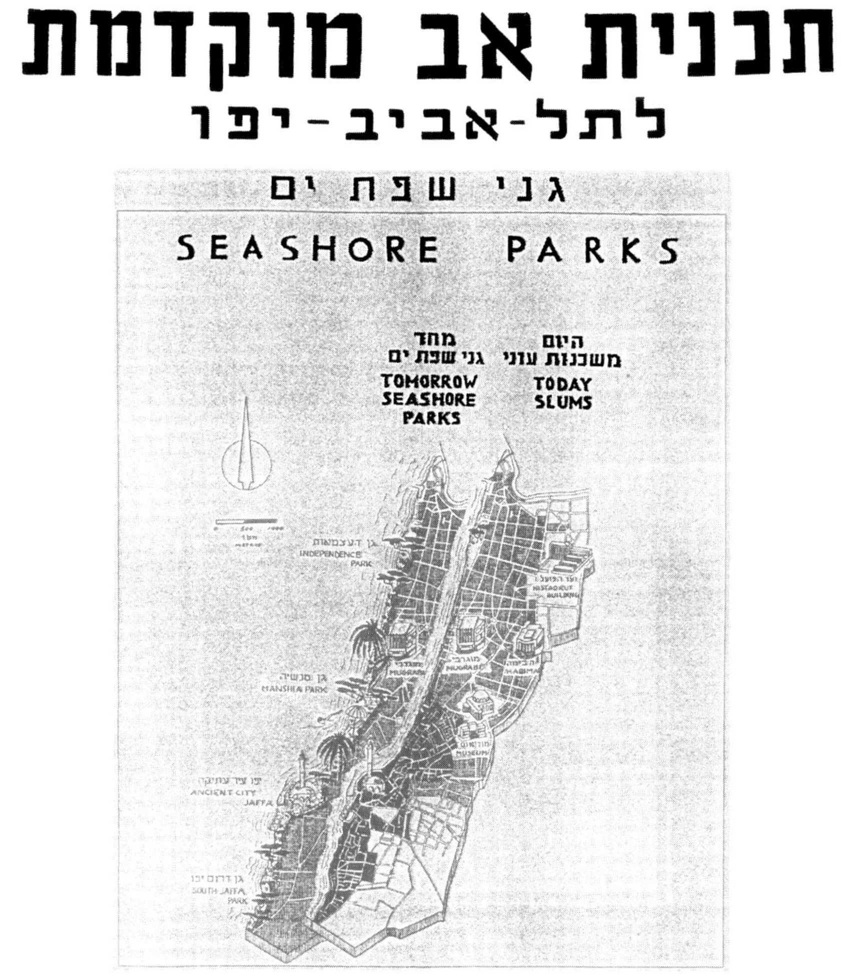

Figure 9.3. Cover of 1954 issue of the *Tel Aviv Town Gazette* (*Yediot Tel Aviv*) featuring plan for the rehabilitation of 'slum' areas, most of them formerly Arab neighbourhoods, into 'sea shore parks.' (Courtesy of *Yediot Tel Aviv*.)

on the architectural map. Neve Tzedek featured much older buildings that attracted a bohemian crowd trying to escape both austere International Style architecture and what Tel Aviv University geographer Juval Portugali has described as a postmodern fetishization of consumption, which had recently taken the ironic form of an easily identifiable, uniform 'postmodern style'.[55]

In fact, there has been something of a rebellion by many residents, and even some architects, against the consumer-driven architecture of the 1970s, as symbolized by the numerous tall residential and office towers in or near Neve Tzedek.[56] This indicated a major change from the time in the early 1970s when the municipality had bragged about how the 'leap up into the skies . . . improved the appearance of the city, adding an extra beauty to its landscape.'[57]

Such renewed appreciation for the city's older architecture can be interpreted as part of a general trend in 'postmodern architecture' against modernism's clean break with the past. Postmodern architectural sentiment has tried to employ a type of 'historicism; historical quotation; an architecture of memory and monuments . . . a search for "character", unique features, visual references.'[58] Yet it can also be explained as part of the process by which architecture, and art in general, has become ever more commodified to cater for consumer tastes (an ironic development in light of the desire to move away from a visually consumerist lived environment). Viewed in this light, the 'renewal' of neighbourhoods like Neve Tzedek can be understood not as preserving the past, but rather as rewriting or inventing it. Thus, buildings and districts have been renovated, restored or rehabilitated to correspond to ideal visions of the past, and at the same time to satisfy contemporary needs and tastes by incorporating new technologies and designs.[59]

If the gentrification of Tel Aviv's older neighbourhoods has generated and reflected contradictory impulses and desires, the process has been even more complicated in Jaffa, which despite being officially part of Tel Aviv, is heavily invested with symbolism as Tel Aviv's alter ego. How has this separation been mediated? The answer becomes clearer if one considers how through the various Zionist/Israeli visions of 'ancient' Jaffa, the neighbourhood has become 'a discursive object created by Israelis as part of turning Israel . . . into particular socio-political spaces.'[60] If Jaffa is seen as a frontier region, it further becomes clear how the spatial policies of the municipality have been used as a powerful tool – much like the power of Orientalist discourse as described by Edward Said – to exert territorial control over, and physically shape, this discursive yet material space.[61]

In the resulting process of cognitive and physical boundary demarcation between Tel Aviv and Jaffa, Jewish 'yuppies' moving to Jaffa 'see residential exclusivity and the redeeming modernizing impact of Zionism as simply engendering a demarcation between two types of territory.'[62] In this vision, Jaffa has served as the historical 'other' of Tel Aviv, thus Tel Aviv has used the historic Jaffa to justify itself. At the same time, having been liberated from its Arab identity, and united with its daughter city, Jaffa has been presented as continuously undergoing a process of renewed youth and progress, the life blood of which is the architectural and planning policies of the municipality. Nevertheless, the neighbourhood's renewal has been dependent upon its permanent fixture in time and space as 'ancient' or 'quaint' – the ideal site for tourist and elite development.

In fact, if a fear of building imaginatively has led to an architectural 'tragedy' in Tel Aviv, Jaffa has become a space where the imagination, although remaining under government supervision, has had freer reign.[63] In other words, as 'picturesque' has become the architectural fashion, the government has realized that 'old, dilapidated Arab neighborhoods have an "oriental" potential.' Thus, the function of the numerous rehabilitation projects of the past two decades has been to expand commerce, tourism and hotels in line with the 'specific character' of the area.[64] More specifically, 'today the slogan is, "gentrify!" As land becomes available, it is sold on stringent conditions that only the wealthy can meet.'[65] As one architectural critic put it, the current style among the Jewish architects practising in Jaffa is to build with arches, 'thousands of arches, wholesale'[66] (figure 9.4).

As one Israeli architect has explained, the end result of this process has been expressed in 'the systematic erasure of the identity of the city of Jaffa as an Arab city.'[67] This may seem ironic given the 'Oriental' feel of current building styles; but in fact Jaffa has had to be emptied of its Arab past, and its Arab inhabitants, in order for architects to be able to re-envision it as a 'typical Middle Eastern city', and construct new buildings based on this imagined space.[68]

It is within this framework that Peter Kook has explained why recent attempts to 'preserve' Jaffa cannot be taken at face value:

This is not 'preservation' [*shimor*], this is Disneyland. The old city and the new projects that attempt to preserve the Arab architecture are cheap imitations,

Figure 9.4. 'Thousands of arches, wholesale' featured in new construction in Jaffa. An article in the Tel Aviv newspaper *Ha'ir*. (Courtesy of *Ha'ir*.)

more decorative, intended for tourists . . . It's for entertainment or amusement (*msha'sha'a*), so why not?[69]

Tourism and the New Market Discipline

When the world economy and the peace process faltered during the tenure of former Prime Minister Benjamin Netanyahu, Israel's finance minister explained that the key to the country's continued economic growth was the real estate market, of which the Tel Aviv metropolitan region is the centre. Such a view has clear implications for current 'renewal' efforts in Jaffa. Obviously, it signals that the municipality will have even less freedom, or incentive, to commit its resources to a poor minority community sitting on valuable land.[70] The influence of such market discourse is readily apparent in current planning in Jaffa. Thus, on the one hand, current policy guidelines have declared that the new regional plan for Tel Aviv must involve residents in planning and work to increase housing for young couples.[71] Yet when Arab community leaders have complained that most young Arab couples cannot afford to live in Jaffa, officials have responded by explaining that 'the market is the market'[72]; and that 'selling some apartments more cheaply would hurt profits.'[73]

The most important impact of the marketization of planning in Jaffa has been the partial or total privatization of several of the bodies directly responsible for the rehabilitation of the quarter since the mid-1990s. Until then as many as 90 per cent of the housing units in Jaffa were partly owned by the government, and a large part of the real estate in Jaffa was in the hands of quasi-governmental companies such as Amidar and Halmish.[74] Since then, however, the transfer of development projects to private developers has been described by Jaffa's Arab councilman (in the same language, it is worth noting, used by the Jaffa newspaper *al-Jam'iah al-Islamiyyah* in 1932 to describe burgeoning land conflicts in Jaffa-Tel Aviv, and in Palestine as a whole) as a major turning point for the quarter.[75]

One such project that has been partially transferred to private developers involves the redevelopment of Jaffa's port, home to a fishing industry supporting 250 families. The stated aim of this project is to 'resurrect and develop old Jaffa's harbour as an area of tourism, recreation and sea sport.'[76] This is to be done by linking the port directly to the lived area of the old city through the construction of as many as 4000 elite residence and hotel units.[77] The symbolism surrounding the port gives a clue to how such a project will be realized. Thus, the official Tel Aviv-Jaffa guide of the Ministry of Tourism explains how

. . . the old city today is alive, her buildings and alleys restored amidst cobbled streets and green parks as a thriving artist's colony . . . great care has been given to developing Old Jaffa as a cultural and historical center while preserving its Mediterranean flavor . . . Jaffa Marina [part of the development project] has been established in the heart of the ancient port . . . and offers all a sailor could desire . . . Visit Old Jaffa anytime. By sunlight and starlight, it is the 'jewel' of Tel Aviv.[78]

Project Shikum [Rehabilitation] is another such project ostensibly designed to 'develop and rehabilitate Jaffa.' It was turned over by the municipality to a private developer, Yoram Gadish, in 1996. But when mismanagement and concerted local opposition led the government and the Tel Aviv Municipality to terminate Gadish's contract, a new private company headed by former Tel Aviv Mayor Shlomo Lahat was awarded the contract to continue the neighbourhood's gentrification.[79] One might note how this relationship – involving the Tel Aviv Municipality, a historic tourist landmark inhabited by Arabs, and a private development company headed by a former mayor – is identical to that in East Jerusalem *vis-à-vis* the City of David project, which is headed by former Mayor Teddy Kolek.[80]

An interview with representatives of the Gadish company while it was administering Project Shikum revealed the thinking underlying both the Jaffa and Jerusalem projects – and thus the discourse governing Israeli planning on both sides of the Green Line. According to Gadish, the goal of the project was

. . . to develop Jaffa because Jaffa is not developed . . . that is, develop infrastructure, sewers, streets, schools, etc., and to develop the empty lands in Jaffa. We want to revolutionize Jaffa [lhafoh et Yafo], to change Jaffa from a neighborhood with so many problems to a tourist city – there's lots of potential for development into a tourist city . . . But you need to have a plan, and like New York or anywhere, sometimes you have to destroy a building as part of development for public needs, and we're working with a committee of architects and the Municipality . . . However, the residents want to keep the status quo because development increases prices, and their children won't be able to live and buy apartments there; also Arabs won't go to other cities like Bat Yam, Herzliyya because there are no services for them. They can go to Lod and Ramle, but they're not ready to go and don't want to develop . . . but with Jews [Jaffa] becomes more beautiful and develops.[81]

The Andromeda Hill Project

The paradigmatic example of the intersection of new global, market-based, postmodern architectural discourse in Jaffa with the almost century-long Zionist/Israeli imagination of the city is the Andromeda Hill project, where basic units were advertised at well over US$300,000 (figure 9.5). Constructed on property at the top of the Ajami Hill, with a commanding view of the port and ocean below, Andromeda Hill has billed itself as 'the incomparable Jaffa . . . the New-Old Jaffa.'

Figure 9.5. Publicity advertisement for Andromeda Hill development which appeared in numerous Hebrew and English publications. (Pamphelet of Andromeda Hill development.)

To help orient prospective customers on its website, the Andromeda Hill virtual brochure explains that 'historic Jaffa' lies to the north of the development, the 'picturesque fishermen's wharf of Jaffa' to the west, and the 'renewed Ajami district, where the rich and famous come to live' to the south. The Hebrew version stresses the architecture of the place even more, in line with the greater importance of architectural discourse in Israeli culture.[82] Moreover, the section of the website entitled 'The Legendary Jaffa' recounts the Greek legend of Andromeda, which supposedly took place on a large rock facing the city outside of Jaffa's port. It explains how 'Andromeda became a symbol of awakening and renewal, and it is not by chance that the project was named "Andromeda Hill", expressing the rebirth of old Jaffa.'

When asked why and how such an architectural design and advertising campaign was chosen for Andromeda Hill, one former employee explained:

> The municipality decided on the style – the windows, the columns, the materials – after going around Jaffa and looking at the buildings . . . The style was very eclectic – Arabic from the beginning of the century influenced by European (specifically Italian) architecture. . . . Arches were a main symbol in a project of this size . . . We didn't use real stone (except in a few places), but rather a manmade material called 'GRC', which is fake stone. In terms of the ads, you have to think about who's going to buy there . . . they expected people from abroad to buy it. Jaffa today is not a nice place, you have to think about the future, what will be attractive. People aren't living there because of the sea, because there's sea all over Israel, they're living there because of the nostalgia, the atmosphere.[83]

The Andromeda Hill discourse, like that of Gadish, exemplifies the conflation of architecture and planning, market forces and government control, that comprise the forces at play in the continuing 'war over land' in Ajami.[84] In fact, visitors to the complex are shown a short video before their tour, whose narration concludes by explaining how 'Andromeda Hill is, in essence, a city within a city' – within Jaffa. This is almost identical to the language used by the founders of Tel Aviv to describe the Jewish position in Jaffa almost one hundred years ago, when they celebrated having created 'a state within a state in Jaffa'. The social, political, and spatial implications of such a discourse are also identical – that is, in each case Jaffa is the object of 'economic conquest' (as Arthur Ruppin described it ninety years ago) by Jewish residents from Tel Aviv.[85]

Such imagery takes on added significance if one recalls Peter Kook's equation of Jaffa with Disneyland and the belief by proponents of global America that 'the wretched of the earth just want to go to Disneyland if given the chance.'[86] Like Disneyland for most of the world's poor, the virtual reality that increasingly cohabits the space of contemporary Jaffa can only be viewed from beyond a 'secured gate' by most residents of Ajami.[87] In Isra-Disney, Jaffa, as symbolized by Andromeda Hill, becomes an 'urban masterpiece', a site of 'artistic renaissance', and 'a museum of magnificent architecturally designed buildings'[88] – a carnival of sites, sights, and sounds that excludes those who cannot afford the entrance fee.

Conclusion: Spatializing Arab Jaffa

More than a century ago Theodor Herzl explained what was necessary to create a Jewish state in Palestine: 'If I wish to substitute a new building for an old one, I must demolish before I construct.'[89] Five decades later, at the height of the era of modernist planning, the French architect and city planner Le Corbusier – several of whose disciples became prominent Zionist planners and architects – quoted a famous Turkish proverb to epitomize the modernist ethic: 'Where one builds one plants trees. We root them up.'[90] From a similar but more critical perspective, Henri Lefebvre has explained how 'the "plan" does not remain innocently on paper. On the ground, the bulldozer realizes "plans".'[91]

This chapter has tried to demonstrate that Jaffa can be understood as a space of both negation and identification for Tel Aviv, and that such ambivalence reflects the larger relationship of the Israeli state toward the Palestinian communities living within its pre-1967 borders. In fact, the entry of Arabs into the Israeli 'national self', or even the self-definition of the Israeli state, is both ambivalent and paradoxical – precisely because of the primacy and power of planning as a vehicle for such articulation. Arab inclusion into the project of modern Israel is ambivalent in that postmodernist architectural sensitivity towards Jaffa's Arab heritage has remained 'superficial' and economic in orientation. It is paradoxical in that the aim of current place-oriented postmodern architecture has been to entice a 'global' (and implicitly, non-Arab) elite, and disallow the potential of political identification from Jaffa's Arab community. The double economy of fixing Jaffa for the Orientalist gaze and developing it along the lines of a market economy implies both the commodification and de-politicization of the Arab community.

The contested space of Jaffa and Tel Aviv further epitomize the complex manner in which architectural movements have been inscribed in the politics of national identity in Israel: first erasing 'tradition' (through International Style), and then reclaiming it (through discourses of heritage promoted by postmodernist architecture). Both movements have been expressed in economic as well as political idioms in the process of constructing the political identity of the nation-state. Such is the dynamic governing the politics of urban design in contemporary Jaffa.

Such a linkage of the metaphors of erasure and rebuilding, and their function as the ideological underpinnings of the Jewish state, has, however, long been recognized by the country's indigenous inhabitants.[92] Thus, community leaders objected to a 1985 development project by explaining that the development policies of both the Tel Aviv Municipality and national-government agencies had generally involved using 'legal' and 'planning' mechanisms to destroy homes and expropriate land from Arabs.[93]

Indeed, the concerted efforts to 'preserve'/erase Jaffa's Arab character or heritage have had a profound effect on the way residents experience the city. On the one hand, while residents attempt to reclaim the city by

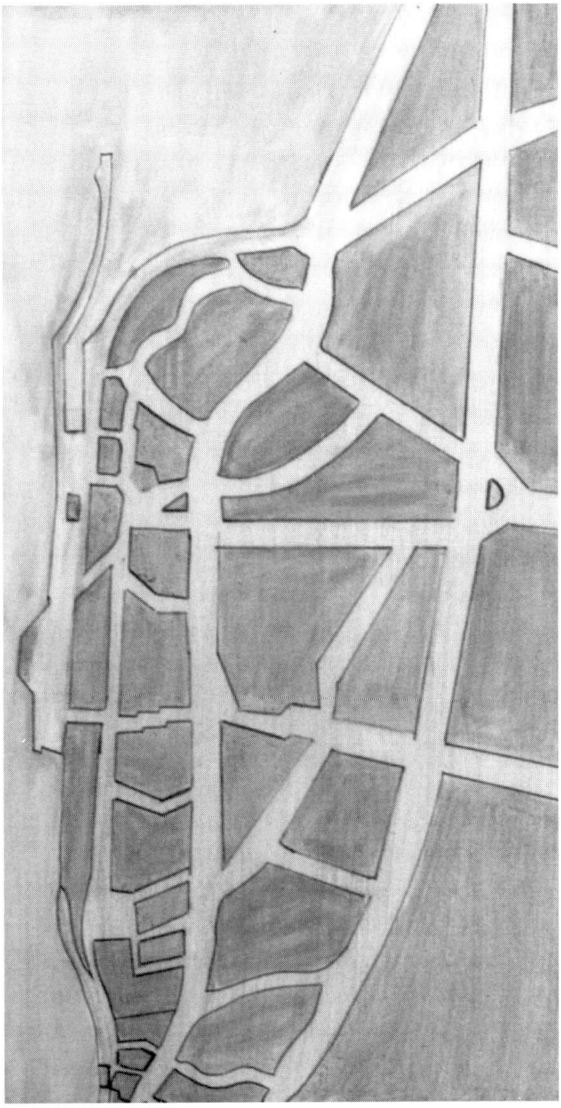

Figure 9.6. Painting by Jaffan artist Suheir Riffi, exhibited in the 'To Live Within a Picture Exhibition,' Jaffa, 1997. (Courtesy of artist.)

referring to streets by their original Arabic names, artists paint Jaffa as empty and vacant (figure 9.6). On the other hand, many residents,

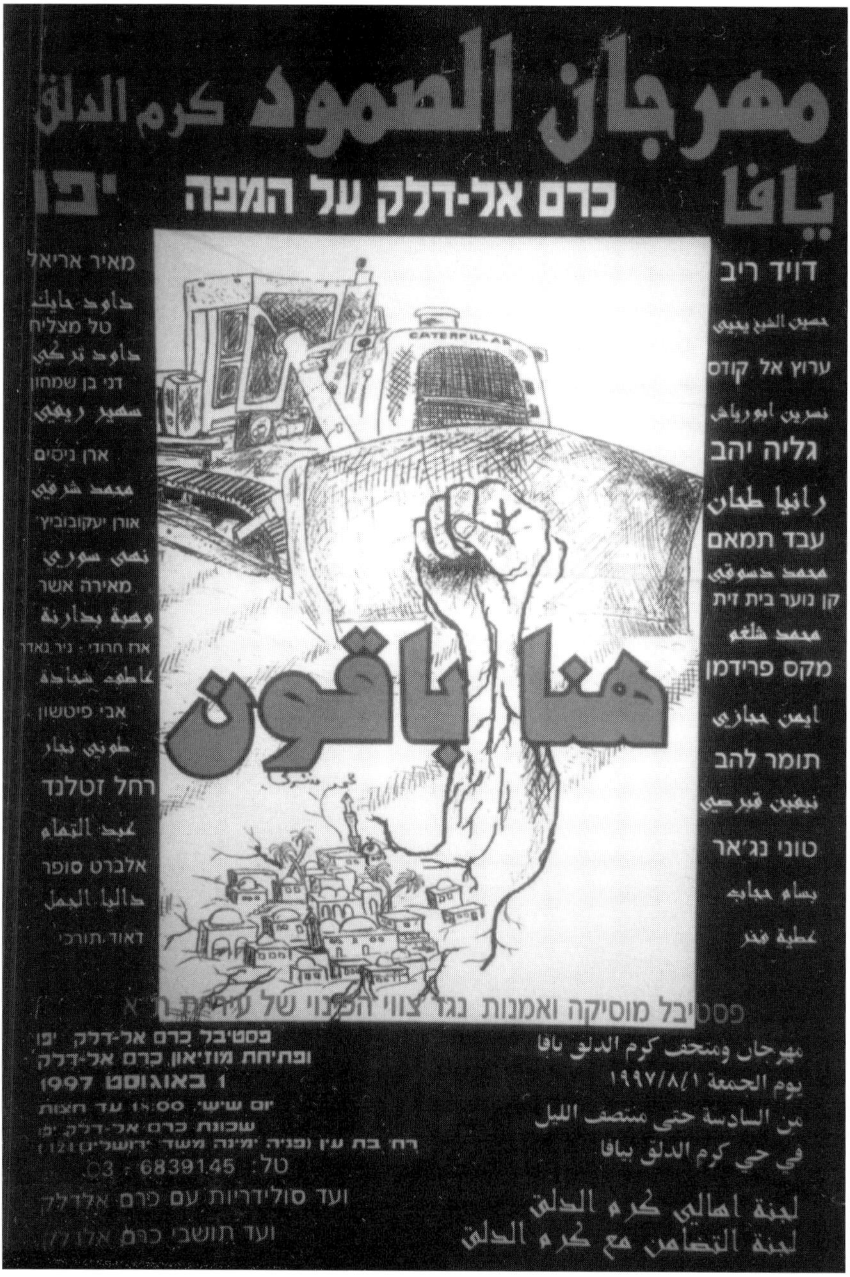

Figure 9.7. Poster for 'Jaffa Festival' on one of the main streets in Tel Aviv. (Photo by author.)

including the former head of al-Rabita,[94] the local Arab community organization, have expressed their belief that the policies of the Tel Aviv Municipality have only strengthened the ties of most of the Arab community to Jaffa and its Arab identity.[95]

Prevented from expressing its identity through design and planning of its lived environment, Jaffa's Arab population has articulated its identity through '*spatializing* social activity'.[96] This has included art festivals, original theatre, organized protests (which became violent in 1994 and 1996), and the fight to return original Arabic street names – or, barring that, appropriating the language of Ajami's luxury developments and deploying it to document its consequences.[97] Thus, a 1997 festival jointly sponsored by local Jewish and Arab grassroots organizations in support of a large group of families threatened with eviction from their land was called the 'Sumud Festival' – *sumud* being the well-known Palestinian slogan for remaining rooted on the land. The festival featured a poster of a bulldozer confronted by a fist rooted in the earth. Its caption read 'Here we Remain . . . We are not alone'[98] (figure 9.7).

These activities should be seen as a form of architecture – indeed, the only form of architecture available to Arab residents, who are prohibited from planning or building their own lived environment.[99] By constructing an alternative landscape, a 'poetic geography' in opposition to that of Zionist/Israeli Tel Aviv, the Arab community has 'cognitively redefined the borders of Jaffa' to include parts of Tel Aviv, such as Neve Tzedek, that historically lay outside Jaffa's borders.[100] This has provided the impetus for the recently intensified calls for 'autonomous' municipal independence from Tel Aviv.[101]

Jaffa can also be understood in terms of Henri Lefebvre's concept of 'representational spaces' – that is, spaces that are linked to the clandestine or underground side of social life: the space of inhabitants, as opposed to the space of planners and political authorities.[102] Lefebvre has characterized this dimension of space as 'imagined'; and it is here that possibility of 're-imagining' the spaces of Jaffa and Tel-Aviv can most felicitously be entertained.[103] Lefebvre's analysis helps show that the spatial system of Ajami and Arab Jaffa is characterized 'not by one social space but by many . . . The worldwide does not abolish the local,' however much it might want to.[104]

Notes

1. The four-year Project on Cities and Urban Knowledges of New York University's International Center for Advanced Studies (at which I was a fellow during 1997-1998, and where many of the ideas presented here germinated) has produced significant new empirical and theoretical research to support this contention. For further analyses of the role of the city in the modern and 'global' periods, see *inter*

alia the works of Henri Lefebvre, David Harvey, Manuel Castells, Saskia Sassen, Janet Abu-Lughod, Nezar AlSayyad, Gwendolyn Wright, Nan Ellis, David Gregory, John Urry, Peter Hall, Anthony King, and James Holston.

2. For a discussion of the implication of the fading away of the state on the contemporary dynamics of identity formation, see M. Castells, *The Power of Identity*, Oxford, Blackwell Publishers, 1997.

3. Cf. D. Tzafrir, *Jaffa: A Glance at Adjami: An Architectural Profile*, Tel Aviv, City Engineer's Department, 1995; and M. LeVine, 'A Nation from the Sands?' *National Identities*, vol. 1, no. 1, 1999, pp. 15-38.

4. M. LeVine, 'Conquest through Town-Planning: The Case of Tel Aviv,' *Journal of Palestine Studies*, Summer, 1998; and LeVine, 'A Nation from the Sands?'

5. O. Stendel, *The Arabs in Israel*, Brighton, Sussex Academic Press, 1996, p. 54.

6. Quoted in T. Segev, *1949: The First Israelis*, New York, Free Press, 1986, p. 75.

7. Quoted in *Ha'ir*, May 2, 1997, p. 24. However, newly arriving immigrants quickly moved into the abandoned houses, and the plan was not carried out. The five other villages surrounding Tel Aviv, much of whose land had already been incorporated into the town-planning area of Tel Aviv in the previous decade, were similarly emptied of their Arab inhabitants only to be re-inhabited by Jews.

8. As Gideon Levi explained in an editorial in *Ha'aretz* ('The Time to Count has Arrived,' May 25, 1997), in the eyes of many veterans who fought in the battle for Jaffa there was no alternative to emptying the city of its Arab residents. Writing in the voice of one of those veterans, Levi rhetorically asked: 'What would have happened to the Jewish state, if, for example, the residents of Jaffa had remained? How would the first Jewish city be seen that is next door?'

9. Amos, 9:14.

10. See *Yediot Yafo*, 10/62, p. 11. For an analysis of the name changes, see A. Mazawi, 'The Chosen Street Names in Jaffa Before and After 1948: Ideological Contents and Political Meanings,' unpublished article in al-Rabita Archive [RA], General ['am] File. Most side streets in Jaffa are still only numbered. However, the local population does not refer to them as such, preferring to use local landmarks such as corner stores or mosques/churches to navigate through the city. Today the Arab community is still greatly concerned about the problem of street names in Jaffa. 'The street names do not express the character of the area and the names relate to Jewish Rabbis and events without connection to the cultural and historic life of the Arabs of Jaffa.' Thus, they want both Arabic street names and road signage in Arabic (RA, General File, 'The Society for Jaffa's Arabs, Agenda for meeting with Tel Aviv Mayor Roni Milo,' April 18, 1997, pp. 13, 15; cf. February 21, 1995, letter from al-Rabita to Tel Aviv mayor, included in appendix to Agenda).

11. A. Golan, 'The Demarcation of Tel Aviv-Jaffa's Municipal Boundaries following the 1948 War: Political Conflicts and Spatial Outcome,' *Planning Perspectives*, vol. 10, 1995, p. 391.

12. *Ibid.*, pp. 393-94; and Tel Aviv Municipality, *Tel Aviv: People and their City*, Tel Aviv, 1974, pp. 5-6.

13. Y. Schnell, 'The Formation of an Urbanite Life Style in Central Tel Aviv,' in D. Nachmias and G. Menahem (eds.), *Social Processes and Public Policy in Tel Aviv-*

Yafo, vol. I, Tel Aviv, Ramot Publishing House, 1993, p. 41. The term 'yuppy/dinkie' describes young urban professionals with double incomes and no children.

14. Source: *Statistical Yearbook of Tel Aviv Municipality*, 1997, p. 59. The proportion of Muslims to Christians increased from 51.5 per cent to 42.9 per cent in 1961, to 62 per cent to 37.3 per cent in 1995.

15. M. Semyonov, N. Lewin-Epstein, and H. Mendel, 'The Labor Market Position of Arab Residents of Tel Aviv-Yafo: A Comparative Perspective,' (Heb.), in Nachmias and Menahem (eds.), *Social Processes and Public Policy in Tel Aviv-Yafo*, vol .II, 1997, p. 195.

16. Position paper drawn up by leading activists, sociologists and educators for Jaffa Councilman Nassim Shaker to use in the debates over the 1997 budget (summarized in A. Waked, 'Milo Worries about Jaffa. Here are some examples,' *Ha'ir*, December 27, 1996, p. 30).

17. Source: M. Semyonov, N. Levin-Epstein, and H. Mendel, Tel Aviv University report on the economic situation of Jaffa's Arab community, reprinted in *Ha'ir*, July 11, 1997, p. 34. The report noted that the number of Arabs employed in the public sector diminished from 23 per cent in 1983 to 4.6 per cent in 1993 – an 80 per cent reduction. The researchers could not include figures relating to budgetary allocations from the Tel Aviv Municipality to Jaffa because the municipality refused to turn over the relevant data. Furthermore, it should be noted that more than 50 per cent of the Arab population of Jaffa is 19 years of age or younger.

18. H. Lefebvre, *Writings on Cities*, E. Kofman and E. Lebas (ed. and trans.), Oxford, Blackwell Publishers, 1996, p. 51.

19. *New York Times*, April 30, 1998, p. A19. Another report called Tel Aviv 'unabashedly sybaritic' (S. Schmemann, 'What's Doing in Tel Aviv,' *New York Times*, December 21, 1997).

20. C. Kummer, 'Tel Aviv: Secular City, Where Israel Meets the Modern World,' *The Atlantic Monthly*, December 1995. Cf. 'Survey of Israel at 50,' *Economist*, April 25, 1998, p. 18; and Schmemann, 'What's Doing in Tel Aviv.' Kummer continues: 'If you want to stroll down an Israeli street lined with quirky, forward-looking shops, or sit back and enjoy a relaxed meal in a restaurant that cares about elegance and service, Tel Aviv is the place.' Cf. *Economist*, 'Survey of Israel at 50'; and *Le Monde*, April 25, 1998.

21. As a 1941 article described it (Tel Aviv Municipal Archive [TAMA], 4/3565, 8/41 article by L.V. Beltner entitled 'City of the Jews'). I offer a critique of this portrayal of Tel Aviv as having been built away from, and thus not infringing upon, Arab-inhabited land in my 'Conquest through Town-Planning.'

22. Cf. J. Holston, *The Modernist City: An Anthropological Critique of Brasilia*, Chicago, University of Chicago Press, 1989, p. 5.

23. *Economist*, 'Survey of Israel at 50,' p. 18.

24. Cf. J. Gorny, 'Utopian Elements in Zionist Thought,' *Studies in Zionism*, vol. 5, no. 1, 1984, pp. 19-27.

25. Kummer, 'Tel Aviv: Secular City.'

26. A. Mazawi, 'Film Production and Jaffa's Predicament,' *Jaffa Diaries*

(compilation of email communications from former and current residents of Jaffa contained at Jaffa Website www.yafa.org), February 9, 1998. These shows have included the popular 'Jaffa Portraits' crime series, which aired on the 'Family Channel' (H. Kaye, 'A "Portrait" of Aviva Marks,' *Jerusalem Post*, March 25, 1997, p. 7).

27. Official 'Tel Aviv-Yafo' guide of the Ministry of Tourism and Tel Aviv Hotel Association, 1997.

28. State of Israel, Ministry of Defense, Museums Unit, *Brochure of the Museum of the I.Z.L*; and Eretz Israel Museum, *Guide to Yafo, for Self-Touring*, Tel Aviv, 1988.

29. Ministry of Tourism, *Official Guide to Tel Aviv*, 1997, pp. 23, 32. This theme of the beauty and quaintness of Jaffa at night was already being used in articles in the local Hebrew Jaffa paper *Yediot Yafo* as far back as 1963. See 'There's Nothing Like Yafo at Night,' *Yediot Yafo*, 7/63, p. 4.

30. Tel Aviv Municipality, *Tel Aviv: People and their City*, pp. 3, 5-6.

31. *Al-Ayyam*, May 19, 1997. Story on the Jaffa Internet discussion group.

32. *Falastin*, May 9, 1946, p. 2.

33. S. Tamari and R. Hammami, 'Virtual Returns to Jaffa,' *Journal of Palestine Studies*, vol. 27, no. 4, Summer 1998, p. 73.

34. April 1998 meeting sponsored by 'Face to Face.' Other bourgeois Palestinians with the ability to travel freely in Israel also have told me of their fondness for visiting Tel Aviv.

35. Elie Rekhess reaches a similar conclusion of increased 'Palestinianization' of Israeli Arabs in the wake of the Six Day War. See his *The Arab Minority in Israel: Between Communism and Arab Nationalism, 1965-1991*, Tel Aviv, Tel Aviv University, 1993, p. 84. Cf. Y. Schnell, *Perceptions of Israeli Arabs: Territoriality and Identity*, Brookfield, VT, Avebury Press, 1994, p. 2.

36. As Yitzhak Schnell writes: 'The substance of nationalism is a combination of ethnic identity with the right to territorial sovereignty' (*Perceptions of Israeli Arabs*, p. 26). Cf. A. Bishara, 'On the Question of the Palestinian Minority in Israel,' *Theory and Criticism*, vol. 3, 1993, pp. 7-20. Also see his 'The Arab-Israeli: Readings in an Incomplete Political Discourse,' *Journal of Palestine Studies*, vol. 24 1995, pp. 26-54.

37. RA, public declaration of al-Rabita, June 24, 1997, entitled 'We Will Not Leave Because We Are Planted Here' ['lan nuraha innna huna manzura'una'].

38. The community has also established ties with the Palestinian National Authority, corresponding with PNA President Yassir Arafat and hosting visiting PNA ministers. Moreover, several delegations of Jaffans have visited Amman to meet with the local Jaffa Society (many bourgeois Jaffans, especially Christians, who fled to Amman in 1948) and put on a music and art festival about Jaffa, at which a senior al-Rabita member rhapsodized the phrase 'Jaffa is Palestine, Jaffa is Palestine' in a speech to the assembled guests. In fact, in internal meetings of al-Rabita that I attended, some younger members, in particular, were clearly anti-Zionist (although by no means anti-Jewish), and heated discussions took place surrounding the relationship with the Tel Aviv Municipality, Arab identity, education, and how much to remain within the Jewish cultural and educational system of Tel Aviv.

39. T. Margolit, 'Cities of the World, Twin Identity,' *Ha'aretz*, May 2, 1997, p. B6.

40. In terms of the entrance into international markets, see Nachmias and Menahem, *Social Processes and Public Policy in Tel Aviv-Yafo*, vol. I, p. 294; and A. Shachar, 'The Planning of Urbanized Areas: A Metropolitan Approach and the Case of the Tel Aviv Region,' in *ibid.*, vol. II, p. 312. In terms of the 'marketization' of social services, see *Ha'ir*, February 28, 1997, p. 31; September 26, 1997, p. 37; and June 6, 1997, p. 30. For a revealing personal account of the life of African workers in Tel Aviv, see *Ha'ir*, September 19, 1997, p. 54.

41. 'Yafo is Disneyland, the North is a Tragedy,' *Ha'ir*, June 12, 1997, p. 24. Most new Israeli apartments, especially luxury models, have special 'security rooms' that are bomb proof and are provisioned in case of chemical attacks.

42. N. Ellis, *Postmodern Urbanism*, London, Blackwell, 1996, p. 251.

43. The word *milhama* ('war' in Hebrew) was used by Tel Aviv's leaders to describe the struggle to expand the city's borders during the Mandate period, and if one considers that the Arab Revolt of 1936-39 began in Jaffa, the martial imagery is not surprising (cf. Central Zionist Archives [CZA], S25/5936).

44. R. Bilski (ed.), *Can Planning Replace Politics? The Israeli Experience*, Boston, Martinus Nijhoff Publishers, 1980. The closest they come is to point out that 'it may safely be assumed, for instance, that the political leadership will not approve any national plan that fails to point out how the primary objectives of the Zionist ideology are to be achieved, such as settlement of the country's arid areas.' The deployment of the geography of 'arid areas', i.e., uninhabited areas, is not by chance, as 'making the desert bloom' was from the start a central justification for Zionist colonization enterprise, while the use of such visually evocative examples elides the reality that hundreds of built-up Arab villages were 'emptied' as part of the 'settlement' of the country. To take another example, in a chapter on 'Urban and Regional Planning in Israel', the discussion begins in the post-1948 period, and thereby has a clean slate, since almost all major Arab towns, like Jaffa, no longer had an Arab presence after the war (M. Hill, 'Urban and Regional Planning in Israel,' in Bilski, *Can Planning Replace Politics?*, p. 259).

45. A. Churchman and R. Alterman, 'Conflict Management in Urban Planning in Tel Aviv-Yafo,' in Nachmias and Menahem, *Social Processes and Public Policy in Tel Aviv-Yafo*, vol.II, 1997. Also in this volume see Shachar's 'The Planning of Urbanized Areas.'

46. B. Yoscovitz, in Nachmias and Menahem, *Social Processes and Public Policy in Tel Aviv-Yafo*, vol.II, pp. 352, 354-356.

47. P. Geddes, *1925 Town Plan for Tel Aviv*, TAMA Library, p. 1.

48. While the term 'frontier' usually refers to sparsely populated national or international border regions, it can also be used felicitously to describe 'internal' regions such as minority neighbourhoods within cities in which the state is attempting to expand its control (O. Yiftachel, 'The Internal Frontier: Territorial Control and Ethnic Relations in Israel,' *Regional Studies*, vol. 30, no. 5, 1995, pp. 494, 496.) Along these lines, Ian Lustick has lucidly explained how majority-minority relations in Israel are characterized by 'control and exclusion' (I. Lustick, *Arabs in the Jewish State: Israel's Control of a National Minority*, Austin, University of Texas Press, 1980, pp. 69, 77, 84, 88). More specifically, Lustick has

detailed an elaborate system of control (which he has stated existed at least through the 1970s, but which I would argue still exists in a different form today), composed of three mutually reinforcing components – segmentation, co-optation, and dependency – and according to which the state has attempted to achieve the quiescence of Palestinians and the exploitation of their resources to further its goals (cf. S. Jirvis, *The Arabs in Israel*, I. Bushnaq (trans.), New York, Monthly Review Press, 1976). Similarly, Aziz Haidar has pointed out that 'the simultaneous non-recognition of the status of the Arabs as a national minority, together with unwillingness to absorb them into Israeli society (because of the definition of the boundaries of the society) is . . . contradictory' (A. Haidar, *On the Margins: The Arab Population in the Israeli Economy*, New York, St. Martins, 1995, p. 4.). S.N. Eisenstadt has defined this complex of attitudes as 'semi-colonial paternalism,' as opposed to 'colonial,' because the Arabs have been officially accorded civil rights. Moreover, 'since it is the large Arab localities which are potentially more capable of developing a modern economy, their exclusion from top priority development status has denied them the appropriate physical infrastructure on which to base economic project, deterring potential investors' (in Haidar, *On the Margins*, p. 32). While Haidar was speaking of Arab municipalities, the same could be said about Jaffa (see M.A. Ramadan, 'La Minorité Palestinienne de l'État d'Israèl,' *L'Observateur des Nations Unies*, 1997, no. 3).

49. Yiftachel, 'The Internal Frontier,' p. 498.

50. Oren Yiftachel has charted the Israeli planning system in such a manner as to demonstrate the interrelationship between official, semi-governmental, and pseudo-autonomous planning, supervisatory, and ownership organizations (see O. Yiftachel, *Watching over the Vineyard: The Example of Majd el-Krum*, Raanana, The Institute for Israeli Arab Studies, 1997, p. 116).

51. Thus, for example, the pseudo-governmental Jewish National Fund (an agency that since 1901 has used donations from Jews around the world to purchase land in Palestine/Israel that, once in its possession, can never be sold to non-Jews) announced in November 1998 that it was severing ties with the Israel Land Authority, the semi-governmental agency that administers both state and JNF-owned lands (and which heretofore has been composed of both government and JNF representatives). Precisely because it was going 'private', it could buck the legal trend toward equality between Jews and Arabs in the government sector and ensure that its huge reserves of land remained 'in the hands of the Jewish people' (*Ha'aretz*, November 6, 1998, p. A1).

52. Interview with Oren Yiftachel, November 20, 1998. Haifa University law professor Sandy Kedar has added that 'privatization is the "in" thing today in Israel . . . the Kibbutzim, Moshavim, and real estate developers all realize that it is no longer as easy to discriminate against Arabs through the state, so they are trying to find new ways to pursue the ideological, economic and psychological goals of continued judaization' (interview with author, November 19, 1998). At the time this chapter was being written, a new government was being formed by Prime Minister-elect Ehud Barak, who during his campaign promised to work for greater equality for Israel's Palestinian citizens.

53. Tzafrir, *A Glance at Ajami*; and LeVine, 'A Nation from the Sands?'

54. See cover of and articles in *Yediot Tel Aviv*, 1954, pp. 8-9. For a description of

the development in Jaffa through 1965, see Tel Aviv Worker's Council, 'Development Activities in Jaffa and the Neighborhoods (1960-65),' TAMA Library 8-10. Some of the neighbourhoods that have received specific attention and planning for renewal are Kfar Shalem, Neve Eliezar, Hatikva and Florentin.

55. Y. Ginsberg, 'Revitalization of Two Urban Neighborhoods in Tel Aviv: Neve Tzedek and Lev Tel Aviv,' in Nachmias and Menahem, *Social Processes and Public Policy in Tel Aviv-Yafo*, vol. I, 1993, p. 151; and J. Portugali, 'The Taming of the Shrew Environment,' *Science in Context*, vol. 7, no. 2, 1994, p. 312. Yet as Portugali points out, the words 'uniform' and 'style' represent the very opposite of postmodernism.

56. *Ha'ir*, May 30, 1997, pp. 32-33; June 6, 1997, p. 32; and July 4, 1997, p. 30. For a discussion of the opposition to other larger towers in Tel Aviv, see *Ha'ir*, May 2, 1997, p. 12.

57. Tel Aviv Municipality, *Tel Aviv: People and their City*, p. 9.

58. Ellis, *Postmodern Urbanism*, pp. 91-92.

59. *Ibid.*

60. D. Rabinowitz, *Overlooking Nazareth: The Ethnography of Exclusion in Galilee*, Boston, Cambridge University Press, 1997, p. 15. Also see his 'An Acre is an Acre is an Acre? Differentiated Attitudes to Social Space and territory on the Jewish-Arab Urban Frontier in Israel,' *Urban Anthropology*, vol. 21, no. 1, 1992, pp. 67-89.

61. Yiftachel, 'The Internal Frontier,' pp. 494, 496, 498. For a more detailed analysis of this dynamic in the country at large, see O. Yiftachel and A. Meir (eds.), *Ethnic Frontiers and Peripheries: Landscapes of Development and Inequality in Israel*, Boulder, CO, Westview Press, 1998.

62. Portugali, 'The Taming of the Shrew Environment,' p. 312. This remark was made regarding the Occupied Territories, but it is equally relevant in this context.

63. 'Yafo is Disneyland, the North is a Tragedy,' *Ha'ir*, June 12, 1997, p. 24.

64. A. Mazawi, 'Spatial Expansion and Building Styles in Jaffa: Past and Present,' in Mazawi (ed.), *Art and Building in the View of the Paintbrush*, Jaffa, the Center for Arabic Culture, 1988.

65. *Challenge*, May-June 1998, pp. 12-13, 18.

66. *Ha'ir*, June 20, 1997, p. 32.

67. Quoted in Mazawi, 'Spatial Expansion and Building Styles in Jaffa,' p. 4.

68. A. Mazawi and M.K. Machool, 'Spatial Policies in Jaffa, 1948-1990,' in H. Liski (ed.), *City and Utopia*, Tel Aviv, The Israeli Society for Publishing, 1991, p. 66.

69. 'Yafo is Disneyland, the North is a Tragedy,' *Ha'ir*, June 12, 1997, p. 24.

70. Architect Yosi Tager, quoted in *Ha'ir*, April 18, 1997, p. 43.

71. 'Regional Descriptive Plan for the Tel Aviv Region,' Interim Report no.3, March 16, 1998, published by Hebrew University under direction of Professor A. Shakhar, pp. 2-4.

72. Quoted in A. Waked, 'Place for Worry,' *Ha'ir*, September 20, 1996, p. 1.

73. *Ha'ir*, August 15, 1997, p. 34.

74. N. Shachar, *Jaffa: At a Fork in the Road*, Jaffa, al-Rabita Publications, 1997, p. 37. Often, when new building is allowed, it is only on the roofs of existing structures in order to keep as much vacant land as possible available for development.

75. *Al-Jam'iah al-Islamiyyah*, December 18, 1932, p. 7. The phrase was 'a fork in the road' ['muftarik fi al-turuq']. Cf. Shachar, *Jaffa: At a Fork in the Road.*

76. *Ibid.*; cf. *al-Sabar*, December 24, 1997, p. 8. For more information on the continuing battle between the fishermen and the municipality, see *al-Sabar*, January 7, 1998, p. 4; and January 12, 1998, p. 9. For another description of the situation, see 'The Fishermen are Furious at the Closure of the Port,' *al-Sabar*, March 18, 1997, p. 9.

77. April 17, 1997, letter to Schmuel Laskar from Dan Darin of Tel Aviv Planning Commission with attached plan. Copy given to author by local journalist. Also see *Ha'ir*, August 15, 1997, p. 22, for reporting on the plan.

78. Ministry of Tourism, *Official Guide to Tel Aviv*, 1997, pp. 23, 32.

79. During his mayoralty, Lahat was a vocal proponent of continued 'judaization' in Jaffa as well as vigilance against attempts by the local Arab community to gain more control over the neighbourhood. For a description of the declared goals of the company, named Ariel Real Estate-Yafo, see their November 1998 publication in Hebrew and Arabic, 'Yediot Yafo.'

80. For details on this project, see project brochure 'Emek Hamelekh,' produced for the Jerusalem 3000 celebration. Also see the English-language 'City of David' brochure, whose cover declares that 'In 1996 the City of David will be 3000 years old. In 1996 the City of David can be ours again.'

81. Author's interview with several representatives of Gadish, July 13, 1997.

82. All quotes from the 1998-99 Andromeda Hill Website: www.andromeda.co.il/home.html.

83. Interview with author. GRC is made out of silicon. It is less expensive, lighter, and easier to use. The video shown to prospective customers misleadingly implies that the whole project as being made out of stone.

84. As the new Arab councilman from Jaffa described it (A. Waked and R. Zartzki, 'I, Rifa'at Turk,' Ha'ir, May 9, 1997, p. 14).

85. For a discussion of the earlier example of this discourse, see LeVine, 'Conquest Through Town-Planning,' and 'A Nation From the Sands?'

86. T. Friedman, *New York Times*, August 15, 1998, Op/Ed page.

87. Cf. Tamari and Hammami, 'Virtual Returns to Jaffa.' A similar architectural and design aesthetic, described as a 'neo village, sans mosque or church, with stone walls with electronic access gates,' is appearing in Palestinian-controlled areas such as Ramallah, where several new developments for wealthy diaspora Palestinians are under construction (E. Hecht, 'Homeward Bound,' *Jerusalem Post* magazine, November 20, 1998).

88. 'Andromeda Hill – The New Old Jaffa: Living an Original,' 1998 brochure for project.

89. T. Herzl, *The Jewish State*, New York, Scopus Publishing Co., 1943, p. 84.

90. Le Corbusier, *The City of Tomorrow and its Planning*, F. Etchells (trans.), London, Architectural Press, 1947, p. 82.

91. H. Lefebvre, *The Production of Space*, New York, Blackwell, 1991, p. 191.

92. Cf. S. Slyomovics, *The Object of Memory: Arab and Jew Narrate the Palestinian Village*, Philadelphia, University of Pennsylvania Press, 1998, p. 29.

93. Labor Archives, V/329/2, November 20, 1985, report by al-Rabita. In response to the plan as proposed, al-Rabita suggested the immediate cessation of house demolitions and working to preserve more buildings.

94. Al-Rabita was formed in 1979 'to protect the Arab Jaffan essence of the Ajami and Jebaliyyah quarters [against] the plans of the authorities whose goal is to transfer us off our land' (flyer from al-Rabita dated January 20, 1986, RA, General File).

95. Interview with author, May 1997.

96. Lefebvre, *Writings on Cities*, p. 188.

97. A perfect example of this was the use of the phrase 'to live within a picture' ['*lagur btokh tziur*'], which spearheaded an ad campaign for the 'Jaffa Village' development, as the title for the Arab art festival in Jaffa from which several paintings have been included in this chapter.

98. For Arabic reporting on the festival, see *al-Sabar*, July 25 and August 8, 1997.

99. In the sense given to it by Lefebvre as 'a social practice among others' (*Writings on Cities*, p. 189).

100. That is, one that is based on the entry of past events into the present as materials with which to imagine and construct the present and future' (J. Portugali, *Implicate Relations: Society and Space in the Israeli-Palestinian Conflict*, Boston, Kluwer Academic Publishers, 1993, p. 140). Geography is poetic in the sense that the past is brought to the present in a new, imaginative configuration, and thus creates a new reality (p. 62).

101. Interviews with leaders of al-Rabita in 1997 and 1998. Cf. A. Waked, 'Politely, Quietly, Jaffa Keeps its Distance (Mitraheket),' *Ha'ir*, April, 1997, p. 42; and R. Tzaror, 'We are Autonomous, from Today,' *Ha'ir*, May 9, 1997, p. 19.

102. Lefebvre, *The Production of Space*, pp. 33, 39.

103. David Harvey has described the process as 'mental inventions,' or spatial discourses that imagine new meanings or possibilities for spatial practices (*The Condition of Postmodernity*, Berkeley, University of California Press, 1989, p. 39).

104. Lefebvre, *The Production of Space*, p. 86.

Bibliography

Abu-Lughod, Janet, *Rabat*, Princeton, Princeton University Press, 1980.

Bilski, Raphaella (ed.), *Can Planning Replace Politics? The Israeli Experience*, Boston, Martinus Nijhoff Publishers, 1980.

Bishara, Azmi, 'On the Question of the Palestinian Minority in Israel' (Heb.), *Theory and Criticism*, vol. 3, 1993, pp. 7-20.

Bishara, Azmi, 'The Arab-Israeli: Readings in an Incomplete Political Discourse,' *Journal of Palestine Studies* (Arabic), vol. 24, 1995, pp. 26-54.
Castells, Manuel, *The Urban Question*, translated by Alan Sheridan, Cambridge, MIT Press, 1977.
Castells, Manuel, *The City and the Grass Roots*, Berkeley, University of California Press, 1983.
Castells, Manuel, *The Power of Identity*, Oxford, Blackwell Publishers, 1997.
Churchman, Arza and Rachelle Alterman, 'Conflict Management n Urban Planning in Tel Aviv-Yafo' (Heb.), in Nachmias and Menahem, (eds.), *Social Processes and Public Policy in Tel Aviv-Yafo*, vol. II, 1997.
Ellis, Nan, *Postmodern Urbanism*, London, Blackwell, 1996.
Geddes, Patrick, *1925 Town Plan for Tel Aviv*, TAMA Library.
Ginsberg, Yona, 'Revitalization of Two Urban Neighborhoods in Tel Aviv: Neve Tzedek and Lev Tel Aviv' (Heb.), in Nachmias and Menahem (eds.), *Social Processes and Public Policy in Tel Aviv-Yafo*, vol. I, 1993.
Golan, Arnon, 'The Demarcation of Tel Aviv-Jaffa's Municipal Boundaries following the 1948 War: Political Conflicts and Spatial Outcome,' *Planning Perspectives*, vol. 10, 1995.
Gregory, David and John Urry, (eds.), *Social Relations and Spatial Relations*, London, Macmillan, 1994.
Haidar, Aziz, *On the Margins: The Arab Population in the Israeli Economy*, New York, St. Martins Press, 1995.
Hall, Peter, *Cities of Tomorrow: An Intellectual History of Urban Planning and Design in the Twentieth Century*, New York, Basil Blackwell, 1988.
Harvey, David, *Social Justice and the City*, Baltimore, Johns Hopkins University Press, 1973.
Harvey, David, 'The Urban Process under Capitalism: A Framework for Analysis,' in M. Dear and A. Scott (eds.), *Urbanization and Urban Planning in Capitalist Society*, London, Methuen, 1981.
Harvey, David, *The Condition of Postmodernity*, Cambridge, MA, Blackwell, 1989.
Harvey, David, *The Urban Experience*, Baltimore, John Hopkins Press, 1989.
Harvey, David, *Justice, Nature and the Geography of Difference*, Cambridge, MA, Blackwell, 1996.
Herzl, Theodor, *The Jewish State*, New York, Scopus Publishing Co., 1943.
Holston, James, *The Modernist City: An Anthropological Critique of Brasilia*, Chicago, University of Chicago Press, 1989.
Jirvis, Sabri, *The Arabs in Israel*, Inea Bushnaq (trans.), New York, Monthly Review Press, 1976.
King, Anthony, *Urbanism, Colonialism, and the World-Economy*, New York, Routledge, 1991.
Le Corbusier, *The City of Tomorrow and its Planning*, F. Etchells (trans.), London, Architectural Press, 1947.
Lefebvre, Henri, *The Production of Space*, New York, Blackwell, 1991.
Lefebvre, Henri, *Writings on Cities*, Eleanor Kofman and Elizabeth Lebas (ed. and trans.), Oxford, Blackwell, 1996.
LeVine, Mark, 'Conquest Through Town-Planning: The Case of Tel Aviv,' *Journal of Palestine Studies*, Summer 1998, pp. 36-52.
LeVine, Mark, 'A Nation from the Sands?' *National Identities*, vol. 1, no. 1, 1999, pp. 15-38.

Lustick, Ian, *Arabs in the Jewish State: Israel's Control of a National Minority*, Austin, University of Texas Press, 1980.
Mazawi, Andre, 'Spatial Expansion and Building Styles in Jaffa: Past and Present,' in Mazawi (ed.), *Art and building in the View of the Paintbrush* (Heb.), Jaffa, the Center for Arabic Culture, 1988.
Mazawi, Andre and Makram Khoury Machool, 'Spatial Policies in Jaffa, 1948-1990,' in Haim Liski (ed.), *City and Utopia*, (Heb.), Tel Aviv, the Israeli Society for Publishing, 1991.
Nachmias, David and Gila Menahem (eds.), *Social Processes and Public Policy in Tel Aviv-Yafo* (Heb.), vols. I, II, Tel Aviv, Ramot Publishing House, 1993, 1997.
Portugali, Juval, *Implicate Relations: Society and Space in the Israeli-Palestinian Conflict*, Boston, Kluwer Academic Publishers, 1993.
Portugali, Juval, 'The Taming of the Shrew Environment,' *Science in Context*, vol.7, no.2, 1994, pp. 307-326.
Rabinowitz, Dan, 'An Acre is an Acre is an Acre? Differentiated Attitudes to Social Space and territory on the Jewish-Arab Urban Frontier in Israel,' *Urban Anthropology*, vol. 21, no. 1, 1992, pp. 67-89.
Rabinowitz, Dan, *Overlooking Nazareth: The Ethnography of Exclusion in Galilee*, Boston, Cambridge University Press, 1997.
Rekhess, Elie, *The Arab Minority in Israel: Between Communism and Arab Nationalism, 1965-1991* (Heb.), Tel Aviv, University of Tel Aviv, 1993.
Sassen, Saskia, *Globalization and Its Discontents: Essays on the New Mobility of People and Money*, New York, W.W. Norton, 1998.
Schnell, Yitzhak, 'The Formation of an Urbanite Life Style in Central Tel Aviv' (Heb.), in Nachmias and Menahem (eds.), *Social Processes and Public Policy in Tel Aviv-Yafo*, vol.I, 1993.
Schnell, Yitzhak, *Perceptions of Israeli Arabs: Territoriality and Identity*, Brookfield, VT, Avebury Press, 1994.
Segev, Tom, *1949: The First Israelis*, New York, Free Press, 1986.
Semyonov, Moshe, Noah Lewin-Epstein and Hadas Mendel, 'The Labor Market Position of Arab Residents of Tel Aviv-Yafo: A Comparative Perspective' (Heb.), in Nachmias and Menahem (eds.), *Social Processes and Public Policy in Tel Aviv-Yafo*, vol. II, 1997.
Shachar, Arieh, 'The Planning of Urbanized Areas: A Metropolitan Approach and the Case of the Tel Aviv Region' (Heb.), in Nachmias and Menahem (eds.), *Social Processes and Public Policy in Tel Aviv-Yafo*, vol. II, 1997.
Shachar, Nissim, *Jaffa: At a Fork in the Road* (Arabic), Jaffa, Al-Rabita Publications, 1997.
Slyomovics, Susan, *The Object of Memory: Arab and Jew Narrate the Palestinian Village*, Philadelphia, University of Pennsylvania Press, 1998.
Stendel, Ori, *The Arabs in Israel*, Brighton, Sussex Academic Press, 1996.
Tamari, Salim with Rema Hammami, 'Virtual Returns to Jaffa,' *Journal of Palestine Studies*, vol. 27, no. 4, 1998, pp. 64-79.
Tel Aviv Municipality, *Tel Aviv: People and their City*, Tel Aviv, 1974.
Tzafrir, Daron, *Jaffa: A Glance at Adjami, An Architectural Profile* (Heb.), Tel Aviv, City Engineer's Department, 1995.
Wright, Gwendolyn, *The Politics of Design in French Colonial Urbanism*, Chicago, University of Chicago Press, 1991.
Yiftachel, Oren, 'The Internal Frontier: Territorial Control and Ethnic Relations in Israel,' *Regional Studies*, vol. 30, no. 5, 1995.

Yiftachel, Oren, *Watching Over the Vineyard: The Example of Majd el-Krum* (Heb.), Raanana, Institute for Israeli Arab Studies, 1997.
Yiftachel, Oren and Avinoam Meir (eds.), *Ethnic Frontiers and Peripheries: Landscapes of Development and Inequality in Israel*, Boulder, Colorado, Westview Press, 1998.
Yoskovitz, Baruch, 'Urban Planning in Tel Aviv-Yafo: Past, Present, and Future' (Heb.), in Nachmias and Menahem (eds.), *Social Processes and Public Policy in Tel Aviv-Yafo*, vol. II, 1997.

Chapter 10

Image Making, City Marketing, and the Aesthetization of Social Inequality in Rio de Janeiro

ANNE-MARIE BROUDEHOUX

City marketing and image making were key features of urban governance in the late twentieth century. With growing interurban competition for global flows of capital and visitors, city managers in search of increased tax revenues and new sources of employment have increasingly been pressured to develop a distinctive urban image to advertise their locales on the world market. However, despite this strong economic rationale there has also been a social logic to the practice of selling places. Urban image construction through public works and marketing campaigns has often served as a tool of social control, as dominant groups have used visual and spatial strategies to impose their views and set the terms for membership in society, sanctioning some actors as participants in urban life, while ignoring, segregating, and making others increasingly invisible.

While most of the literature on the social dimensions of image making and city marketing has referred to the experience of First World cities, the impact of such practices has been felt with even greater magnitude in cities of the developing world.[1] Third World cities have been faced with specific urban realities which demand that different priorities be given for the use of scarce public funds. Recent global restructuring and economic instability have contributed to a widening of income disparities and an increase in social conflicts in most cities of the developing world. This chapter examines recent urban beautification efforts in Rio de Janeiro, Brazil, to understand the mechanisms of urban image construction and the relationship between space, power, and social justice in the practice of selling places in a developing economy.

Image Matters

Observers of the contemporary city have described the late capitalist urban condition as characterized by a trend toward aestheticization, where the primacy of the visual and the centrality of the image have reduced the city to a landscape of visual consumption, an object to be gazed upon, or a spectacle.[2] Current urban design practices are said to nourish this appeal for the embellishment of the material world by giving precedence to appearance over substance, and by establishing the primacy of the facade in the creation of urbane disguises, thereby reducing the effect of much architecture to two dimensions.[3]

The growing predilection for the production and dissemination of urban images has generally been justified based on economic imperatives. According to this logic, in the course of the last few decades the changing configuration of the global political and economic order has forced cities worldwide to undergo major restructuring to become more competitive in the international market. In their struggle for economic survival, and in search of new sources of employment and revenue, cities have turned to marketing, image making, and in some cases urban 'imagineering' – the Disney expression for the engineering of imaginary places – to boost local distinctiveness and attract visitors and capital. With a growing awareness of their city's position in the global hierarchy, city officials and local entrepreneurs have collaborated to exploit city images and 'sell' localities by harnessing actual or perceived attributes.

As one of the fastest-growing global industries and the largest employer in the world, tourism has played an important role in the development of this urban image-construction process. Tourism can be seen as one of the most concrete and pervasive forms of globalization, reaching out to the most remote regions on earth and bringing people from distant places face-to-face on an everyday basis. Since the 1980s large cities have arguably become the most important type of tourist destination, and tourism has grown as a source of revenue for many metropolitan areas, thus greatly influencing their economic and physical development. To tap into this attractive source of foreign exchange and employment, urban managers have tried to refashion their municipalities to match the expectations of potential investors and tourists.[4]

It would be a mistake, however, to consider image making to be a purely economic phenomenon and to overlook its social implications. Urban images are constructed both through discourse – as in marketing campaigns, promotional brochures, and tourist advertising – and by more concrete means, including the transformation of the built environment through public works, historical preservation, and redevelopment programmes. Such interventions may alter a community's material and symbolic capital, and thus have an impact upon collective representation.

In order to grasp fully the social implications of urban image construction, it is first necessary to consider the relationship between space and society. The built environment is intimately related to the social construction of meaning, and it plays a central role in the formation of collective consciousness and self-perception. In fact, society and the built environment can be conceived as reciprocally constitutive of one another. Thus, while human intent and actions inscribe meaning and transform space into experienced places, places in turn structure human values and actions. As the physical embodiment of specific ideologies, and of social, cultural, political and economic relations and practices, the built environment not only represents but also constrains and enables these same relations and practices.[5]

The recognition of position within both society and space is fundamental for identity formation.[6] People's associations with and consciousness of the places where they live constitute vital sources for cultural identity construction, a point of departure from which people orient themselves in the world. Local history and collective memory, embodied in the walls and streets of the city, play a central role in the construction of meaning.[7] Alteration of the urban environment, through rhetorical or physical interventions, is thus bound to have an impact on both individual and collective identities.

Place image – often equated with place identity – also represents a source of symbolic capital that may be exploited in city-promoting activities. But since no two people experience space the same way, there are at least as many place identities as there are people. The imposition of single-stranded images onto urban diversity in the process of city marketing and the reduction of place identity to constricted and easily packaged urban 'products' thus represent necessarily exclusionary processes that often privilege the views of one group over another. More specifically, the ready-made identities assigned by city boosters and disseminated through the mass media generally serve the aspirations of dominant groups in a city, and, as such, reflect the values, lifestyles and expectations of potential investors and tourists.[8]

Recent studies of the late-twentieth-century city have raised issues about social justice and the right to representation of certain urban populations who have been made increasingly invisible by such processes of urban image formation.[9] Central to the problem of image making is thus the question of what is to be promoted and valued, and in whose interest. As urban-growth entrepreneurs and property investors have used spatial and visual strategies of social differentiation to set the terms of membership in society and symbolize 'who belongs where', major urban centres have become the sites of battles over the right to the city.

As indicated above, the aestheticization of the urban environment has also served an ideological purpose, not only acting to legitimize consumerism and social acceptance of the imperatives of capitalism, but also to help depoliticize the city. With the city thus reduced to a surface assumed to be transparent and unproblematic, aestheticization has distracted attention from real social and economic injustice.[10] Urban image construction can also be understood as a means of social control where urban managers have attempted to manipulate cultural forms and symbols to engineer consensus among city residents, either to insure social stability and unity, or to reinforce political allegiances. Similarly, image-making projects have enabled ruling minorities to use the power of visual imagery and mental and emotional associations to determine who will dominate, use, live in, and profit from urban spaces. Finally, image construction has also been used to boost the confidence of local commercial interests, encourage civic pride, and (as in the case of capital cities) foster national sentiment. Overall, urban images must thus be read as ideological and historical products, behind whose unified appearance may exist struggles between various organized groups and disputes over use and design.[11]

Image Making in Rio de Janeiro

Contemporary Rio de Janeiro constitutes both a unique and a typical case for the study of city marketing and image making in a Third World context. As Brazil's capital of culture and tourism, Rio is known as the *cidade maravilhosa*, the marvellous city. But the city is also marked by some of the highest income inequalities in the world, and due to Rio's unique geography, rich and poor have come to live and work in very close proximity to one another (figure 10.1). The 1990s were considered Rio's 'urban renaissance', a decade devoted to the restoration of the city's reputation, both locally and abroad. One reason was that since 1990 Rio's city government has embarked on a massive image-making programme in an attempt to revamp the city's economy.

Such intense preoccupation with international image has not been without precedent in the city's history. Brazilian ruling elites have long been image conscious, and they have repeatedly tried to use architecture to advertise Brazil's progress to the more advanced nations of the world.[12] Most importantly, at the turn of the nineteenth century the public health and beautification campaign of the Passos Reforms, which led to Rio's Haussmannization and the construction of imitation Parisian buildings and boulevards, was conceived as a means of attracting foreign capital and competing with other more modern trading centres, such as Buenos Aires. By providing a familiar image to the international elite, Rio's reformers hoped to convince an international audience that Brazil was a serious and deserving participant in the European economic order.[13]

However, with the shift of the national capital to Brasilia in 1960, Rio's economy lost one of its main driving forces. And the city was subsequently

Figure 10.1. Face-to-face but worlds apart: social inequalities are clearly inscribed in Rio's landscape, as rich and poor literally live side by side. A golf course, a *favela*, upper-class residences, and beachfront luxury condominiums are juxtaposed in the exclusive São Conrado. (Photo by author.)

hit hard by the economic crisis of the 1980s, when a lack of public funds led to massive disinvestment, creating serious urban problems.[14] As a result, the late 1980s and early 1990s were characterized by rising trends toward poverty, criminality and insecurity, which tarnished the national and international image of Rio and threatened its identity as *cidade maravilhosa*.[15] During this period tourism suffered terribly, and there was a nearly 50 per cent drop in the number of international tourists entering Brazil between 1986 and 1990. Public safety was cited as the main cause of dissatisfaction among visitors to Rio.[16]

In preparation for the 1992 United Nations Conference on the Environment and Development, however, the city government initiated a major clean-up operation. Parks were weeded, main thoroughfares were resurfaced, and street children were rounded up and put out of sight.[17] As part of the Rio Orla project, the city's main beachfronts were also remodelled with new paving, lighting systems, urban furnishing, and uniform food kiosks. Such interventions did little to improve the citywide quality of life, however. And a few months after the Rio Summit, the city's most disenfranchised expressed their discontent in a series of riots on the affluent beaches of Copacabana and Ipanema, creating panic among the sunbathers.[18] The city's international image was further tarnished when, within the next year, a group of street children were killed by the police and the national army occupied several of the city's *favelas* (squatter settlements) in an attempt to curb endemic crime.

It was under such circumstances that after 1993 the city government resolved to restore the city's image in hopes of restraining investment flight and making the city competitive again on the world tourism market. As part of this process, Rio's entry in the competition for hosting the 2004 Olympic Games led to the drafting, in the mid-1990s, of the Strategic Plan for the City of Rio de Janeiro.[19] This plan was strongly influenced by such foreign urban interventions as the revitalization of Barcelona for the 1992 Olympics (and was, in fact, developed with the help of an advisory board from that city).[20] The launching of the strategic plan was also stimulated by Brazil's 1988 constitutional reform, which returned central government revenues and political control to local governments. Significantly, this constitution guaranteed that matters of urban development would be placed under the control of municipal governments, and that master planning would be mandatory for all Brazilian cities with populations of more than 20,000.[21]

Three Public Programmes

Since 1993 a series of public-works programmes have significantly transformed the image of Rio de Janeiro. Closer analysis of three of these programmes as they have been recently implemented by the Rio's city government provides a better understanding of the mechanisms of image construction in the city and their effects on social relations. The programmes illustrate how space, power, and social justice are closely intertwined in the process of image making in cities of the developing world.

Rio Cidade

First among the city's recent renovation programmes was Rio Cidade, or Rio City, an urban design project said to represent the most comprehensive urban intervention programme implemented in Rio in decades. The project was initiated in late 1993 by Cesar Maia, Rio's Mayor at the time, with the help of his then Municipal Secretary of Urbanism and now current Mayor, Luiz Paolo Conde. In collaboration with the Institute of Planning of the City of Rio de Janeiro (IPLANRIO) and of the Institute of Architects of Brazil, intervention sites were chosen and a design competition was held to select architects who would work in teams on different neighbourhoods. According to official sources, the project affected fifteen neighbourhoods, with 60 per cent of the $227 million budget spent on infrastructure.[22]

The first phase of Rio Cidade focused on commercial districts of the city, many located in Rio's Zona Sul (South Zone) along the beaches, where most tourist attractions and accommodations are concentrated. According to official publications, the project sought to improve the image of these neighbourhoods and heighten their sense of identity by giving them a unique visual character. Much emphasis was placed on the improvement of aesthetics, security, circulation and parking. Focused interventions included the redesign of public spaces, the integration of different circulation systems, tree planting, landscaping, and the installation of new signage and urban furnishing. Rio Cidade also aimed to replace century-old sewers, fractured pavement, and street lights. The project gained widespread recognition abroad and was selected by the United Nations as part of its Best Practices Database for urban planning.[23]

As a response to the pressing demands of residents of other areas of the city for a similar level of public improvements in their neighbourhoods, and to address criticisms that the first phase was elitist, a second phase of Rio Cidade was initiated in March 1998. Its aim has been to improve twenty-seven additional districts located mainly in the city's lower-income, less glamorous West and North Zones.

Favela-Bairro

A second series of urban improvements was initiated by the administration of former Mayor Cesar Maia under the Favela-Bairro programme, which has sought generally to turn *favelas* (informal settlements) into *bairros* (formal neighbourhoods). The Favela-Bairro programme was implemented almost two years after the beginning of Rio Cidade, in an apparent attempt to appease social discontent in neighbourhoods which did not benefit from the first Rio Cidade programme (although this has been denied by representatives of the city government). One out of ten people in Rio lives in a *favela*, which are generally lacking in basic services and infrastructure such as access to running water, garbage collection, sewerage, mail distribution, and proper drainage. Residents of such neighbourhoods also lack the security of tenure enjoyed by the residents of formal neighbourhoods.

The idea behind the Favela-Bairro programme was to move away from previous attempts to eradicate *favelas* by finding new ways to upgrade informal settlements and integrate them into the larger urban community. Thus, in June 1995 a competition was launched to select design professionals who could develop strategies to improve the conditions of the *favelas*. Thirty-four teams of architects and urbanists contributed ideas, of whom fifteen signed contracts with the city to try out their proposed methodologies. According to official sources, in the following four years interventions would be extended to ninety *favelas*, in order to improve the conditions of 300,000 inhabitants, less than a third of Rio's *favelados*.[24]

Although conducted under the supervision of the city's housing authorities, this programme has not, however, been concerned with housing provision. Rather, it has focused on integrating *favela* communities into the city through the provision of such infrastructure as street paving, electricity, sewerage, canalization of rivers, tree planting, and landscaping. It has brought such new services to the *favelas* as day-care centres, community centres, and sport and leisure facilities. The programme has also sought to improve the development of communities through participation in sanitation programmes, education, job training, and cultural and educational activities. Much has been done to remove the stigma of *favelas* as dangerous, marginal areas. It has also been claimed that 'officialization' of *favelas* would reduce crime, especially drug-trafficking.[25]

The first phase of Favela-Bairro was implemented at a cost of $300 million, 60 per cent of which came from a loan from the Inter-American Development Bank (IDB). Much of the work has been carried out through paid community work. In 1996 the IDB approved another $300 million loan to benefit ninety-two more *favelas* and 300,000 more residents. The programme has been widely publicized and is gaining international recognition, and the IDB recently approved a similar project in Argentina. The World Bank has also been studying the Favela-Bairro methodology for use in future projects.[26] Favela-Bairro has also appeared on the United Nations' Best Practices Database.[27]

Rio Incomparavel

In September 1998 the city government of Rio de Janeiro initiated a third programme, aimed at improving the image of the city on both a national and international level. Baptized Rio Incomparavel, or Rio Incomparable, this programme has combined public works with a major marketing campaign in the interests of making Rio more attractive and convenient to tourists. Rio is Brazil's primary tourist destination, attracting almost half the country's tourists, and the tourism industry represents one of Rio's most important sources of revenue.[28] Rio Incomparavel was expected to increase the annual number of national and international tourists visiting the city from 4.5 million to 7 million before the year 2000, which would represent a 55 per cent increase.[29]

The marketing aspects of Rio Incomparavel have built upon earlier marketing activities, including in 1996 a $3.3 million international advertising campaign initiated on the Cable News Network (CNN), which helped create a notable increase in foreign tourists.[30] Today, as Rio's image to foreign tourists has been recognized as a valuable asset in its development, city government has stressed development of a Cooperative Image Plan and a Marketing Working Plan for Rio, both aimed at promoting a 'Rio Product'.[31]

Meanwhile, under the name RioMar, the public-works portion of Rio Incomparavel has been designed to include major interventions along the waterfront in the Zona Sul, with new tourist attractions, restaurants, urban furniture, and street lighting. The project is to include a reinforcement of security along the waterfront, with the installation of one hundred video cameras and three police kiosks for the municipal guard and military police. Before the end of 1999 two new bus lines were also to have been implemented to serve the special needs of tourists, with more than forty stops at strategic tourist sites and major hotels. The city government has been preparing a new system of signalization in the city, one that indicates major tourist attractions and the best ways to reach them. The plan calls for these interventions to be financed mainly by the private sector, especially through the leasing of commercial concessions in newly developed leisure areas. Petrobras, the Brazilian national oil company, will also contribute $7.5 million for the right to build service stations in these areas.[32]

One portion of the Rio Incomparavel marketing campaign has been the distribution of propaganda material in hotels and airports of the city. Major advertising campaigns will also be carried out in five other Brazilian cities and in seventeen cities around the world. Eight international marketing offices will also be opened in such foreign centres as New York, Frankfurt, Madrid and Paris. This marketing campaign, which is to be extended through the year 2000, has been estimated to cost $36 million, with the hope of raising city tourism income from its present $1.18 billion to $2.65 billion. Rio's secretary of tourism has stated that one of the goals of the campaign is to make Rio the favourite city of South America for international tourism.[33] A special campaign will also be directed at Rio residents in the hope of stimulating civic pride. Toward this end, advertising on the backs of public buses already asks 'Aren't you lucky to live in a city where you'd love to spend your holidays?'[34]

Urban Image versus Social Reality

Although all three of the programmes described above have been received with much enthusiasm in a city which undeniably has a great need for physical improvement, the programmes have also been the object of severe criticism as being motivated less by commitment to civic improvement than by desire to serve particular social, political and economic interests. And while most criticisms have been aimed at Rio Cidade – the first phase of which has already been completed – many negative comments have also been directed to the other two programmes.

One of the most common concerns about the programmes has been that they have placed too much emphasis on aesthetics, serving as little more than beautification schemes. Although some observers have emphasized the important role of aesthetic surroundings in promoting civic pride, public satisfaction, community self-esteem, and social well-being, many others have questioned the propriety of such superficial urban design interventions in the context of contemporary Brazil.[35] In particular, critics have denounced the decision to give priority to cosmetic interventions when many more pressing issues in the areas of employment, education, public transport, and health have been neglected.[36] For example, the first phase of Rio Cidade included the burying of more than 210 kilometres of phone and electric lines considered to be visually obtrusive in subterranean ducts when a large portion of the city's population did not even have access to such facilities. Critics have also denounced the fact that private interests have often been allowed to benefit from such public undertakings. For example, the powerful Global media network has taken advantage of Rio Cidade interventions to install cable television lines in the exclusive Zona Sul.

The projects have also been severely criticized for serving the needs and interests of only a small fraction of the local population. Thus, a great number of the neighbourhoods affected by the first phase of Rio Cidade were concentrated in the central district and in the Zona Sul, where the city's small middle and upper classes reside, and where the main tourist attractions and accommodations are found. Indeed, this has been a pattern for most of the twentieth century, as the Zona Sul has traditionally been one of the main beneficiaries of major public interventions.[37]

The decision of the Secretary of Tourism to privilege the Zona Sul in the interventions related to Rio Incomparavel, especially in terms of increased security, has also triggered bitter debate. Despite claims by the Commandant of the military police that efficient crime prevention requires an even level of police surveillance throughout the city, the Secretary maintained that since the Zona Sul attracted the most tourists and generated the most revenues, it deserved more security (figure 10.2).[38]

Extravagant spending and uneven resource allocation among different city neighbourhoods have also given rise to public debate. For example, the first phase of Rio Cidade clearly privileged chic Zona Sul neighbourhoods like Leblon, Ipanema and Copacabana. Here, urban furnishings were custom designed, and bus-stop shelters averaged more than $10,000 apiece, twice the cost of shelters in lower-income neighbourhoods such as Meier. Residents have been outraged to discover that their otherwise needy city now has some of the world's most expensive bus shelters – ones which have, additionally, been more successful at making a design statement than at actually sheltering people from sun and rain (figure 10.3).[39] Indeed, some of this expensive urban furniture has already been replaced, since it did not fulfil its intended function.

Furthermore, many critics have argued that Rio Cidade is a money pit because of the extremely high maintenance costs associated with many of

Figure 10.2. Pretense of security: casually dressed members of the *policia militar* walk along the affluent beaches of Ipanema and Leblon to reassure beachgoers and tourists who fear attacks from *pivetes* (street kids). Note in the upper left corner the sprawling Vidigal *favela* which looms over the Sheraton hotel. (Photo by author.)

Figure 10.3. City of contrasts: cart pullers walking down the street in the upscale neighbourhood of Leblon with its sleek bus stop. (Photo by author.)

its improvements. In fact, many of Rio Cidade's interventions have suffered from low quality and fast deterioration, due in part to the poor execution of sophisticated designs which did not take into account the limitations of local workmanship and construction technology. Moreover, the extreme diversity of urban furnishing – with each neighbourhood having its own custom-designed elements – has led to such difficulties as complicated inventories, costly storage of spare units, and over-dependence on certain suppliers.[40] In the second phase of Rio Cidade, which has concentrated on lower-income areas, efforts have been made to keep costs low and avoid some of these excesses for which the first phase is now notorious.

To add to these difficulties, popular discontent has been growing among groups who feel neglected by recent public-works campaigns, or who resent their lack of involvement in the transformation of their neighbourhoods. Many of these groups have come to despise interventions meant to 'restore their neighbourhood's identity', which have only resulted in attempts by urban designers to leave their personal architectural signatures on the city. Expressions of public dissatisfaction have ranged from open opposition, to derision, to vandalism. Protests have been particularly strong in Copacabana and Ipanema, where many of the design interventions have become the object of public ridicule. In Ipanema, residents objected to the construction of a landmark arched pedestrian overpass, which they considered not only wasteful but intrusive since it would block views and allow people to peer into nearby apartments. After the protests, the arch was built without the stairs that would have allowed it to function as an overpass, so it now fulfils only a visual role.

People have also derided the city's attempts to slow down traffic by painting the pavement at intersections with colourful patterns. In practice, no one pays attention, and the paint has already worn off as drivers continue to speed through red lights. New lamp posts in Ipanema have likewise become the object of ridicule because they lean at a curious angle toward the street. Local people have joked that they are drunken lamp posts – a characteristic they find quite fitting for Ipanema, a neighbourhood known for its bohemian culture. Many of these projects have also been the object of vandalism, since kids from the *favelas* have taken to destroying the new things, because their neighbourhoods have not received similar treatment.

The improvement programmes have also been denounced for the lack of fit between the image they project and the reality of everyday life in Rio. The Rio which is portrayed in promotional material and reinforced by recent design interventions has often had more to do with the aspirations of the ruling class than with the daily experiences of most of the city's residents. The image of easy living, prosperity and modern comfort, summarized as sea, sun and modernity, is embodied in the logo of the Rio Incomparavel campaign, where a wave, a high-rise building, and a sun spell out the letters R-I-O on a sandy background. But this symbolism does not reflect the extreme poverty in which the vast majority of Rio residents live. It rather constructs an image of the city where the poor and their informal activities clearly have no place.

Many observers have equated recent revitalization efforts in Rio with social cleansing, keeping certain elements of society at bay, out of sight of foreign visitors. One of the central goals of Rio Cidade, they have claimed, has been to prevent the informal sectors from infringing on Rio's public spaces. Thus, while sidewalks have been repaved with dynamic patterns said to represent the *joie de vivre* central to the city's identity, they have also been cleared of the informal activities that used to enliven them and provide an important source of income for the urban poor. One official publication even insisted that the uncontrolled growth of street vendors 'threatened the identity of Copacabana, symbol of the city of Rio for both locals and tourists.'[41]

But despite the absence of visual and verbal references to the *favelas* in the official tourism literature, foreign visitors often romanticize this aspect of Rio's urban life and deplore recent sanitizing efforts. The most popular tourist items sold at Rio's craft fairs depict life in the *favelas*, and a few private agencies have started to promote so-called exotic tours of Rio, taking tourists on safaris through some of the most famous *favelas*, like Rocinha, advertised as Latin America's largest squatter settlement.

All these recent urban interventions have raised questions about the constitution of citizenship in contemporary Rio. The rhetoric used by certain city officials has suggested a narrow vision of those considered as deserving and legitimate citizens of Rio. For example, some of the often-stated goals of Rio Cidade have been 'to recuperate strategic commercial areas of the city'; 'to take back the use of public spaces'; and 'to recuperate the city's image for its residents and for the international community.'[42] The repeated use of the verbs 'recuperate' and 'take back' has implied that these spaces were appropriated by people who did not have a legitimate right to do so. It has portrayed the poor, involved in informal activities often not by choice but by desperate need, as intruders who have threatened the comfort and security of the more deserving. As a result, street vendors, whose activities have been condemned as 'predatory privatization of the public realm', have been restricted to a few designated areas.[43] Ironically, some of these very 'predators' have been hired by the city to work as street sweepers to make sure the city's sidewalks remain clean and unencumbered.

Also notable in the justification for all three projects has been a strong rhetorical reference to issues of public order, rationality and modernity. Official documents have talked of 'urban rationalization' and 'regularization' as necessary measures to 'stop urban disorder' and 'reduce visual chaos'. Such clear references to notions of 'norms and forms' have not only been inspired by the modernist tradition which continues to be the trademark of Brazilian urbanism, but they suggest attempts by the city's elites to impose their own conception of order on society. There has been a strong sense among the city's ruling class that it is their role to 'educate' people on 'urban values' – that is, to teach them how to comply with upper-class ideas about 'proper' urbanity.[44] This notion of 'civilizing' Rio has been a running theme in the city's history, with the upper classes attempting to control the activities, dress code, and behaviour of the masses to serve their interests.[45] Local elites have long fantasized about Rio being a First World city, and they have worked hard to maintain this illusion.

Official Favela-Bairro publications today illustrate how the ruling class has tried to impose its norms upon society. These documents state that *favelas* could be transformed into so-called 'real' or 'normal' working neighbourhoods, and that their residents should be allowed to access the city, but in an orderly way.[46] Such statements suggest a notion of *favelas* as being pathological environments which need to be brought up to the norms of modern society.[47]

The programmes have also been criticized for resting on the belief that the city's problems can be solved through design. For example, Favela-Bairro – which claimed to be the first municipal programme to deal actively with the problems of Rio's *favelas*, rather than just ignoring or demolishing them (as was the common practice in the past) – was initiated by means of a design competition, and the teams in charge of implementing the programme were subsequently dominated by architects, with a minority of social workers. One result was that there was a near total absence of sociological studies prior to project implementation, and few of the teams reporting on their work in the *favelas* have ever mentioned the widespread social problems, such as drug-trafficking and alcoholism, that exist there. Rather, they have focused on infrastructural and landscaping interventions.[48] One official Favela-Bairro publication has expressed its faith in the power of design in this rhetorical question: 'What better way to integrate informal settlements into the formal city than by giving them the same access to high quality architectural, landscape and urbanistic projects?'[49]

Some critics have suggested that the recent interventions only attempt to cure the symptoms, while they fail to address the real causes of the city's problems, which, especially in the case of Favela-Bairro, have less to do with design and infrastructure provision than with more complex social issues.[50] For example, many of the circulation problems Rio Cidade claimed to resolve – such as lack of respect for traffic lights, speed limits, and pedestrians – can be considered more problems of control, management, and rule enforcement than of urban design. The conviction with which design has been proposed as a solution to all urban problems has often been blamed on the fact that many city officials, including present Mayor Luiz Paolo Conde, are architects, and thus they overestimate the social power of architecture, or simply seek to reproduce their own profession.

They seem to forget that more than a century of Brazilian urbanism has not been able to change Brazil's reality as one of the world's most unequal societies.

Beyond such faith in simple environmental determinism, there has also been an obvious lack of real commitment on the part of the administration to resolving Rio's core problems. In many recent interventions, social-welfare concerns have appeared to be only rhetorical, while beautification efforts have been clearly motivated by a desire to serve the political and economic interests of the leading minority. For example, those *favelas* that were made eligible for Favela-Bairro interventions were those that were already in an advanced state of consolidation, with social structures well in place, and whose configuration prefigured a smooth implementation. As a result, rather than focusing on the poorest *favelas* and the most precarious sites, the programme concentrated on those requiring minimum intervention, but those which also ensured quick results and photo opportunities for upcoming elections.

Similarly, the bulk of Favela-Bairro interventions have taken place around entry points to the *favelas*, most visible to outsiders. According to a publication of the Inter-American Development Bank, the provision of public services and leisure spaces at the border between *favelas* and formal neighbourhoods – with urban furniture, sports facilities, and landscaping that could be enjoyed by residents of both sides – would not only promote social integration, but would also 'alleviate the visual shock of the transition between *favela* and better neighbourhoods.'[51] Such a statement suggests a concern not only with improving the living conditions of the *favelas*' residents but for concealing visual eyesores to please formal-sector residents (figure 10.4).

Figure 10.4. Tower of progress: this landmark tower stands at the heart of the new plaza built to mark the entrance of Fernando Cardim, one of the flagship projects of the programme Favela-Bairro. (Photo by author.)

Furthermore, the implementation of urban programmes such as Favela-Bairro and Rio Cidade has been highly politically motivated, and the projects have served as showcases for election campaigns (figure 10.5). Despite Brazil's electoral law which bars mayors from seeking instant re-election to higher office, the position of mayor is often used as a trampoline for access to higher positions in Brazilian politics. Most mayors therefore try to initiate as many projects as they can in a four-year mandate. And to insure succession to their allies, many of these projects are initiated at the end of their terms.[52] Some have even suggested that Rio Cidade and Favela-Bairro – both carried out near the end of Cesar Maia's term, during a political campaign, were used to blackmail the public into electing his candidate, Luiz Paulo Conde, so that the projects would be carried to completion.[53] As a result, many individual public works suffered from the short time span of campaign politics, and were built with a lack of concern for long-term maintenance.

Today Rio Cidade and Favela-Bairro have become so associated with ex-Mayor Cesar Maia that they have become key tools in his recent campaign for Governor of the State of Rio de Janeiro. They have also been the object of fierce attacks by his political opponents. Central to Mr. Maia's October 1998 campaign was a new project called Re-Cidade which follows the same lines as Rio Cidade to upgrade infrastructure and urban image in other large and medium-sized cities in the state.[54]

A final source of criticism has been the growing involvement of the private sector in public works due to limited public funds to support urban projects in Rio. With the increased use of urban furniture for commercial

Figure 10.5. Showcase project: a new community center created as part of Favela-Bairro interventions in São Sebastian is used as backdrop for political propaganda by Cesar Maia and his followers during the fall 1998 election campaign. (Photo by author.)

advertising, some now fear the corporatization of the city's public spaces. Corporate sponsors have already set the conditions under which some public spaces are used in Rio. For example, in Ipanema the Praça Nossa Senhora da Paz, a key public park, has recently been restored with funding from Citibank. The bank's corporate logo now appears at each entrance to the park, and free *tai chi chuan* classes are offered to local residents provided they purchase and wear a Citibank t-shirt for the activity.

Concluding Thoughts

Recent image-making efforts in Rio have exemplified the relationship between space, power, and social justice in a society inundated with free-market ideology and intensified social polarization. Today, as the public has been increasingly removed from the decision-making process, and as elites and large corporations have taken control of urban space and its representation, urban image construction has become a means of manipulating public opinion and controlling social behaviour to serve particular social, political or economic interests. Decision-makers have used the built environment to manipulate consciousness, and they have disguised this manipulation in order to reproduce their political ideology and naturalize their power. As the concretization of elite aspirations, the image of the city has become an essential tool in the consolidation of elite hegemony and the perpetuation of social inequality.

But the example of Rio de Janeiro also demonstrates how such attempts at social control and exclusion have not always gone unchallenged. People excluded from the dominant image have often resorted to diverse forms of popular reworking to express symbolically their resistance to this system of growing inequity. In their attempts to contest elite control of spatial use and meaning, they have used the built environment and everyday practices to subvert the agendas of dominant groups and derail their systems of representation through ridicule, appropriation, symbolic boycott, unintended use, and unruly activities such as graffiti and vandalism. The example of Rio also suggests that the current use of cosmetic solutions as a form of popular pacification is highly unsustainable, and may actually be counterproductive. By camouflaging the deep social problems of Brazilian society, and denying representation to certain population groups, such actions may in reality only exacerbate social instability and conflict, and ultimately tarnish Rio's international image.

Notes

1. For a survey of the literature related to city marketing and image making, see G.J. Ashworth and H. Voogd, *Selling The City: Marketing Approaches in Public Sector Urban Planning*, London, Belhaven Press, 1990; H. Madsen, 'Place Marketing in

Liverpool: A Review,' *Journal Of Urban And Regional Research*, vol. 16, 1992, pp. 633-640; P. Kotler *et al.*, *Marketing Places: Attracting Investment, Industry and Tourism to Cities, States and Nations*, New York, Free Press, 1993; G. Kearns and C. Philo, *Selling Places: The City as Cultural Capital, Past and Present*, Oxford, Pergamon, 1993; and B. Erickson and M. Roberts, 'Marketing Local Identity: The Importance of the Physical,' paper presented at the 17th International Making Cities Livable Conference, Freiburg, Germany, 1995; S.V. Ward, *Selling Places: The Marketing and Promotion of Towns and Cities 1850-2000*, London, E & FN Spon, 1998.

2. D. Harvey, *The Condition of Postmodernity: An Inquiry Into The Origins of Cultural Change*, Cambridge, MA, Blackwell, 1989; H. Lefebvre, *La production de l'espace*, Paris, Editions Anthropos, 1986; M. Sorkin (ed.), *Variations on a Theme Park: The New American City and the End of Public Space*, New York, Noonday Press, 1992; and J. Urry, *Consuming Places*, New York, Routledge, 1995.

3. M.C. Boyer, *The City of Collective Memory: Its Historical Imagery and Architectural Entertainments*, Cambridge, MA, MIT Press, 1994; and S. Zukin, *Landscapes of Power: From Detroit to Disney World*, Berkeley, University of California Press, 1991.

4. On tourism and its economic and spatial implications, see S. Page, *Urban Tourism*, New York, Routledge, 1995; R.E. Wood, 'Tourism and The State: Ethnic Options and Construction of Otherness,' in M. Picard and E. Wood (eds.), *Tourism, Ethnicity and the State in Asian and Pacific Societes*, Honolulu, University of Hawaii Press, 1997, pp. 1-34; C.M. Law, *Urban Tourism: Attracting Visitors to Large Cities*, New York, Mansell, 1993; and R. Britton, 'The Image of the Third World in Tourism Marketing,' *Annals of Tourism Research*, vol. 6, 1991, pp. 318-329.

5. On the relationship between space and society, see J.A. Agnew and J.S. Duncan, *The Power of Place: Bringing Together Geographical and Sociological Imaginations*, Boston, Unwin Hyman, 1989; M. Gottdiener and A.P. Lagopoulos, *The City and the Sign: An Introduction to Urban Semiotics*, New York, Columbia University Press, 1986; A.D. King, *Re-presenting the City: Ethnicity, Capital, and Culture in the 21st-Century Metropolis*, New York, New York University Press, 1996; and Lefebvre, *La production de l'espace.*

6. For discussions of the role of the built environment in the formation of self- and collective identity, see K. Lynch, *The Image of the City*, Cambridge, MA, MIT Press, 1960; E.C. Relph, *Place and Placelessness*, London, Pion, 1976; and D.J. Walmsley, *Urban Living: The Individual and the City*, London, Longmans, 1988.

7. P. Connerton, *How Societies Remember*, Cambridge, Cambridge University Press, 1989; M. Halbwachs, *On Collective Memory*, Chicago, University of Chicago Press, 1992; and D. Lowenthal, *The Past is a Foreign Country*, New York, Cambridge University Press, 1985.

8. Visitor's expectations are at times dissonant with local reality, and attempts at image construction to fit those expectations can be highly disruptive for local population groups. For more on the socio-cultural impacts of tourism development, see F. Ascher, *Tourism: Transnational Corporations and Cultural Identities*, Paris,

UNESCO Press, 1985; D. Hayden, *The Power of Place: Claiming Urban Landscapes as People's History*, Cambridge, MA, MIT Press, 1995; D. MacCannell, *The Tourist: A New Theory of The Leisure Class*, New York, Schocken Books, 1976; J.E. Turnbridge and G.J. Ashworth, *Dissonant Heritage: The Management of the Past as a Resource in Conflict*, New York, J. Wiley, 1996; and G.J. Ashworth and A.G.J. Dietvorst, *Tourism and Spatial Transformations*, Wallingford, CAB International, 1995.

9. S. Sassen, 'Whose City is it? Globalization and the Formulation of New Claims,' *Public Culture*, vol. 8, 1996, pp. 205-223; S. Zukin, *The Cultures of Cities*, Cambridge, MA, and Oxford, Blackwell, 1995; S. Fainstein, I. Gordon, and I and M. Harloe, *Divided Cities*, Oxford, Blackwell, 1992; and Boyer, *The City of Collective Memory*.

10. Place-promoting, money-making activities such as World Exhibitions and Olympic Games often serve as a means of pacification by directing attention away from internal conflicts in a form of social control reminiscent of the 'bread-and-circus' tactics of the late Roman empire. See, for example, Boyer, *The City of Collective Memory*; Sorkin (ed.), *Variations on a Theme Park*; and Zukin, *Landscapes of Power*.

11. Erickson and Roberts, *Marketing Local Identity*; and Kearns and Philo, *Selling Places*.

12. J. Holston, *The Modernist City: An Anthropological Critique of Brasilia*, Chicago, University of Chicago Press, 1989.

13. D. Underwood, 'Civilizing Rio de Janeiro: Four Centuries of Conquest through Architecture,' *Art Journal*, vol. 51, no. 4, 1992, pp. 48-57; and J.D. Needell, *A Tropical Belle Époque: Elite Culture and Society in Turn of The Century Rio de Janeiro*, New York, Cambridge Press, 1987.

14. C. Vainer and M.O. Smolka, 'Em tempos de liberalismo: Tendências e desafios do planejamento urbano no Brasil,' in R. Piquet and A.C.T. Ribeiro (eds.), *Brasil: Território da Desigualdade*, Rio de Janeiro, J. Zahar Ed., Fundação Universitária Jose Bonifacio, 1991.

15. A. Garotinho, *Violência e criminalidade no Estado do Rio de Janeiro: diagnóstico e propostas para uma política democrática de segurança pública*, Rio de Janeiro, Hama, 1998.

16. *Brazil in Figures*, Rio de Janeiro, Instituto Brasileiro de Geografia e Estatistica (IBGE), 1997.

17. M. Nichols, 'Unruly Rio Tidies up: Rio de Janeiro, Brazil, Hosts Earth Summit Conference,' *Maclean's*, June 15, 1992, p. 42.

18. 'Arruaça na Areia,' *Veija*, October 28, 1992, pp. 18-22.

19. IPLANRIO, *Strategic Plan for the City of Rio de Janeiro: Rio Forever Rio*, Rio de Janeiro, Iplanrio-Empresa Municipal de informatica e planjamento S.A., 1996.

20. V. del Rio, 'A requalificação urbana e a imagem da cidade o projeto Rio Cidade no Rio de Janeiro,' draft for publication in *ÓCULUM*, June 1998.

21. 'The City of God, or Someone,' *The Economist*, vol. 340, no. 7985, 1996, p. 56.

22. Prefeitura da Cidade do Rio de Janeiro, *Riocidade: Linhas gerais do projeto urbanistico,* Rio de Janeiro, Iplanrio – Empresa Municipal de informatica e planjamento S.A., 1996; V. del Rio, 'Reconquistando A Imagem Urbana E O Espaço Dos Pedestres O Projeto Rio-Cidade No Centro Funcional Do Méier, Rio De Janeiro,' paper presented at the VIth Meeting of the National Association of Graduate Research in Urban Planning, University of Brasilia, May 1995; and V. del Rio, 'Restructuring Inner-City Areas in Rio de Janeiro: Urban Design for a Pluralistic Downtown,' *Journal of Architectural and Planning Research*, vol. 14, no. 1, 1997, pp. 20-34.

23. Organization of the United Nations, 'Rio Cidade (Urbanism back to the Streets),' *Best Practices Database*, <http://www.bestpractices.org>, 1998.

24. Prefeitura da Cidade do Rio de Janeiro, *Favela Bairro: Integrating Slums in Rio de Janeiro,* Rio de Janeiro, Iplanrio – Empresa Municipal de informatica e planjamento S.A., 1997.

25. C.R. Duarte, O.L. Silva, and A. Brasileiro, *Favela, un bairro: Propostas metodologicas para intervençao publica em favelas do Rio de Janeiro*, São Paulo, Pro-Editores, 1996.

26. D. Magurian, 'Novo futuro para favelas do Rio,' *Bidextra* Supplement to *O BID*, publication of the Interamerican Bank of Development, 1997.

27. Organization of the United Nations, 'Favela-Bairro Program, Rio,' *Best Practices* database, <http://www.bestpractices.org>, 1998.

28. IBGE (Instituto Brasileiro de Geografia e Estatistica), *Brazil in Figures*, Rio de Janeiro, 1997.

29. D. Matta and S. Schmidt, 'Que venham os turistas: Campanha pretende elevar em 55% o numero de visitantes do Rio até o ano 2000,' *O Globo*, August 28, 1998, p. 13.

30. A. Wiederhecker, 'Turismo do Rio gahou fôlego em 97,' *Jornal do Brasil*, March 5, 1998.

31. IPLANRIO, *Strategic Plan.*

32. *Rio Incomparavel*, Rio de Janeiro, Riotur (Rio de Janeiro Tourism Authority), 1998.

33. *Ibid.*

34. 'Sorte sua vider numa cidade em que vocé adoraria passar ferias,' (author's translation).

35. Professor D. Machado, interviewed in *Trabalho sobre Rio Cidade*, video edited by Roberto Dyer, Rio de Janeiro, Universidad Federal de Rio de Janeiro, 1997.

36. L.C. Ribeiro *et al.*, *Observatorio de politicos urbanas e gestao municipal. Projeto plano de avaliação do programa Favela-Bairro,* Rio de Janeiro, IPPUR, 1997.

37. Underwood, 'Civilizing Rio de Janeiro'; and M. Abreu, *Evolução Urbana do Rio de Janeiro,* 3rd ed., Rio de Janeiro, Iplanrio, 1997.

38. Matta and Schmidt, 'Que venham os turistas,' p. 13.

39. 'Quanto mais longe, mais barato: Nos bairros da Zona Norte, obras custam

bem menos,' *Jornal Do Brasil*, May 18, 1996, p. 20. In the second phase of Rio Cidade, in which most projects have been located in lower-income areas far from tourist sites, efforts have been made to keep costs low and avoid some of the excesses for which the first phase of Rio Cidade has become notorious. Most designers have attempted to come up with original designs using urban furniture already in storage in the city's warehouses. See del Rio, 'A requalificação urbana.'

40. Professor W. Bittar, interviewed in *Trabalho sobre Rio Cidade*, video edited by Roberto Dyer, Rio de Janeiro, Universidad Federal de Rio de Janeiro, 1997.

41. Prefeitura da Cidade do Rio de Janeiro, *Riocidade*, p. 19. Ironically, it is those same street vendors that Norma Evenson once identified as being at the heart of Copacabana's identity. See N. Evenson, *Two Brazilian Capitals: Architecture and Urbanism in Rio de Janeiro and Brasilia*, New Haven, Yale University Press, 1973.

42. Prefeitura da Cidade do Rio de Janeiro, *Riocidade*, pp. 3-5.

43. *Ibid.*, p. 7.

44. This language was used by the architect from IPLANRIO who guided me and other visitors through a few of the Favela-Bairro projects, and it is also found in most of the literature put out by the prefectura on Favela-Bairro.

45. See, for example, S. Chalhoub, *Trabalho, lar e botequim: o cotidiano dos trabalhadores no Rio de Janeiro da Belle Epoque*, Sao Paulo, Brasiliense, 1986; T.A. Meade, *'Civilizing' Rio: Reform and Resistance in a Brazilian City 1889-1930*, University Park, Pennsylvania State University Press, 1997; and Underwood, 'Civilizing Rio de Janeiro.'

46. Prefeitura da Cidade do Rio de Janeiro, *Favela Bairro*.

47. For an earlier but similar criticism of elite perceptions of the *favelas* of Rio de Janeiro, see J.E. Perlman, *The Myth Of Marginality: Urban Poverty And Politics in Rio De Janeiro*, Berkeley, University of California Press, 1976.

48. Duarte, Silva, and Brasileiro, *Favela, un bairro*.

49. *Ibid.* (author's translation).

50. E. Bessa, 'Avaliação crítica de um programa municipal de urbanização de favelas na cidade do Rio de Janeiro,' paper presented at the meeting of the Latin American Studies Association in Guadalajara, Mexico, April 1997.

51. Magurian, 'Novo futuro para favelas do Rio.'

52. 'The City of God, or Someone,' p. 56.

53. Bittar, as interviewed in *Trabalho sobre Rio Cidade*.

54. M. Dias, 'Informe,' *Jornal Do Brasil*, June 7, 1998, p. 6.

Bibliography

Anon, 'Arruaça na Areia,' *Veija*, October 28, 1992, pp. 18-22.

Anon, 'Quanto mais longe, mais barato: Nos bairros da Zona Norte, obras custam bem menos,' *Jornal Do Brasil*, May 18, 1996, p. 20.

Anon, 'The City of God, or Someone,' *Economist*, vol. 340, no. 7985, 1996, p. 56.

Anon, *Brazil in Figures*, Rio de Janeiro, Instituto Brasileiro de Geografia e Estatistica (IBGE), 1997.

Anon, *Trabalho sobre Rio Cidade*, video edited by Roberto Dyer, Rio de Janeiro, UFRJ, n.d.

Abreu, M., *Evolução Urbana do Rio de Janeiro*, 3rd ed., Rio de Janeiro, Iplanrio, 1997.

Agnew, John A. and James S. Duncan, *The Power of Place: Bringing Together Geographical and Sociological Imaginations*, Boston, Unwin Hyman, 1989.

Amin, Ash, *Post-Fordism: A Reader*, Oxford, Blackwell, 1994.

Ascher, François, *Tourism: Transnational Corporations and Cultural Identities*, Paris, UNESCO Press, 1985.

Ashworth, Gregory J. and A.G.J. Dietvorst, *Tourism and Spatial Transformations*, Wallingford, CAB International, 1995.

Ashworth, Gregory J. and Henk Voogd, *Selling the City: Marketing Approaches in Public Sector Urban Planning*, London, Belhaven Press, 1990.

Bessa, Eliane, 'Avaliação crítica de um programa municipal de urbanização de favelas na cidade do Rio de Janeiro,' paper presented at the meeting of the Latin American Studies Association, Guadalajara, Mexico, April, 1997.

Boyer, M. Christine, *The City of Collective Memory: Its Historical Imagery and Architectural Entertainments*, Cambridge, MA, MIT Press, 1994.

Britton, R., 'The Image of the Third World in Tourism Marketing,' *Annals of Tourism Research*, vol. 6. 1991, pp. 318-329.

Chalhoub, S., *Trabalho, lar e botequim: o cotidiano dos trabalhadores no Rio de Janeiro da Belle Epoque*, Sao Paulo, Brasiliense, 1986.

Connerton, Paul, *How Societies Remember*, Cambridge, Cambridge University Press, 1989.

del Rio, Vicente, 'A requalificação urbana e a imagem da cidade o projeto Rio Cidade no Rio de Janeiro,' draft for publication in *ÓCULUM*, June, 1998.

del Rio, Vicente, 'Reconquistando A Imagem Urbana E O Espaço Dos Pedestres O Projeto Rio-Cidade No Centro Funcional Do Méier, Rio De Janeiro,' paper presented at the VIth Meeting of the National Association of Graduate Research in Urban Planning, University of Brasilia, May, 1995.

del Rio, Vicente, 'Um Balaço Do Urbanismo Contemporâneo No Brasil: De Curitiba Ao Rio De Janeiro,' paper presented at the VIII Congresso Iberoamericano de Urbanismo, Porto, Portugal, September, 1998.

del Rio, Vicente, 1997, 'Restructuring Inner-City Areas in Rio de Janeiro: Urban Design for a Pluralistic Downtown,' *Journal of Architectural and Planning Research*, vol. 14, no. 1, pp. 20-34.

Dias, Maurício, 'Informe,' *Jornal Do Brasil*, June 7, 1998, p. 6.

Duarte, C.R., O.L. Silva and A. Brasileiro, *Favela, un bairro: Propostas metodologicas para intervençao publica em favelas do Rio de Janeiro*, São Paulo, Pro-Editores, 1996.

Eco, Umberto, 'Function and Sign: Semiotics of Architecture,' in Mark Gottdiener and Alexandros Lagopoulos (ed.), *The City and the Sign: An Introduction to Urban Semiotics*, New York, Columbia University Press, 1991.

Erickson, Bill and Marion Roberts, 'Marketing Local Identity: The Importance of the Physical,' paper presented at the 17th International Making Cities Liveable Conference, Freiburg, Germany, 1995.

Evenson, Norma, *Two Brazilian Capitals: Architecture and Urbanism in Rio de Janeiro and Brasilia*, New Haven, Yale University Press, 1973.

Fainstein, Susan S., I. Gordon, and M. Harloe, *Divided Cities*, Oxford, Blackwell, 1992.

Fainstein, Susan S., *Restructuring the City: The Political Economy of Urban Redevelopment*, New York, Longman, 1986.

Garotinho, A., *Violência e criminalidade no Estado do Rio de Janeiro: diagnóstico e propostas para uma política democrática de segurança pública*, Rio de Janeiro, Hama, 1998.

Gottdiener, Mark and Alexandros P. Lagopoulos, *The City and the Sign: An Introduction to Urban Semiotics*, New York, Columbia University Press, 1986.

Halbwachs, Maurice, *On Collective Memory*, Chicago, University of Chicago Press, 1992.

Harvey, David, *Consciousness and the Urban Experience: Studies in the History and Theory of Capitalist Urbanization*, Baltimore, Johns Hopkins University Press, 1985.

Harvey, David, *The Condition of Postmodernity: An Enquiry into the Origins of Cultural Change*, Cambridge, MA, Blackwell, 1989.

Hayden, Dolores, *The Power of Place: Claiming Urban Landscapes as People's History*, Cambridge, MA, MIT Press, 1995.

Holston, James, *The Modernist City: An Anthropological Critique of Brasilia*, Chicago, University of Chicago Press, 1989.

IBGE (Instituto Brasileiro de Geografia e Estatistica), *Brazil in Figures*, Rio de Janeiro, 1997.

IPLANRIO, *Strategic Plan for the City of Rio de Janeiro: Rio Forever Rio*, Rio de Janeiro, Iplanrio – Empresa Municipal de informatica e planjamento S.A., 1996.

Jameson, Fredric, *Postmodernism, or, the Cultural Logic of Late Capitalism*, Durham, Duke University Press, 1991.

Kearns, Gerry and Chris Philo, *Selling Places: The City as Cultural Capital, Past and Present*, Oxford, Pergamon, 1993.

King, Anthony D., *Re-presenting the City: Ethnicity, Capital, and Culture in the 21st-Century Metropolis*, New York, New York University Press, 1996.

Kotler, P. et al., *Marketing Places: Attracting Investment, Industry and Tourism to Cities, States and Nations*, New York, Free Press, 1993.

Law, Christopher M., *Urban Tourism: Attracting Visitors to Large Cities*, New York, Mansell, 1993.

Lefebvre, Henri, *La production de l'espace*, Paris, Editions Anthropos, 1986.

Lowenthal, David, *The Past Is a Foreign Country*, New York, Cambridge University Press, 1985.

Lynch, Kevin, *The Image of the City*, Cambridge, MA, MIT Press, 1960.

MacCannell, Dean, *The Tourist: A New Theory of the Leisure Class*, New York, Schocken Books, 1976.

Madsen, H., 'Place Marketing in Liverpool: A Review,' *Journal of Urban and Regional Research*, vol. 16, 1992, pp. 633-640.

Magurian, David, 'Novo futuro para favelas do Rio,' *Bidextra* Supplement to *O BID*, publication of the Interamerican Bank of Development, 1997.

Matta, Daniela and Selma Schmidt, 'Que venham os turistas: Campanha pretende elevar em 55% o numero de visitantes do Rio até o ano 2000,' *O Globo*, August 28, 1998, p. 13.

Meade, T.A., *'Civilizing' Rio: Reform and Resistance in a Brazilian City 1889-1930*, University Park, Pennsylvania State University Press, 1997.

Needell, J.D., *A Tropical Belle Époque: Elite Culture and Society in Turn of The Century Rio de Janeiro*, New York, Cambridge Press, 1987.

Nichols, Mark, 'Unruly Rio tidies up (Rio de Janeiro, Brazil hosts Earth Summit Conference),' *Maclean's*, June 15, 1992, p. 42.

Organization of the United Nations, 'Favela-Bairro Program, Rio,' *Best Practices Database*, <http://www.bestpractices.org>, 1998.

Organization of the United Nations, 'Rio Cidade (Urbanism back to the Streets),' *Best Practices* database, <http://www.bestpractices.org>, 1998.

Page, Stephen, *Urban tourism*, New York, Routledge, 1995.

PCRJ (Prefeitura da Cidade do Rio de Janeiro), *Favela Bairro: Integrating slums in Rio de Janeiro*. Rio de Janeiro: Iplanrio – Empresa Municipal de informatica e planjamento S.A., 1997.

PCRJ (Prefeitura da Cidade do Rio de Janeiro), *Riocidade: Linhas gerais do projeto urbanistico*, Rio de Janeiro, Iplanrio – Empresa Municipal de informatica e planjamento S.A., 1996.

Perlman, Janice E., *The Myth Of Marginality: Urban Poverty And Politics In Rio De Janeiro*, Berkeley, University of California Press, 1976.

Pred, Allan Richard, *Recognizing European Modernities: A Montage of the Present*, New York, Routledge, 1995.

Relph, E.C., *Place and Placelessness*, London, Pion, 1976.

Ribeiro, Ana Clara Torres, 'O espetáculo urbano no Rio de Janeiro: comunicação e promoção cultural,' *Cardenos IPPUR/UFRJ*, 1995.

Ribeiro, Luiz Cesar de Q. *et al.*, *Observatorio de politicos urbanas e gestao municipal. Projeto plano de avaliação do programa Favela-Bairro*, Rio de Janeiro, IPPUR, 1997.

Rio Incomparavel, Rio de Janeiro, Riotur (Rio de Janeiro Tourism Authority), 1998.

Sassen, Saskia, 'Whose city is it? Globalization and the Formation of New Claims,' *Public Culture*, vol. 8, 1996, pp. 205-223.

Sciorra, Joseph, 'Return to the Future: Puerto Rican Vernacular Architecture in New York City,' in A.D. King (ed.), *Re-presenting the City: Ethnicity, Capital and Culture in the 21st Century Metropolis*, New York, New York University Press, 1996.

Sorkin, Michael (ed.), *Variations on a Theme Park: The New American City and the End of Public Space*, New York, Noonday Press, 1992.

Turnbridge, J.E. and Gregory J. Ashworth, *Dissonant Heritage: The Management of the Past as a Resource in Conflict*, New York, J. Wiley, 1996.

Underwood, D., 'Civilizing Rio de Janeiro: Four Centuries of Conquest through Architecture,' *Art Journal*, vol. 51, no. 4, 1992, pp. 48-57.

Urry, John, *Consuming Places*, New York, Routledge, 1995.

Vainer C. and M.O. Smolka,'Em tempos de liberalismo: Tendências e desafios do planejamento urbano no Brasil,' in R. Piquet and A.C.T. Ribeiro (eds.), *Brasil: Território da Desigualdade*, Rio de Janeiro, J. Zahar Ed., Fundação Universitária Jose Bonifacio, 1991.

Walmsley, D.J., *Urban Living: The Individual and the City*. London, Longmans, 1988.

Ward, Stephen V. *Selling Places: The Marketing and Promotion of Towns and Cities 1850-2000*, London, E & FN Spon, 1998

Watts, Michael, 'Development II: The Privatization of Everything,' *Progress in Human Geography*, vol. 18, no. 3, 1994, pp. 371-384.

Wiederhecker, Angelica, 'Turismo do Rio gahou fôlego em 97,' *Jornal do Brasil*, March 5, 1998.

Wood, Robert E., 'Tourism and the State: Ethnic Options and Construction of Otherness,' in Michel Picard and Robert E. Wood (eds.), *Tourism, Ethnicity and the State in Asian and Pacific Societes*, Honolulu, University of Hawaii Press, 1997.

Zukin, Sharon, *Landscapes of Power: From Detroit to Disney World*, Berkeley, University of California Press, 1991.

Zukin, Sharon, *The Cultures of Cities*, Cambridge, MA, and Oxford, Blackwell, 1995.

Chapter 11

'Authentic' Anxieties

DELL UPTON

Better than anyone else before or since, Marx and Engels articulated the central theme of modernity. 'All that is solid melts into air, all that is holy is profaned': for better or for worse, all the old certainties of self, society, and culture dissolved in the face of capitalism. They went on to explain that 'The need of a constantly expanding market for its products chases the bourgeoisie over the whole surface of the globe. It must nestle everywhere, settle everywhere, establish connections everywhere.' As a result, 'The bourgeoisie has through its exploitation of the world market given a cosmopolitan character to production and consumption in every country.'[1]

Consuming Tradition, Manufacturing Heritage elaborates on Marx and Engels' insights. Capitalism no longer seeks raw materials and markets for its industrial goods alone, but cultural raw materials that can be transformed into hard cash through the conservation, restoration, and outright fabrication of indigenous landscapes and traditional cultural practices for the amusement of metropolitan consumers. In this light, the rise of heritage and cultural tourism stands as another episode in the two-century history of modernity, though perhaps one given particular urgency in the face of the increased, and increasingly global, scale of tourism and the transformation of local cultures and societies that it has seemed to engender.

At the same time, the significance of tourism transcends the vast profits it generates or the idle amusement it offers. There would be no market for these kinds of activities, no crowds willing to undertake expensive, sometimes arduous journeys of discovery, if the contemplation of the indigenous and the traditional were not so central to the experience of the modern. The adjectives *traditional* and *modern* are themselves artefacts of modernity: tradition did not exist until it was imagined as the defining complement of modernity. As the chapters included in this volume reveal, the root pair can be elaborated into a whole lexicon of dichotomous adjectives: ancient and modern, indigenous and cosmopolitan, hidden and transparent, mysterious and known, obscure and legible, pure and impure, substantial and ephemeral, and most of all authentic and inauthentic. In this sense, the concept of tradition serves as a mirror of the anxieties to which Marx and Engels gave voice, that Raymond Williams explored in *The Country and the City*, and that has been the single great theme of Euro-American and much non-Euro-American literary and cultural writing for two centuries: the anxiety that arises from the fear that modern life is by its nature inauthentic – even counterfeit or spurious.[2]

The articulation of tradition and modernity is a global phenomenon. Tradition was sorted out from modernity in the Euro-American core at least as early as it was in its peripheries and dependencies, and the manufacture of heritage operates as industriously in the developed as in the developing world.[3] Yet the essays here imply that there is a significant difference when the practice is asymmetrical, as it often is in tourism to developing nations. It is the nature of that difference that is at issue: how can we assess the impact of heritage and cultural tourism on its destinations? To what extent do the residents of tourist regions accept or even participate in the construction of a visitable heritage for visitors? When do concepts of heritage and its wealth-making potential become tools of power for indigenous elites, as they have in Timothy Mitchell's New Gurna or Anne-Marie Broudehoux's Rio de Janeiro? In what ways can residents use the concepts of tradition and heritage as a rubric within which to resist that power, as in Mark LeVine's Jaffa?

These analyses of heritage are worthy additions to the contemporary discussion of modernity, but many also share some of the troubling assumptions of that discussion. These are evident in the title of this book: *Consuming Tradition, Manufacturing Heritage*. How can heritage be manufactured or tradition consumed? 'Manufactured' is often a term of abuse, implying the false, as in a manufactured excuse or situation. Even more, something that is manufactured is artificial, the antithesis of the natural that is so valorized in Euro-American thought. 'Consumption' is a using up – ingestion followed by the excretion of waste – and by analogy a wasting disease (tuberculosis). In the folklore of capitalism, consumption is hedonistic indulgence, the opposite of the productive self-discipline that Max Weber called 'the Protestant Ethic'.[4]

Manufactured traditions are not inherently pernicious, as Nezar AlSayyad notes in his introduction. Eric Hobsbawm and Terence Ranger argued in their classic work *The Invention of Tradition* that 'authentic' and 'manufactured' traditions serve nationalist and ethnic goals equally well.[5] In many cases, it is frustrating and possibly counterproductive even to try to make the distinction. The outdoor museums that Paul Oliver discusses in this volume are collections of moved, restored, conserved, reconstructed, and newly built structures that defy the attempt to certify or deny authenticity.

Despite these ambiguities, scholars of heritage and tradition too often leave unchallenged the notions of tradition as a definable, hence commodifiable, entity, and of authenticity as a testable and desirable quality of tradition. Implicitly or explicitly, they assume the defence of tradition and authenticity against modernity and artificiality. Like Dean MacCannell's tourist, they search for some 'true' experience of difference, one created for insiders, available only to the most sympathetic and persistent outsiders.[6] Yet – again – 'all that is solid melts into air': every appearance of authenticity turns out to be a mirage, a simulacrum manufactured only to be consumed. Thus the attempt to deconstruct the discourse of modernity becomes simply an example of that discourse, betraying a continued allegiance to modernity that flourishes in the so-called postmodern age.

Our loyalty to modernity's concept of authenticity has much to do with our own insecurity about what we do. As architects, planners, historians, anthropologists, and preservationists, we are part of the mechanism that manufactures heritage. As relatively prosperous, privileged people, we are consumers of tradition. As postmodern intellectuals, we understand that authenticity is an elusive, perhaps non-existent quality. At the same time, for most of us at least, our political or emotional commitments incline us to side with those who seem to be the victims of modernity, and we search for some objective grounds for valorizing their position over that of Marx's cosmopolitan bourgeoisie. In our heart of hearts, we want to be Gramscian organic intellectuals.

To reconcile our emotional investment in authenticity with our intellectual scepticism, we commonly locate authenticity in the realm of *identity*, defined by *difference* and validated by *culture*. *Tradition* is evidence of the continuity of identity through time. *Heritage* is the visible product of tradition, calcified and commodifiable. By this reasoning, identity conveys an authority to local forms that is lacking in cosmopolitan ones, and to 'traditional' forms that is lacking in 'manufactured' ones.

But culture is a slippery concept, and identity and difference even more so. Consequently, it might be more fruitful to understand heritage, tradition, and modernity as strategic political positions, rather than as fixed or essential qualities of sites or cultural practices, much less of individual identities. Individuals routinely shift from one cultural position to another, adopt one identity or another, as occasion demands. Sites likewise seem to take on varying colourations according to the angle from which one views them. To account for this, recent theorists have begun to speak of 'hybridity', a term that retains the essentialist qualities of the original dichotomies. A hybrid is a third thing created by the amalgamation of two fixed, because genetically coded, entities.

A fresh start requires a subtler understanding of identity. We might begin with the anthropologist Anthony Cohen's useful distinction between selfhood and personhood. The former denotes the 'substance of me', while the latter refers to my social being.[7] Selfhood and personhood are both contingent and fragmented, but they are not the same. In a classic discussion of the varied, often contradictory, aspects of selfhood, the psychologist Ulric Neisser subsumed personhood under the heading of the 'conceptual self', one of five aspects of the self or 'kinds of self-knowledge' that also include data derived from direct perception of the physical environment, from species social signals, memory, and the sense of singularity created by awareness of the uniqueness of one's personal experience.[8] In the context of these other modes of self-knowledge, the powerful structures of personhood, though influential, are not determining of our identity.

Analyses of heritage and tradition often conflate selfhood and personhood, personal and cultural identity. This is a relic of nationalist projects, which treated the state and the individual as macro- and microcosms of one another.[9] Nationalist rhetoric reduced the person to a socio-political identity; cultural nationalism reduces the same person to a common body of beliefs and practices – to a cultural artefact, that is. The alternative is the centreless inauthenticity ascribed to modern selfhood in Nelson Graburn's developmental metaphor for understanding heritage and tradition. Evocative as it is, his construct fundamentally conflates selfhood and personhood. If we respect the distinction, however, we should locate cultural variability and change in the realm of personhood rather than in a hybridity grounded in an essentialized self. Individual social identities are fluid even within apparently homogenous societies; cross-cultural contact introduces more possibilities for fluidity, rather than fluidity itself.

The distinction between selfhood and personhood, then, may help to understand cultural fluidity in a nuanced way. Host people can engage as persons, playing shifting cultural roles as the cultural, economic, or political occasion seems to demand, without necessarily understanding themselves as compromising essential selfhood.[10]

Something similar might be said about landscapes. That is, they have no inherent 'traditional' or modern identities or meanings. They are products of multiple, only partially overlapping, circles or realms of knowledge, practice, and significance. A single building, for example, stands at the intersection of realms of craft, finance, commerce, social practice, and sometimes cosmological or intellectual endeavour. It might be constructed by locally trained craftworkers, using materials manufactured half a world away, to house a colonial governor, decorated with classical columns, and oriented geomantically.[11] The 'foreign' materials might represent modernity to one observer, making do to another. To a third, they might have no significance at all. The building has no legible or unified identity. Instead, the attachment of a label such as 'traditional', 'indigenous', or 'modern' *conveys* identity on the building, and by extension on its makers and users, rather than makers or users conveying an identity on their building. In that sense, all traditions are manufactured.

This is not to say that all such categories are imposed from outside. The same kinds of distinctions are recognized locally. An example is the cross-cultural phenomenon of the ancestral homeland, an imaginary assemblage of natural, historic, and sometimes newly created monuments, edited, altered, added to, and mythically knit together by bits and pieces of historical fact and fantasy to give it a distinctive, characteristic, legible form capable of grounding claims to territory, cultural identity, or nationhood. The Navajo homeland, Dinétah, defined by four sacred mountains, architectural ruins left behind by previous occupants of the territory (drafted as ancestors by the Navajo), and a body of sacred stories, is one. Like the Anglo-Saxon 'colonial' landscape of the East Coast of the United States or the Confederate landscape of the South, Dinétah shares space with the differently imagined homelands of competing indigenous and Euro-American populations.[12]

In a tourist setting, craft workers understand that they and their products have very different significance for local and visiting publics. African-American basket makers on the South Carolina coast work 'traditional' materials in 'traditional' ways, but they make 'traditional' basket forms only for tourists, to whom they offer business cards, so that their customers can recommend the maker to their friends.[13] On the other hand, the New Mexico *santero* (image carver) George López sold the same objects to tourists and local buyers. He told folklorist Charles L. Briggs that 'The block of wood is nothing more than wood . . . The difference is not in the wood or in the way in which one carves it. If you were baptized Catholic . . . then you can take your carving to the father and he will bless it. It does not do anything to pray to the carvings, because they are not holy, they have not been blessed.' Once the image was blessed, it was different, holy, but so was a car or a rosary that had similarly been blessed. 'He who sells it is responsible' for its proper disposition.[14]

Selfhood and personhood can offer a different way of looking at the visitors' side of the tourist-host relationship, as well. Tourists sometimes provoke dramas of social dominance (perhaps less brazenly in the present era of mass tourism and shifting political and cultural values than in Gregory's Egypt), and more often, probably, indulge a romantic mystification of the other. Cultural tourism is both competitive – tourists vie to visit the most, or the most authentic, or the most obscure sites – and integrative – they seek self-improvement through seeing the sights one 'must' see to be a cultivated or educated member of one's home society.[15] In this respect, touristic consumption is outward looking, an aspect of personhood.

But it is also a function of selfhood, which is more complex. Touristic consumption of 'heritage' and 'tradition', which from one perspective appears merely as appropriation of the cultural 'property' of others, is also an act of bricolage through which consumers manufacture selves in ways that cannot be reduced to simple formulas.[16] As in all consumption, tourists seek pleasure, a kind of 'mentalistic hedonism' in which the cultural meanings sought may be less messages – about what it is to be modern, to be privileged, to be a member of one nationality or ethnicity and not another – than less articulated and less articulatable experiences of selfhood in all the varieties Neisser outlined.[17] However commodified the product may be, in the act of consumption the purchaser or user discovers and internalizes the associated human and psychological meanings. This process has been linked by the anthropologist Daniel Miller to Hegel's account of the process of objectification and reconciliation of Spirit.[18] Since analogous experiences of cultural difference are available to members of the host population, the effects of heritage and cultural tourism are much more difficult to assess than they appear to be at first.

All this is by way of arguing that the issue is not the authenticity of particular modes of engagement, or the legitimacy of the much-derided touristic gaze. Nor is it the mere tourist presence. To frame heritage and tradition in terms of authenticity – of the choice of traditional values, authentic forms, undiluted identities – is to miss the point. The problem is one of conflicts among values, all of which have some legitimacy. The problem, in other words, is political, as Mitchell, LeVine, and Broudehoux all make plain.

The rhetoric of heritage, identity, and authenticity are typically evoked in times of great political and economic changes – in Europe in the mid-eighteenth century, globally in the late nineteenth century and again in recent decades. This rhetoric is a way of claiming or challenging power where traditional political-economic authorities and ideologies close off more direct routes. One anthropological interpretation of ethnicity, for example, is as a device to draw boundaries that can help decide the distribution of scarce resources. This is one way to understand Mark LeVine's story of Tel Aviv-Jaffa, where ethnic difference codified as landscape 'heritage' articulated competing claims to real estate in a tight market. In the tourist world, exaggerated and commodified difference, as in the 'manufactured' traditional architecture of the many 'Chinatowns' found throughout the South Chinese diaspora, or the 'folk' festivals springing up in the American South, can divert needed income from more 'authentic', but often less conspicuous, competitors.[19]

Viewed in this light, the manufacture of heritage and the consumption of tradition become more difficult to define and equally difficult to distinguish in their effects from authentic and indigenous practices. The dichotomies of modernity lose their persuasiveness in the process. The focus of critical analysis begins to drift far away from cultural effects and to move toward political-economic causes.

Notes

1. Karl Marx and Friedrich Engels, *The Communist Manifesto*, ed. Samuel H. Beer, New York, Appleton-Century-Crofts, 1955, p. 13.

2. Raymond Williams, *The Country and the City*, New York, Oxford University Press, 1973.

3. Stuart Piggott, *Ruins in a Landscape: Essays in Antiquarianism*, Edinburgh, Edinburgh University Press, 1976; Peter Collins, *Changing Ideals in Modern Architecture, 1750-1950*, Montreal: McGill University Press, 1967; Richard Handler and Eric Gable, *The New History in an Old Museum: Creating the Past at Colonial Williamsburg*, Durham, N.C., Duke University Press, 1997; and see the chapters by Phil Gruen and Paul Oliver in this volume.

4. Colin Campbell, *The Romantic Ethic and the Spirit of Modern Consumerism*, Oxford, Blackwell, 1986, pp. 60-76; Max Weber, *The Protestant Ethic and the Spirit of Capitalism*, trans. Talcott Parsons, New York, Scribners, 1930.

5. Eric Hobsbawm and Terence Ranger (eds.), *The Invention of Tradition*, Cambridge, Cambridge University Press, 1983.

6. Dean MacCannell, *The Tourist: A New Theory of the Leisure Class*, New York, Schocken, 1989, pp. 91-104.

7. Anthony P. Cohen, *Self-Consciousness: An Alternative Anthropology of Identity*, London, Routledge, 1994, pp. 56-57.

8. Ulric Neisser, 'Five Kinds of Self-Knowledge,' *Philosophical Psychology*, vol. 1, no. 1, 1988, pp. 35-59. The other four are the ecological self, the interpersonal self, the extended self, and the private self, respectively.

9. Cohen, *Self-Consciousness*, p. 157.

10. Peter Lunt, 'Psychological Approaches to Consumption: Varieties of Research – Past, Present and Future,' in Daniel Miller (ed.), *Acknowledging Consumption: A Review of New Studies*, London, Routledge, 1995, p. 249.

11. Dell Upton, *Architecture in the United States*, Oxford, Oxford University Press, 1998, pp. 282-283. The historian of science Peter Galison has made a similar point in arguing that 'progress' in science is usually the product of the efforts of actors (e.g., instrument makers and theoretical physicists) whose interests, values, and circles of reference overlap only partially. In his metaphor, what appears to be a single strand of rope (change in science) is, on closer inspection, a series of short, intercalated fibres, none of which has an exact counterpart in any other. (Galison, *Image and Logic: A Material Culture of Microphysics*, Chicago, University of Chicago Press, 1997).

12. Upton, *Architecture in the United States*, pp. 78-86; Mike Wallace, *Mickey Mouse History and Other Essays in American Memory*, Philadelphia, Temple University Press, 1996.

13. On the differentiation of goods for local and tourist markets, see Nelson H. H. Graburn (ed.), *Ethnic and Tourist Arts: Cultural Expressions from the Fourth World*, Berkeley, University of California Press, 1976, pp. 1-32.

14. Charles L. Briggs, *The Wood Carvers of Córdova, New Mexico: Social Dimensions of an Artistic 'Revival'*, Knoxville, University of Tennessee Press, 1980, p. 193.

15. MacCannell, *Tourist*, pp. 91-104; Lunt, 'Psychological Approaches to Consumption,' p. 248.

16. Colin Campbell, 'The Sociology of Consumption,' in *Acknowledging Consumption*, p. 117.

17. Campbell, 'Sociology of Consumption,' p. 118.

18. Daniel Miller, *Material Culture and Mass Consumption*, Oxford, Blackwell, 1986, pp. 19-33, 178-180.

19. Rodger Lyle Brown, *Ghost Dancing on the Cracker Circuit: The Culture of Festivals in the American South*, Jackson, University Press of Mississippi, 1997.

Bibliography

Briggs, Charles L, *The Wood Carvers of Córdova, New Mexico: Social Dimensions of an Artistic 'Revival'*, Knoxville, University of Tennessee Press, 1980.

Brown, Rodger Lyle, *Ghost Dancing on the Cracker Circuit: The Culture of Festivals in the American South*, Jackson, University Press of Mississippi, 1997.

Campbell, Colin, *The Romantic Ethic and the Spirit of Modern Consumerism*, Oxford, Blackwell, 1986.

Campbell, Colin, 'The Sociology of Consumption,' in Daniel Miller (ed.), *Acknowledging Consumption: A Review of New Studies*, London, Routledge, 1995, pp. 96-126.

Cohen, Anthony P, *Self-Consciousness: An Alternative Anthropology of Identity*, London, Routledge, 1994.

Collins, Peter, *Changing Ideals in Modern Architecture, 1750-1950*, Montreal, McGill University Press, 1967.

Galison, Peter, *Image and Logic: A Material Culture of Microphysics*, Chicago, University of Chicago Press, 1997.

Graburn, Nelson H. H. (ed.), *Ethnic and Tourist Arts: Cultural Expressions from the Fourth World*, Berkeley, University of California Press, 1976.

Handler, Richard and Eric Gable, *The New History in an Old Museum: Creating the Past at Colonial Williamsburg*, Durham, N.C., Duke University Press, 1997.

Hobsbawn, Eric and Terence Ranger (eds.), *The Invention of Tradition*, Cambridge, Cambridge University Press, 1983.

Lunt, Peter, 'Psychological Approaches to Consumption: Varieties of Research – Past, Present and Future,' in Daniel Miller (ed.), *Acknowledging Consumption: A Review of New Studies*, London, Routledge, 1995, pp. 238-263.

MacCannell, Dean, *The Tourist: A New Theory of the Leisure Class*, New York, Schocken Books, 1976, 1989.

Marx, Karl and Friedrich Engels, *The Communist Manifesto*, ed. Samuel H. Beer, New York, Appleton-Century-Crofts, 1955.

Miller, Daniel, *Material Culture and Mass Consumption*, Oxford, Blackwell, 1986.

Neisser, Ulric, 'Five Kinds of Self-Knowledge,' *Philosophical Psychology*, vol. 1 no. 1, 1988, pp. 35-59.

Piggott, Stuart, *Ruins in a Landscape: Essays in Antiquarianism*, Edinburgh, Edinburgh University Press, 1976.

Upton, Dell, *Architecture in the United States*, Oxford, Oxford University Press, 1998.

Wallace, Mike, *Mickey Mouse History and Other Essays in American Memory*, Philadelphia, Temple University Press, 1996.

Weber, Max, *The Protestant Ethic and the Spirit of Capitalism*, trans. Talcott Parsons, New York, Scribners, 1930.
Williams, Raymond, *The Country and the City*, New York, Oxford University Press, 1973.

Index

aborigines 41, 42, 91
Alexandria 119, 123, 124, 128, 130, 133, 137, 145
American West 21, 152-190
Andromeda Hill Project 255-258
Arab Jaffa 240-272
Arab Palestine 240, 241-243, 246
Aswan 117, 121, 123, 124, 127, 133, 134, 135, 136, 219
Aswan Dam 134-136, 145, 226
Australia 41-42, 45, 91, 104
authenticity 4-5, 8-12, 18, 21, 25-26, 43, 54, 97, 98, 112, 114, 115, 117, 123, 141, 181, 196, 201, 206, 298-306

Bali 4, 58, 59, 60, 206
birthright 69, 71
Brazil 74, 273-297

Cairo 12, 22, 113, 117, 123-124, 127, 128, 128, 130-134, 218, 219, 220, 222, 223
capitalism 2, 5, 8, 20, 36, 44, 53, 90, 92, 95, 96, 98, 114, 115-116, 276, 298-299,
Carter, Howard 213, 215
Celebration, Florida 11, 12-14
Chinatown, San Francisco 161, 164
cidade maravilhosa 272-277
clan 71
colonial nostalgia 20, 111-151
colonial urbanism 3
Colonial Williamsburg 9-10, 202
colonialism 3, 4, 5, 35, 45, 55, 93, 112-114, 116, 140-141, 240
communism 74, 79, 80, 201
community heritage 70
Conde, Luiz Paolo 278, 287, 289
consumption 2-3, 5, 8-25, 34, 36, 38, 45, 47, 51, 52, 68-69, 71, 78, 96, 178, 231-233, 246, 251, 298-299, 302-303
Copacabana 277, 282, 284, 285

dahabeah 116-118, 121, 123, 124-127, 131-133, 136,
Dearborn, Michigan 191
Delhi 206-207
developing country 36, 37, 38, 206
difference 3, 5, 7, 8, 16, 90, 99-100, 102, 104-105, 158, 300
Disney Corporation 9, 13, 14, 274
Disney World 98
Disneyland 4, 9, 25, 253, 257-258
dragoman 117, 122, 128, 133, 135, 138

Egyptian nationalism 213-215
Engels 298, 299
EPCOT 13, 25

familial relationships 74
Favela-Bairro 279-280, 286, 287, 288
favelas 278, 284-288
Fathy, Hassan 12, 22, 216-222, 223, 228
First World 4-5, 12, 17, 36, 38, 42, 44, 45, 47, 53, 55, 273, 286
Flanders 195
Ford, Henry 191
Fort Collins, Colorado 25
Foucault, Michel 18, 20, 68, 115, 249
France 25, 197-198, 202

Geddes, Patrick 249
gentrification 23, 50, 240, 250, 252, 255
Germany 75, 78-79, 193, 195, 197, 200, 229
global domination 93
globalization v, 3, 8, 15-16, 53-54, 240, 247, 274
Greenfield, Michigan 191-192, 200

Habermas, Jurgen 96
Helsinki 192
heritage preservation 2, 3
Hong Kong 15
Hobsbawn, Eric 14, 20, 112
hybrid 116, 300
hybridity 3, 16, 300, 301

identity 7-8, 15-16, 18, 23, 43, 50-55, 91, 98, 99, 100, 105, 212, 213, 214, 215, 300-303
identity and alterity 69-73, 78, 79, 81-82
industrialization 38, 91, 92, 96
inter-cultural conflict 34-67
International Association for the Study of Traditional Environments 99
intra-cultural conflict 34-67
Ipanema 277, 282, 284, 289
Irigaray, Luce 102

Jaffa 23, 240-272, 299, 303
Jakarta 206
John Mason Cook 124, 127-130

Kenya 77, 206
Khedive, the 124, 125, 129
kinship 18, 69-75, 81
Kook, Peter 247, 253, 257

Las Vegas 10, 11, 26
Latour, Bruno 20, 115
Learner, Daniel 99
le Corbusier 93, 95, 258
Lefebvre, Henri 20, 95, 96, 114, 240, 258
Levinas, Emmanuel 102
Luxor 12, 22, 111, 126, 127, 130, 133, 213, 216, 217, 219, 222, 223, 224, 228, 232
Lyotard, Jean Francois 104

machines 91, 92
Maia, Cesar 278, 279, 280
Main Street, Disneyland 25
manufacture 2-3
manufactured traditions 9-15, 18, 20, 21, 25, 159, 167, 178, 216, 233, 299-303
Marceline, Missouri 25
Market St, San Francisco 152, 171, 178
Marx, Karl 298, 299, 300
Marx, Werner 102
marxism 45
Massachusetts 195, 196, 197
matrimony 70
Michigan 191
Mies van der Rohe 93
modernist architecture 92, 96, 105, 242, 252, 258
modernist design 92, 93, 244
modernist environments 90, 95
modernity 90, 96-102, 240, 241, 284, 286
modernization 19, 45, 90, 91, 93, 102
Morocco 25
Mount Vernon 194
Musée des Arts 202
museums, open-air 191-211

Nairobi 206
nationalism 3, 4, 7, 15, 23, 73, 80, 96, 157, 212-215, 240, 301
New Gurna 11, 12, 22-23, 212-239, 299
New Urbanism 11-14

New Zealand 203
Nile, the 20, 111-141, 213, 214, 216, 220, 221, 228, 230
Oedipal 71, 77-78
Olympic Games 278
Orient, the 117, 118, 120, 123, 125, 138
Orientalism 112, 116, 117, 120, 134, 137, 141, 252-253, 258
'other', the 3, 4, 8-19, 21, 68, 70, 79, 90, 95, 100, 102-104, 105, 158, 179, 200, 245, 252

Palestine 23, 74, 240, 241, 244, 246, 254, 258
Paris Exposition, 1887 6
Paris Exposition, 1889 25
Passos Reforms 276
patrimony 69, 70, 71
personhood 301-303
Philadelphia 153, 172, 200
Philae 111, 134-135
postmodernism 240, 244, 251, 252, 258
poststructuralism 90, 102-104, 105
Poundbury 11-12
Prince of Wales 11
Project Shikum 255

rehabilitation 243, 250, 253, 254, 255
restoration 49, 196-197, 199, 201, 203, 298
Rio Cidade 24, 278-179, 281, 282, 284, 285, 286, 288, 289
Rio de Janeiro 24, 273-297
Rio Summit 277
Romania 192, 193, 207
Royal Society of Arts 191

Said, Edward 20, 111, 252
San Francisco 21, 101, 152-190
Santa Fe 11, 25
selfhood 301-303
Shakespeare, William 74, 199
sibling rivalry 73, 74
Singapore 6, 7, 15, 36, 49
skansen 191-194
Suez Canal 113, 114, 124, 141
Sweden 192

Taut, Bruno 93
technological age 19, 90-110
technological development 90, 92
technological modernism 92, 93, 96
Tel Aviv 23, 240-272, 303
Thailand 38
Third World 3, 4 5, 6, 24, 26, 44, 45, 54, 55, 273, 276
Thomas Cook 20, 113, 114, 124-127, 131-141, 156, 158
Toraja 207
tourism development 3, 17, 36, 44, 45, 49, 50, 51, 228
tourism industry 3, 16-17, 19, 35, 36, 39, 40, 40-41, 44, 45, 46, 47, 48, 49, 50, 52, 53, 54, 222, 231-232, 280
touristic gaze 3-4, 20, 23, 38, 39, 42, 46, 48, 50, 53, 71, 117, 135, 139, 159, 257, 303
traditional culture 6, 8, 41, 43-44, 49, 68-89, 95, 98, 298
traditional environments 2, 19, 90, 92, 95, 98-102, 104, 111, 298
Traditional Neighborhood Development [TND] 11
travel 1, 3, 35, 53, 54, 97-98, 111-116, 140-141
Tutankhamun 213

Uluru 41
United Nations 241, 277, 279, 280
urban design 51, 260, 274-280, 282, 284, 287, 288
urban tourism 157, 173
urbanism 3, 6, 11
urbanization 102, 155, 157, 240

vandalism 284, 290
vernacular building traditions 3, 49, 81, 92, 94, 194, 197, 200, 207-208
vernacular environment 92, 198

Weissenhofsiedlung Experimental Housing Project 93
Williams, Raymond 2, 299
World Heritage Site 193
World Trade Organization [WTO] 34, 36, 37, 38

Zona Sul 278, 281, 282